Quantitative X-Ray Diffractometry

Springer
New York
Berlin
Heidelberg
Barcelona
Budapest
Hong Kong
London
Milan
Paris
Santa Clara
Singapore
Tokyo

Lev S. Zevin Giora Kimmel

Quantitative X-Ray Diffractometry

Edited by Inez Mureinik

With 70 Illustrations

Springer

Lev S. Zevin (deceased)
The Institutes for Applied Research and
Department of Materials Engineering
Ben-Gurion University of the Negev
Beer-Sheva 84105, Israel

Giora Kimmel
Department of Materials Engineering
Ben-Gurion University of the Negev
Beer-Sheva 84105, Israel

Edited by
Inez Mureinik
The Institutes for Applied Research
Ben-Gurion University of the Negev
Beer-Sheva 84105, Israel

Library of Congress Cataloging-in-Publication Data
Zevin, Lev S.
Quantitative X-ray diffractometry / Lev S. Zevin, Giora Kimmel; edited by Inez Mureinik.
p. cm.
Includes bibliographical references and index.
ISBN-13:978-1-4613-9537-9 e-ISBN-13:978-1-4613-9535-5
DOI: 10.1007/978-1-4613-9535-5

1. X-rays—Diffraction. 2. X-rays—Diffraction—Industrial applications. I. Kimmel, Giora. II. Mureinik, Inez. III. Title.
QC482.D5Z48 1995
545′.81—dc20 95-19428

Printed on acid-free paper.

Softcover reprint of the hardcover 1st edition 1995

Production managed by Hal Henglein; manufacturing supervised by Jeffrey Taub.
Camera-ready copy prepared from the editor's Microsoft Word files.

9 8 7 6 5 4 3 2 1

ISBN-13:978-1-4613-9537-9

Preface

The first book dealing with quantitative analysis in powder mixtures (if only in one chapter) was that of Klug and Alexander, *X-ray Diffraction Procedures,* which was published in 1954. The only work dedicated entirely to quantitative X-ray powder diffraction (QXRD) that has appeared since that time is the 200-page monograph of Davis, *Reference Intensity Method of QXRD Analysis.* However, this publication, which appeared in 1988, was devoted only to a single aspect of QXRD, as reflected in its title. On the other hand, most books on the characterization of materials by X-ray powder diffraction do include a section or a chapter on QXRD. In contrast to structure determination, in which single crystal methods are favored, quantitative analysis of phases is specific to the powder method, and the lack of a special book dealing with QXRD analysis is inexplicable. The aim of the present book is to fill this void.

Since the terminology pertaining to QXRD is not always standardized in the literature, the reader should pay special attention to descriptions of the particular methodology and to the formulation of each method in order to compare similar procedures described under different names by other workers.

The book is divided into six chapters. The first chapter is introductory, giving the history of X-ray powder diffraction and comparing QXRD to other analytical methods. The second chapter deals with the physical background of X-ray powder diffraction in general and with its application to QXRD, with special emphasis on the attenuation of X-rays and microabsorption. The third chapter is devoted to instrumentation, with particular reference to the Bragg-Brentano diffractometer. Other powder facilities are also reviewed in detail.

The methodology of QXRD is given in the fourth chapter, which constitutes the core of the book. A variety of problems and solutions in the field of quantitative phase analysis are given. The chapter starts with the general case, in which all absorption coefficients are known, and then continues with the internal standard method (including the addition methods of doping and dilution); full-phase analysis (external standard); the reference intensity ratio method; the full diffraction pattern approach; standardless methods; combination of

X-ray diffraction with chemical analysis; crystallinity of polymers; and analysis of low-mass samples.

The fifth chapter is devoted to practical aspects of QXRD: optimal data collection; hints for sample preparation; intensity measurement; absorption; analytical standards; pattern treatment and simulation; correction of preferred orientation; and errors and limitations. The section on preferred orientation was kindly edited by Prof. Matti Järvinen. The final chapter addresses some applications, most of them industrial.

The unique quality of this book lies in the fact that it brings together between one set of covers all the aspects of the multifaceted discipline of QXRD, and we thus trust that it will make a major contribution to promoting QXRD in a wide spectrum of applications.

Giora Kimmel

Editor's preface

It might seem strange to dedicate a book to the author himself, but this book is dedicated to the memory of Lev Zevin, who sadly did not live to see its completion. Zevin, as he was known affectionately to those of us who worked with him at Ben-Gurion University of the Negev, was an outstanding diffractionist—the range of his knowledge and experience is reflected in the scope and depth of this work. Indeed, the concept of the book—of bringing together all the facets of quantitative X-ray analysis, both theoretical and applied—evolved during many years of teaching, research and analytical work, first in the former USSR and later in Israel. I sincerely hope that what stands on these pages will be a fitting tribute to a scientist of his caliber.

Inez Mureinik

Acknowledgments

Prof. Lev Zevin dedicated his life to scientific research. He was a hard-working, active, and creative personality who never hesitated to share his knowledge and inexhaustible sources of original ideas with all who came into contact with him—researchers in different fields, colleagues, and especially the students who were so important to him.

We hope that this book will contribute to progress in the field of quantitative X-ray diffractometry. The methods of QXRD analysis outlined in this book will be of use to researchers and scientists at universities and industrial laboratories who are investigating compounds with special technological properties; to chemists and physicists working on the synthesis and properties of new materials; to geologists studying natural deposits; and to specialists dealing with ore beneficiation. These methods also have important applications in the routine technological processes of the cement industry. If this book will serve as an everyday working manual for all these fields, it will be the best way to preserve the memory of Prof. Zevin.

Prof. Zevin started working on this book a few years before his death. At the time he passed away, the book was almost completely written, except for the last few sections of Chapter 6. The final editing and completion of the book were accomplished after his death. Before he died, Prof. Zevin asked his colleague and friend, Prof. Giora Kimmel, to see the book through to completion. We thank him for his part.

The publication of the book was made possible due to the efforts of the highly qualified and professional editor, Ms. Inez Mureinik. We would like to express our deepest gratitude to her.

Special thanks are due to Prof. M. Järvinen of Lappeenranta University of Technology, Finland, who read and edited the sections on crystallite orientation.

We would also like to thank Ms. Shula Rabinovitz for her highly professional illustrations; Ms. Marion Milner for typing the manuscript; and Ms. Dorot Imber for her understanding and encouragement.

Finally, we would like to thank Dr. Thomas von Foerster, Springer-Verlag's Senior Editor for Physics and Mathematics, for his sympathetic attitude and patience, and the Production staff at

Springer-Verlag, for their investment of time and effort in the production aspects of the book.

If we—his wife and children—have inadvertently omitted any of Prof. Zevin's colleagues who were involved in the writing of this book, we apologize and thank them.

Isabella, Shaul, and Miriam Zevin

Contents

1 Introduction

1.1 Phase analysis—when and why

There are unlimited possibilities of combining two, three, or more elements into chemical compounds, whose specific physical properties depend on the chemical composition, type of bond between the atoms, and structure of the compound. Nature itself is a great creator of chemical compounds: there are thousands of natural substances, including minerals, and the formation of new compounds continues apace. Today, however, nature is being overtaken by mankind, and artificially synthesized compounds outnumber minerals and natural organic compounds. An enormous creative initiative is fueling the search for compounds with special technological properties. The geologist and mineralogist studying natural deposits or the chemist and physicist synthesizing new compounds will need to identify their findings and new chemicals. Thus, methods for phase identification are required. The logical progression of this need is quantitative estimation of the phase composition. A quantitative analysis can be obtained only if the chemical composition of the component phases is known, if the system is sufficiently stable, and if the equilibrium phase diagram is known. All three conditions are rarely met, and computational methods for phase quantification must be developed.

Most of the technologically important materials, such as metal alloys, ceramics, and cements, are multiphase materials, and quantification is essential for the understanding of these materials. The physical and chemical properties of multiphase materials (strength, toughness, corrosion resistance) depend on the quantitative phase composition. An important example is that of Portland cement: the strength of the concrete will increase with an increasing proportion of alite ($3CaO.SiO_2$), and the resistance to sulfate corrosion will decrease with an increase in the proportion of tricalcium aluminate ($3CaO.Al_2O_3$). The analysis of hardened steel is an example of phase quantification in metallurgy. In many cases, it is important to know or even to restrict the amount of retained austenite in hardened steel, because an austenite-martensite phase transition, which is accompanied by a change in specific volume, is unwanted in machined parts. Another technologically important application of quantitative methods is industrial control, especially in beneficiation plants in which the main purpose of the technological process is the separation of a single mineral or a group of useful minerals. In environmental studies, phase analysis has acquired significant importance in the analysis of airborne dust, including, of course, such health hazards as quartz and asbestos. In geology, diffractometry is probably the most important analytical tool. And finally, we come to research and development in industry. In the development of any technology in which materials comprise the final product, a method for quantitative estimation of the phase abundance is, of course, essential. Thus, reliable methods for quantitative phase analysis are required in many facets of today's industrial world.

1.2 Phase analysis as an analytical method

1.2.1 Direct methods of phase analysis

Methods of phase analysis based on the measurement of any inherent property of the phase as a whole, such as magnetic susceptibility, density, or crystal structure, may be classified as direct methods to distinguish them from the widely used indirect methods based on chemical data. Examples of direct methods of phase analysis are given below.

1.2.1.1 Magnetic methods

If one of the phases in a multiphase material is ferro- or ferrimagnetic, then determination of the magnetic susceptibility of the sample might give a measure of the abundance of the magnetic phase. This method is used, for example, to analyze ferromagnetic martensite in steels that also contain paramagnetic austenite. (This method will, however, not work in the technically important case of small amounts of austenite.)

1.2.1.2 Selective dissolution

The rates of dissolution in acids or bases of the phases constituting a multiphase sample might be quite useful for application in phase analysis. For example, the rate at which the calcium silicate phases of Portland cement dissolve in maleic acid is much faster than the rate of dissolution of Al- and Fe-bearing minerals. This effect is used for the partial phase analysis of Portland cement.

1.2.1.3 Density measurements

The density of a crystalline compound, such as polyethylene, might be quite different from that of the same amorphous compound, and this fact is used to determine the crystallinity of the compound. Density differences might also be exploited for physical separation and consequent analysis of the phases in a powder sample.

The three methods discussed above have one feature in common—they can be applied only to materials that contain a phase with a prominent property, such as high magnetic susceptibility or a high dissolution rate. Their use is thus very specialized and not universal. The next group of methods is more general because the techniques are based on an inherent and very specific property of each phase—its atomic structure.

1.2.1.4 Optical (microscope) methods

The optical properties of a mineral (birefringence, optical constants) in combination with its specific morphology form the basis of microscopy techniques for mineral identification. In this technique the area covered by a certain mineral is a measure of its abundance. In a scanning electron microscope equipped with an X-ray microprobe analyzer, morphological observations are supplemented by an approximate chemical analysis. Today, this type of method is greatly enhanced by computer-assisted image analysis.

1.2.1.5 Thermal analysis (differential thermal analysis or differential scanning calorimetry)

Thermal methods take advantage of phase transitions occurring during heating that are specific to each phase of a multiphase sample. For example, two different forms of calcium sulfate—the dihydrate $CaSO_4.2H_2O$ and the hemihydrate $CaSO_4.1/2\ H_2O$—exhibit endothermic effects at different temperatures that can be used to distinguish them from one another and from anhydrous $CaSO_4$. The amount of heat evolved during the transition is a measure of the abundance of a particular phase.

1.2.1.6 X-ray diffraction

X-ray diffraction corresponds directly to the structure of a phase and is the most universal of the structure-sensitive methods for phase analysis of multiphase samples. This technique is discussed in detail in this book.

1.2.2 *Indirect methods of phase analysis*

Indirect methods of phase analysis are based on chemical data from which assumptions are made regarding the phase composition of the analyte. These methods have obvious limitations because of the need to make some preliminary presupposition as to the composition or other properties of the analyzed phases. We should remember that in many element-sensitive analytical methods, such as optical or X-ray spectroscopic techniques, the analyte is dissolved, thus destroying the phase composition. However, elemental composition of the sample may be used to compute the phase composition provided that the elemental composition of the phases is known. Phase analysis from elemental composition will be discussed in detail in Section 4.11.

1.2.3 General approach to phase quantification

Suppose that each phase j in the analyzed material possesses a set of properties (or characteristics) a_{ij}. This set may comprise density, magnetic susceptibility, element abundance, or diffraction intensities. Suppose also that the rule of additivity holds, and a global characteristic A_i of the material is simply described as the weighted sum of the corresponding characteristics a_{ij} of the pure phases:

$$A_i = \sum_{j=1}^{n} a_{ij} c$$

where c_j is the weight fraction of phase j in the analyzed sample, and n is the number of phases.

For a set of m characteristics:

$$\mathbf{A} = \mathbf{a}\,\mathbf{c} \tag{1.1}$$

where **A**, **a**, and **c** are 1 x m, n x m, and 1 x n matrices, respectively. The system of linear equations (1.1) may be solved for c_j provided that we have measured a sufficient number of characteristics and have *a priori* knowledge of the matrix of phase characteristics a_{ij}. The number of characteristics m must be not less than n (i.e., the number of phases) in the general case and not less than n – 1 if we intend to utilize the mass-balance equation:

$$\sum_{j=1}^{n} c_j = 1$$

On the other hand, if m > n – 1, we have more information than we need for a single solution, and the selection of the best result is made by statistical methods.

Although the linear dependency of A_i on c_j has already been assumed, the matrix **a** should be in a form that provides fast and efficient quantitative analysis. In the most favorable case this matrix is diagonal, and the set of linear equations (1.1) transforms into a single simple equation:

$$A_i = c_j a_{ij} \tag{1.2}$$

Sometimes the conditions (1.2) might be met for indirect methods of phase analysis. For example, in a sample containing quartz, calcite, and pyrite, the weight fractions of silicon, calcium, and iron are directly related to the abundance of the corresponding minerals. However, this is a rare example, and a diagonal property matrix might be expected for a direct structure-sensitive analytical method, like X-ray diffraction.

1.3 History of quantitative X-ray phase analysis

1.3.1 Quantification techniques

It seems that the first mention of quantitative phase analysis was made in the paper by Hull, "A new method of chemical analysis," published in 1919 in the *Journal of the American Chemical Society.* Discussing the diffraction pattern of a mixture of substances, he wrote, ". . . the photograph obtained with a mixture is a superimposed sum of photographs that would be obtained by exposing each of the components separately for the same length of time. This law applies quantitatively to the intensities of the lines (provided absorption is negligible for each of the components), as well as to their positions, so that the method is capable of development as a quantitative analysis."

The first attempts at quantitative analysis soon followed (Navias 1925). At this early stage, quantitative X-ray diffraction (QXRD) was based on a comparison of diffraction line intensities for the analyzed sample with a series of mixtures containing known amounts of the analyte. Probable variations of sample absorption were ignored in the first QXRD techniques. Usually, diffraction patterns of the prepared mixtures were obtained first, followed by those of the material to be analyzed. An external standard was not exposed simultaneously with the analyzed sample in order to account for probable variations in experimental conditions. It is thus no surprise that the first analyses were rather inaccurate. After the structure of martensite (Bein 1924) and the shear mechanism of the austenite-martensite transition (Kurjumov and Sachs 1930) had been established, attention was turned to the phase analysis of the technologically important compound hardened steel, in which martensite and austenite are the two major phases. These two phases with well-known and simple structures (austenite of the γ-Fe type and martensite of the tetragonally distorted α-Fe type) constitute a rather simple system with constant sample absorption.

Apparently, Tamaru and Sakito (1931) were the first workers to use X-ray diffraction to analyze retained austenite in hardened steel. They employed a cylindrical Debye-Sherrer camera with a rotating cylindrical sample 5 mm in diameter and with an incident beam tangent to the sample surface. In order to compensate for probable variations in experimental conditions (electrical setting, exposure, developing conditions), a thin strip of gold foil was glued to the sample. The sample material and the gold strip were alternately exposed to the primary beam during the sample rotation. Photometric traces were obtained, and the ratio of austenite to the gold phase was found not to vary with experimental conditions. This ratio was used as a measure of austenite abundance. Tamaru and Sakito used 100% austenitic steel with high manganese content to calibrate the intensity ratios against the austenite content.

Umanski and Khidekel (1941) extended this technique to a powder material, a mixture of tungsten with tungsten carbides WC and W_2C. Cylindrical specimens were pressed from fine-grained powders, and thin Al or brass foil was used as the external standard. The analytical procedure was distinguished by a remarkable detection limit of 0.1%-0.2% and an acceptable precision of a few percent of the analyzed phases.

A modified external standard technique was used by Vozdrigenski and Sergeev (1935) in studying the stress-induced decomposition of austenite in stainless steel. A Pt wire clamped to the exit diaphragm of the X-ray camera was used as an internal standard. The distance between the wire and the film was such that the 111 line of Pt was close to the analytical line of austenite. Very good agreement was obtained between the amount of retained austenite determined and the hardness of the steel. If c_m and c_a are the weight fractions of martensite and austenite, then:

$$c_a = [1 + (I_a)_0/(I_m)_0]^{-1}$$

where $(I_a)_0$ and $(I_m)_0$ are the intensities of the diffraction lines of pure austenite and pure martensite.

The ratio $(I_a)_0/(I_m)_0$ is calculated easily, if the structure of the phases is known [(see equation (3.8)]. In the simplified case of austenite and martensite, the term F^2/V^2 is almost constant, and thus:

$$(I_a)_0/(I_m)_0 = (LP)_0 M_0/(LP)_m M_m$$

where LP and M are Lorentz-polarization and multiplicity factors for chosen diffraction lines. In this technique we can recognize the first application of calculated patterns in phase quantification.

The method of homological pairs and the technique proposed by Tamara and Sakito have a common drawback: they are based on peak intensities rather than integrated peak intensities. Phases in hardened steel, especially martensite, exhibit line-broadening effects, and peak intensities thus become inaccurate measures of phase abundances. In a further development of the phase analysis technique, Averbach and Cohen (1948) took into account all the aspects of absorption correction and intensity measurements and produced an excellent method for the determination of retained austenite, which retains its significance today.

The simplicity of the phase analysis of hardened steel lies in the constant sample absorption, which is not the case for most other materials subjected to quantification. Even in the 1930s it was understood that sample absorption can drastically affect the intensities of the diffraction lines. The simplest remedy to this problem is the internal standard, in which technique the analyzed material (powder) is spiked with a known amount of internal standard. The idea was evidently taken from an analytical spectroscopic technique in which it had been used for some time. The internal standard technique was reported by Clark and Reynolds (1936) for the analysis of mine dusts. The analyzed phase was quartz, and the internal standard, fluorite. The reliability of the internal standard technique was then confirmed for the analysis of quartz in matrices of calcite, zircon, microcline, and titanite. The merits of the internal standard technique soon became appreciated, and the method was extended to the analysis of quartz in various ores and rocks by Ballard et al. (1940). Despite inherent limitations of intensity measurements by the photographic technique, the accuracy of quartz analysis was 5%-10% of the measured amount, and the detection limit was below 1%.

In addition to the obvious effect of sample absorption on the diffracted intensity, the less evident but equally important microabsorption phenomenon was discovered in the early stages of quantitative analysis. Microabsorption is a grain-related effect that becomes significant if the absorption in an individual particle produces an appreciable reduction in the ratio of the diffracted intensities. Studies by Brentano (1935), Shäffer (1933), Taylor (1944), and Brindley (1945) provided the basis for subsequent developments of the appropriate intensity corrections.

Further development and extension of quantitative phase analysis were undoubtedly facilitated by the formidable collection of standard diffraction patterns for minerals and chemical compounds, which began after the development of the powder camera by Debye and Sherrer (1916, 1917) and Hull (1917). The well-known paper of Hanawalt et al. (1938) presenting 1,000 patterns and a classification scheme was the first and most important contribution to the Powder Diffraction File (PDF), which is currently managed by the International Centre for Diffraction Data. A similar study in the former USSR was undertaken by Boldyrev et al. (1939), who published a collection of diffraction patterns for minerals. Here, we should also remember the collections of patterns of intermetallic phases (Pearson 1958, 1967; Villars and Calvert 1985).

The current stage in the history of phase quantification started in the late 1940s after the introduction of two important developments:

1. The design and industrial manufacture of powder diffractometers;
2. The formulation of a theoretical basis for quantitative analysis by Klug and Alexander in a series of papers published in 1948

(Alexander and Klug 1948; Alexander et al. 1948; Klug et al. 1948a,b).

This book is devoted to a critical discussion of this new era of quantitative analysis, in which the simple and elegant formulations of Klug and Alexander have been developed into contemporary schemes either utilizing a considerable portion of the diffraction pattern or combining phase quantification with structure refinement of the component phases.

1.3.2 Instrumentation

The commercial production of X-ray generators with sealed X-ray tubes, which started in the 1930s, undoubtedly facilitated the analytical application of X-ray diffraction. X-ray cameras of different designs were used with flat, bent, or capillary-mounted samples. Descriptions of the cameras can be found in the book of one of the most prominent designers of cameras (Buerger 1942).

The history of the major contemporary instrument for phase analysis—the diffractometer with parafocusing Bragg-Brentano geometry—is rather interesting. The original Bragg ionization spectrometer (Bragg and Bragg 1913) was designed for use with a single crystal. In 1921, Bragg described the focusing X-ray geometry to be used with powder samples, which is equivalent to the well-known Bragg-Brentano geometry shown in Fig. 3.2. The ionization chamber used by Bragg as the detector of diffracted rays has the inherent drawbacks of low sensitivity and instability that make the registration of ionization inferior to the photographic technique. Brentano (1925, 1935, 1937) used the same focusing geometry, but a photographic film instead of an ionization chamber for the detection of the diffracted rays. However, in this scheme the focusing conditions are maintained only for a limited portion of the film. The full pattern can be registered only by moving the flat sample in steps. LeGalley (1935) made diffraction measurements with a Geiger counter, which is a much more efficient detector than an ionization chamber. However, he used a geometry similar to that of the Debye-Sherrer camera, which is inadequate for a "point"-type detector such as a Geiger counter. Due to an unfortunate choice of the absorption factor in the geometry used and the low sensitivity of the Geiger counter to Mo radiation, the results obtained with LeGalley's method were generally poor.

The first real progress was achieved when the Geiger counter was combined with proper focusing geometry, in which an appreciable amount of analyzed material was spread over a relatively large surface to obtain favorable absorption characteristics. This combination of the Bragg-Brentano focusing principle with the then-modern Geiger counter was independently proposed by Lindemann and Trost (1940) in Germany and Friedman (1945) in the USA.

The first diffractometric designs do, however, suffer from relatively low angular precision, broad peaks, and very moderate intensity. It was only in the late 1940s that Parrish (1965) devised the basic geometric features of the contemporary powder diffractometer (tremendous improvement of angular resolution, resulting in a reduction of peak width from 0.35° to ~0.10°, and introduction of scintillation and proportional detectors).

2
Physical basis

2.1 Interaction of X-rays with material

Throughout this book, we will use the phrase "intensity of X-rays," which is usually synonymous with energy flow per unit time (seconds) per unit area normal to the direction of the X-ray beam (wave vector). The intensity of X-rays passing through a layer of any material is attenuated: the degree of attenuation depends on the wavelength of the X-rays, the chemical composition of the material, and the geometric features of the sample. In the derivation of the basic equations, we assume that the attenuation of monochromatic X-rays by a very thin layer of material is proportional to the thickness of the layer:

$$\Delta I = -\mu I \Delta x \tag{2.1}$$

where I is the intensity of the beam at the boundary of the layer facing the X-ray source, Δx is the thickness of the layer, and μ is a proportional coefficient, known as the coefficient of attenuation.

Suppose that the thin layer is situated inside a material block of total thickness t (Fig. 2.1). We will obtain an expression for the attenuation by integration[1] of equation (2.1) to give:

$$I_t = I_0 \exp(-\mu t) \tag{2.2}$$

where I_t is the intensity passing through a material block of thickness t and I_0 is the intensity of the primary beam. The dimensions of the coefficient μ are reciprocal length, usually cm^{-1}. This coefficient is also

---

1 $\int dI/I = -\mu \int dx; \qquad \ln I = -\mu x + \text{const}$

Taking into account that at $x = 0$, $I = I_0$ = the intensity of the primary beam, we define the integration constant as being equal to $\ln I_0$, and finally we obtain equation (2.2).

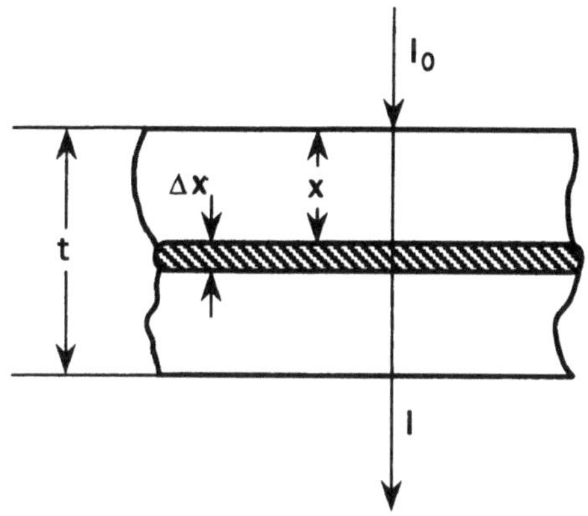

Figure 2.1. Attenuation of an X-ray beam by a block of material.

known as the linear absorption coefficient (see Section 2.1.3). We define the mass attenuation coefficient as $\mu^* = \mu/\rho$, where ρ is the density of the material that attenuates the X-rays. For radiations useful in powder diffraction, the linear absorption coefficient in solids ranges from a few cm^{-1} for light elements to thousands of cm^{-1} for heavy elements. The range of the mass coefficient is narrower, from a few cm^2/g for light elements to hundreds of cm^2/g for heavy elements.

The process of energy dissipation during the attenuation of X-rays is complicated, and a comprehensive discussion of the subject is beyond the scope of this book. The interested reader will find a detailed account elsewhere (Blokhin 1957; Azároff 1963, 1968). Here, we will discuss only two major processes in the interaction of X-rays with materials that are essential for understanding the contents of this book: scattering and absorption.

2.1.1 Scattering of X-rays

Scattering of X-rays means that X-ray photons, after interaction with matter, deviate from their initial direction. Thus, a certain transfer of momentum must be accompanied by this phenomenon. Sometimes there is also energy transfer.

2.1.1.1 Bragg (coherent) scattering

Bragg scattering constitutes the basis of our subject—X-ray diffraction. In this case energy transfer is zero, i.e., the scattering occurs without a change of wavelength. The momentum of the scattered photons changes (with the exception of scattering in the direction of the incident beam), and the corresponding change in the scattering material is accommodated by the average lattice. This type of scattering can be described in the framework of classical electrodynamics, and we will use this approach in calculations of the coefficient of scattering. The alternating electromagnetic field of the X-ray wave causes oscillation of electrons, which become a source of secondary (scattered) waves with the same frequency (wavelength) as the incident wave.

The intensity scattered by one free electron I_e at a scattering angle of 2θ measured at a distance R from the electron is given by the Thomson formula for nonpolarized incident radiation:

$$I_e = I_0 \frac{r_e^2}{R^2} \frac{1 + \cos^2 2\theta}{2} \tag{2.3}$$

where I_0 is the intensity of the incident beam and $r_e = e^2/mc^2$ is the electron radius (where e is the electron charge, m is the electron mass, and c is the velocity of light. Integration of equation (2.3) over the whole space angle produces the total power scattered by one electron W_e (Richtmyer et al. 1955; Cohen 1966):

$$W_e = (8\pi/3)\, I_0\, r_e^2 \tag{2.4}$$

Let us consider an X-ray beam striking a thin layer of any element of area 1 cm^2 and thickness Δx. The number of electrons in our specimen $N = (ZA/M)\, \rho\, \Delta x$, where Z is the atomic number, M is the atomic weight, A is Avogadro's number, and ρ is the density of the element. The intensity removed from the incident beam due to scattering by N free electrons may be expressed as:

$$\Delta I = W_N = (8\,\pi/3) I_0\, r_e^2\, (ZA/M)\, \Delta x = W_e\, (ZA/M)\, \Delta x$$

Taking the definition of the attenuation coefficient given in equation (2.1), we arrive at the following expression for the linear scattering coefficient σ that describes the attenuation of the X-ray beam due to scattering phenomena:

$$\sigma = (8\pi/3) r_e^2\, (ZA/M)\, \rho = W_e\, (ZA/M)\, \rho \tag{2.5}$$

The mass scattering coefficient σ^* can then be expressed as:

$$\sigma^* = \sigma/\rho = (8\,\pi/3) r_e^2\, (ZA/M) = W_e (ZA/M) \tag{2.6}$$

If we suppose that $Z/M \approx 0.5$, then $\sigma^* \approx 0.2\ cm^2/g$ and does not depend on atomic number.

These calculations are too simple to be precise and give only an order of magnitude. The basic concept of free oscillating electrons used in calculations does not hold, especially for medium and heavy elements.

Thus, a real scattering coefficient increases gradually with atomic number and reaches 1 cm^2/g for heavy elements at the end of the periodic table. This value is still no more than a few percent (or even less) of the attenuation coefficient for wavelengths common in X-ray diffraction. Thus, coherent scattering is a minor source of attenuation of X-rays passing through a material, excluding one particular case, that in which the incident angle of the primary beam corresponds to the Bragg angle of the very strong reflection. In this case a double-reflected wave propagating in the direction of the primary wave and acquiring a phase shift of 180° attenuates the primary beam. This effect is known as a primary extinction and will be discussed in Section 2.2.3.

Let us calculate the fraction of the power of the incident wave scattered by a typical sample. Suppose that the sample area is 1 cm^2, the thickness is 10^{-3} cm, and the volume is 10^{-3} cm^{-3}. If the atomic radius is ~$1.5 \cdot 10^{-8}$ cm, then the volume per atom in the solid matter is $\upsilon_a = 27 \cdot 10^{-24}$ cm^3, and there are ~$4 \cdot 10^{19}$ atoms in the sample. The number of electrons in the sample is thus $4 \cdot 10^{19}$ Z. The fraction of the power of the incident wave scattered by the sample is given by:

$$\xi = (W_e/I_0) \times 4 \cdot 10^{19} Z$$

where W_e/I_0 is defined in equation (2.4). Thus, $\xi = 4 \cdot 10^{-5}$ Z, i.e., ξ is equal only to $4 \cdot 10^{-5} - 4 \cdot 10^{-3}$ of the power of the incident wave, depending on the atomic number of the scattering material.

2.1.1.2 Thermal diffuse scattering

In the case of thermal diffuse scattering (TDS), there is a certain, although very low, energy transfer. The value of ΔE is ~10^{-2} to 10^{-1} eV, i.e., the order of magnitude of phonon energy. This value is much less than both the X-ray photon energy (10^4 eV) and the natural width of the spectral line. Thus, this energy transfer is not detected experimentally, i.e., this type of scattering can be regarded as occurring without a change in wavelength. The intensity of TDS increases with rising temperatures, but at ambient conditions it is much lower than the intensity of Bragg scattering.

2.1.1.3 Incoherent (Compton) scattering

Incoherent scattering results from elastic collisions between X-ray photons and electrons. Electrons (recoil electrons) are responsible for the transfer of momentum, and there is a substantial energy transfer of ΔE ~10–10^2 eV to these electrons. The difference in wavelengths between an incoherently scattered wave λ_c and an incident wave λ_0 is given by the following equation:

$$\lambda_c - \lambda_0 = 0.0242(1 - \cos 2\theta) \qquad [\text{Å}] \qquad (2.7)$$

where 2θ is the scattering angle of an X-ray photon.

The width of the incoherently scattered spectral line is θ-dependent and decreases towards smaller scattering angles.

Compton scattering is substantial for free electrons and is negligible for tightly bound electrons. Thus, the contribution of incoherent scattering to the total scattering of X-rays increases for lighter elements and for higher energies of the incident photons. It is quite substantial for organic materials. The difference in coherent scattering between lighter and heavier elements is illustrated by the spectrum of MoK_{α} radiation scattered by $CaCO_3$ and $MgCO_3$ (Fig. 2.2). The ratio of incoherent to coherent scattering is greater for the lighter compound $MgCO_3$.

Incoherent scattering does not produce any diffraction effects, but it does contribute to the background. This contribution decreases with decreasing scattering angle. For the subject discussed in this book, incoherent scattering, like any background source, is mostly deleterious.

Ruland (1964) sought conditions that would eliminate incoherent scattering by monochromatization of the diffracted beam. He found that effective attenuation of Compton scattering is possible only at medium and high scattering angles and only with a monochromator with a narrow spectral band. Due to the uncertainty at low scattering angles, Ruland recommended elimination of Compton scattering by calculation. Further details on the angular behavior of Compton scattering and the calculations required for its elimination are given in Section 5.6.3. There is one useful application of incoherent scattering—the estimation of mass attenuation coefficients; this application will be discussed in Section 5.7.5.

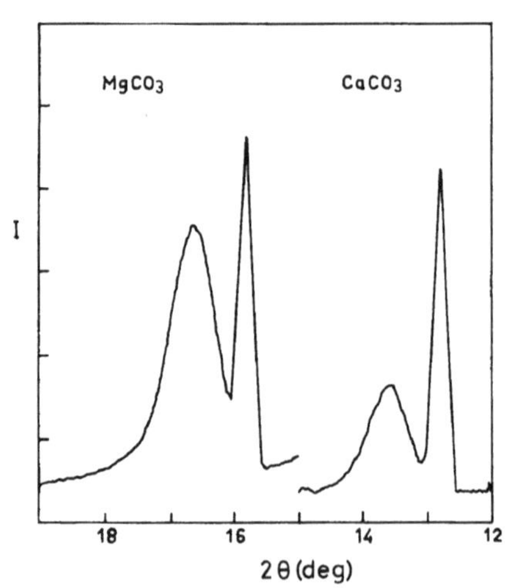

Figure 2.2. Spectrum of MoK_{α} radiation scattered by (a) $CaCO_3$ and (b) $MgCO_3$. The broad peaks are incoherent and the sharp are coherent.

2.1.2 *Absorption of X-rays*

Absorption is essentially a process that differs from scattering because in this case all the energy of the photon is transferred to an electron, and the photon disappears. Of course, this effect is possible only if the photon has sufficient energy to remove the electron from the atom. An atom that has lost an electron is in an excited state and returns to its stable state by emitting energy, which has a characteristic spectrum. This secondary X-ray radiation is called *fluorescence radiation,* in analogy with optical fluorescence. The absorption coefficient increases with increasing wavelength, approximately of the order of λ^3, but it drops suddenly if the photon energy falls below the energy of the electrons in the K, L, and M shells. The dependence of the absorption coefficient on the wavelength may be seen in Fig. 2.3, in which the $\mu(\lambda)$ curve is shown for iron. Fluorescence radiation yields extensive information on the chemical composition of materials and is thus widely used in analytical X-ray spectroscopy. In diffraction experiments, this radiation is often the major contributor to the background and must be eliminated.

We have briefly discussed three primary processes of interaction of X-rays with matter: coherent scattering, incoherent scattering, and absorption. There are numerous complicated complementary processes emitting X-rays and electrons. For example, we touched on the emission of fluorescence radiation as a result of primary absorption. In fact, photons created in any of the three processes might interact with atoms of the material and give rise to tertiary radiation. Moreover, all these processes are accompanied by the emission of secondary and tertiary electrons, and so on. However, attenuation of the incident X-ray beam after passing through a layer of material is due to the three primary processes described above. Comparison of the attenuation and scattering

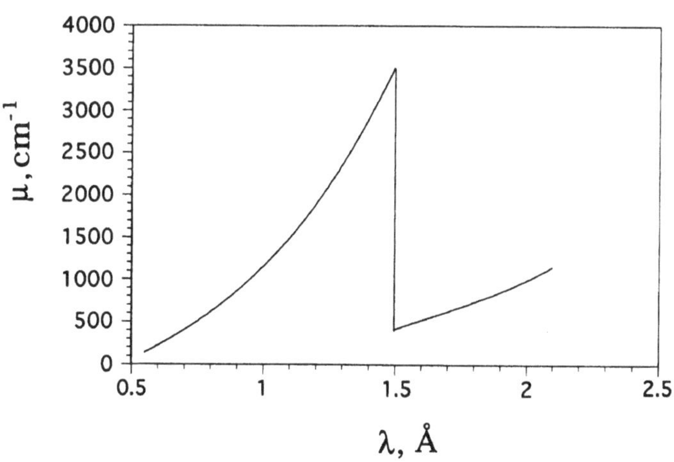

Figure 2.3. Linear absorption coefficient of solid iron.

coefficients leads us to the conclusion that the attenuation of X-rays in the spectral range useful in X-ray diffractometry (wavelength between $MoK_\alpha \sim 0.707$ Å and $CrK_\alpha \sim 2.29$ Å) is due largely to absorption rather than to scattering. This fact is reflected in the professional terminology: instead of the correct term "attenuation coefficient" the term "absorption coefficient," is used. There is nothing we can do about this sloppy practice other than to continue using it.

2.1.3 *Various forms of absorption coefficients*

The linear absorption coefficient μ was defined in equation (2.1). The value of μ depends on the density of the material and will, of course, be different, for example, for liquid water and water vapor. The absorption coefficient of a multiphase sample is expressed through the absorption coefficients of the component phases μ_j.

Suppose $s_j = \sum_i s_{ij}$ is the total path of an X-ray beam through the particles (grains) of phase j (Fig. 2.4); then the attenuation of X-rays after passing through a plate of a multiphase material is:

$$I/I_0 = \exp(-\sum_j \mu_j s_j) = \exp(-\mu t) \tag{2.8}$$

where t is the thickness of the plate. Thus, the absorption coefficient of a multiphase material is expressed as:

$$\mu = \sum_j \mu_j s_j / t = \sum_j \mu_j v_j \tag{2.9}$$

If the grains are small enough and the number of grains of each phase

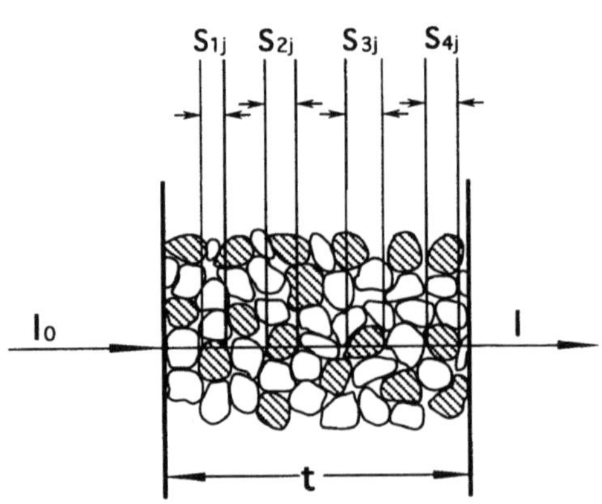

Figure 2.4. Attenuation of an X-ray beam by a multiphase powder specimen.

crossed by the X-ray beam is large enough, then the segment fraction s_j/t is equal to the volume fraction v_j of phase j, and we obtain the final equation (2.9).

In many cases, it is important to characterize absorption by a constant term that does not depend on the physical state of matter but only on its chemical composition. The attenuation law for a plate can thus be rewritten as:

$$I/I_0 = \exp(-\mu^* \rho t) \tag{2.10}$$

or

$$I/I_0 = \exp(-\mu^* G) \tag{2.11}$$

where $G = \rho t$ is the mass of material per 1 cm^2 of its area, and $\mu^* = \mu/\rho$ is the mass absorption coefficient possessing the sought constant characteristic. Indeed, according to its definition, μ^* characterizes the attenuation produced by 1 g of material evenly distributed per 1 cm^2 of its area and is thus independent of the physical state and the porosity of the specimen. Note that ρ is the apparent density of the specimen. For a powder specimen, this density might be much smaller than the theoretical density of a nonporous material.

Expressing the volume fraction of a phase v_j through its weight fraction c_j:

$$v_j = c_j \rho / \rho_j \tag{2.12}$$

where ρ_j is the density of phase j, and substituting this expression for v_j in equation (2.9), we obtain a working expression for the mass absorption coefficient:

$$\mu^* = \sum_j \mu_j^* c_j \tag{2.13}$$

where μ_j^* is the mass absorption coefficient of phase j. The mass absorption coefficient of any phase is easily calculated from its chemical composition:

$$\mu_j^* = \sum_k \mu_k^* x_k \tag{2.14}$$

where μ_k^* is the mass absorption coefficient of a component k and x_k is the weight fraction of component k.

Another, less-common, form of the absorption coefficient is the atomic absorption coefficient μ_a which is defined by the equation:

$$I = I_0 \exp(-\mu_a N) \tag{2.15}$$

where N represents the number of atoms in the absorbing layer per 1 cm^2. The value of N is defined as:

$$N = \frac{G}{M} A \tag{2.16}$$

where A is Avogadro's number, M is the atomic (molecular) weight, and G is the mass per 1 cm^2 of the sample surface.

Comparison of equations (2.8), (2.10), and (2.12) leads to the following ratios:

$$\mu_a = \mu^* M/A = \mu M/(\rho A) \tag{2.17}$$

Atomic and mass absorption coefficients, measured in cm^2/atom and cm^2/g, are tabulated in the International Tables for X-Ray Crystallography, Vol. IV (1974).

2.2 Intensity in powder diffraction

The intensity diffracted by a polycrystalline sample of randomly oriented grains is evenly distributed along the Debye ring. The intensity distribution across the Debye ring is maximal at a scattering angle close to that derived from Bragg's law and descends more or less sharply on both sides of the maximum. The exact shape of cross-sectional distribution depends on numerous sample-, phase-, and instrument-related factors, which are difficult to account for. Theoretical calculations are usually carried out for the intensity integrated over a scattering range including the diffraction peak (integrated intensity). In practical measurements, the chosen angular range is scanned by a detector with a receiving slit of constant width H_c. Thus, integration is automatically carried out over the arc of the Debye ring corresponding to the constant chord H_c. For conventional diffractometers with a diffractometer radius of 150–200 mm and H_c = 10–15 mm, the difference between the arc length and the chord length H_c can be neglected, and the integrated intensity obtained by a powder diffractometer is expressed as:

$$I = \int_{-H_c/2}^{H_c/2} ds \int_{2\theta_1}^{2\theta_2} i(2\theta) \cdot d(2\theta) \tag{2.18}$$

where $i(2\theta)$ is the diffraction peak profile and $2\theta_1$, $2\theta_2$ define the angular range in which $i(2\theta)$ has a detectable value. The first integration is carried out along the Debye ring, and ds is its element. Due to integration over an area in space, the integrated intensity has the meaning of energy flow per second or power of radiation. Note that for the intensity of the primary (incident) beam we retain its usual meaning of energy flow per unit time and unit area given at the beginning of this chapter.

2.2.1 *General expression derived from the kinematic theory of diffraction*

We will start with a small polycrystalline specimen, powdered or coherent, composed of a large number of tiny crystals. The smallness of the specimen has special implications, as does the smallness of the composite crystals. Let us suppose that the absorption path in our specimen is so short that attenuation of the X-ray beam due to absorption can be neglected. The smallness of crystals means that we can also neglect the effects of extinction. If the latter condition is met, then the so-called kinematic theory of X-ray diffraction is applicable. The full development of this theory may be found in a number of excellent books (James 1962; Guinier 1963; Azároff 1963, 1968; Cohen 1966; Warren 1969). According to the theory, we may write the following equation for the integrated intensity of X-rays diffracted per the arc H_c of the Debye ring:

$$I = I_0\left[\frac{MH_c}{8\pi R\sin\theta}\right] QV = I_0\, r_e^2 \lambda^3 \frac{1+\cos^2 2\theta}{\sin^2\theta\cos\theta} \cdot \frac{H_c}{32\pi R} |F|^2\, D\,(1/\vartheta^2)\, MV \tag{2.19}$$

where I is the integrated diffracted intensity per arc H_c of the Debye ring, I_0 is the intensity of the incident beam at the sample, M is the multiplicity factor, R is the distance from the specimen to the detector, V is the specimen volume, ϑ is the volume of a unit cell, λ is the wavelength of X-rays, and $D = \exp[-2B(\sin^2\theta/\lambda^2)]$ is the Debye-Waller (temperature) factor, in which B is defined by mean square deviations of atoms from their equilibrium positions. In equation (2.19) Q is the integrated intensity (reflecting power) of the crystal unit volume, which is equal to:

$$Q = r_e^2 \frac{1+\cos^2 2\theta}{2} \frac{1}{\sin 2\theta} \lambda^3 |F|^2\, D\,(1/\vartheta_c^2)$$

where $r_e^2 = (e^2/mc^2) = 7.9.10^{-26}$ cm^2 is the square of the classic electron radius, θ is the Bragg angle, $\frac{1+\cos^2 2\theta}{2} = P$ is the polarization factor for a nonpolarized incident beam[1], $1/\sin 2\theta = L$ is the Lorentz factor, and

[1]The expression for P will be different if monochromators are used, as will be discussed in Section 2.2.2.

$F_{hkl} = \sum_j f_j \exp[-2\pi i(hx_j + ky_j + lz_j)]$ is a structural factor, in which f_j is the atomic scattering factor for the *jth* atom in the unit cell, x_j, y_j, z_j are the fractional atomic coordinates of the atom j, and h, k, l are the Miller indices of reflection.

Equation (2.19) holds in the limiting case of very thin films or other types of low-mass (μg range) specimens. In the general case, considerable absorption occurs in the specimen, and an absorption factor A should be included in equation (2.19). This factor takes into account the attenuation of the primary and reflected beams passing through the specimen. The exact form of the absorption factor depends on the geometry of the particular instrument and will be discussed in Chapter 3.

In order to compare various diffractometric schemes, it is convenient to rewrite equation (2.19) in the following form:

$$I_i = I_0 E_i AV \tag{2.20}$$

where the term $E_i = QMH_c/8\pi R \sin\theta_i$ is constant for given experimental conditions, A is an absorption correction term, and the subscript i indicates a particular diffraction peak.

The term in brackets in equation (2.19) expresses the fraction of correctly oriented crystallites, i.e., crystallites reflecting X-rays into the segment of the Debye ring of length H_c. The multiplier $(\sin\theta)^{-1}$, expressing the angular dependence of this fraction, may be regarded as the "powder" supplement to the conventional "single-crystal" Lorentz factor $(\sin 2\theta)^{-1}$, bringing the Lorentz factor for a powder to its usual form $(\sin^2\theta\cos\theta)^{-1}$. It was supposed in the derivation of this expression that the extensions of the focal spot and the specimen in the direction normal to the diffraction plane are negligible. In fact, both the focal spot and the specimen have finite dimensions, and thus fewer restrictions are imposed on the orientation of reflecting crystals. Although this fact does not change the value $L_p = (\sin\theta)^{-1}$ for a randomly oriented specimen, it has some importance both for oriented specimens and for the evaluation of intensity fluctuations due to particle sampling. The orientation of reflecting particles in a diffractometer will be discussed in Chapter 5.

If our specimen is a multiphase specimen and if we consider the intensity of a particular diffraction peak i of any phase j, then, of course, only grains of this phase will contribute to I_{ij}. The volume of these grains in our specimen is v_jV (where v_j is the volume fraction), and this product should be used in equation (2.19) instead of V.

Taking these modifications into account, we can rewrite equation (2.19) as:

$$I_{ij} = I_0 r_e^2 \lambda^3 \frac{1+\cos^2 2\theta_i}{\sin^2\theta_i \cos\theta_i} \cdot \frac{H_c}{32\pi R} \cdot |F_{ij}|^2 D_{ij} M_{ij} A_i \vartheta_j^{-2} v_j V \tag{2.21}$$

where the subscripts i and j refer to the diffraction peak i and the phase j, respectively.

The volume fraction v_j (or, more often, the weight fraction c_j) is of prime interest in quantitative X-ray diffractometry (QXRD), and the quantitative phase analysis is based on the dependence of I_{ij} on v_j, as expressed in equation (2.21).

Unfortunately, many other factors influence the intensity of diffraction. Some of them are fundamental well-known constants, and others (I_0, R, λ) are chosen as experimental conditions, which can be kept constant. However, yet another group of factors affecting intensity, such as real crystal structure and X-ray absorption, have to be controlled. Thus, the main task that concerns us in QXRD is to isolate the useful intensity-composition interdependence on the background of various interfering factors.

2.2.2 Polarization factor in X-ray diffractometry

The polarization factor is an outcome of the weakening of the wave component with an electrical vector parallel to the diffraction plane (π-component) as opposed to the wave component with an electrical vector normal to the diffraction plane (σ-component). If the incident beam is not polarized, then the polarization factor for a powder sample is:

$$P = (1 + \cos^2 2\theta)/2 \tag{2.22}$$

Another polarizing element—the monochromator crystal—is present in diffractometers having incident- or reflected-beam monochromators. The general form of the polarization factor is (Azároff 1955):

$$P = (1 + K\cos^2 2\theta)/(1 + K) \tag{2.23}$$

where $K<1$ is the polarization ratio. For ordinary diffractometric geometry, when the sample and monochromator diffraction planes are parallel and the monochromator crystal is considered as an ideal mosaic diffractor, the polarization ratio $K = \cos^2 2\theta_m$, where $2\theta_m$ is the monochromator scattering angle (Whittaker 1953; Azároff 1955).

Under the same conditions, a dynamic theory approach to monochromatic scattering gives us $K = \cos 2\theta_m$. In practical diffractometry, the polarization ratio also depends on beam divergence and can vary between $\cos^2 2\theta_m < K < 1$. Although the scattering angle $2\theta_m$ is usually low and $\cos^2 2\theta_m$ is close to unity, uncertainty in the polarizing ratio K can affect the calculated diffraction intensities beyond modern standards of accuracy. Jennings (1984) surveyed values of K obtained by various experimental techniques. For two important diffractometric settings with quartz and graphite monochromators ($\cos 2\theta_m = 0.894$ for both) and

CuK_α radiation, the mean values of the polarization factor were 0.912 for a graphite monochromator and 0.905 for a quartz monochromator; i.e., the polarization ratio was very close to cos $2\theta_m$. Altree-Williams and Jordan (1980) have proposed a simple technique for the experimental evaluation of K. It is based on measurements of intensity ratios for two reflections carried out with and without a monochromator. Their value of K = 0.93 is close to the average reported by Jennings (1984).

2.2.3 *Extinction of X-rays*

It was shown in Section 2.1.1.1 that only a small portion of the incident intensity is generally removed by the scattering of X-rays. However, there is one particular case, described below, in which that portion may increase to an appreciable amount. Suppose that our specimen is an ideal single-crystal plate reflecting X-rays coherently (Fig. 2.5). If the angle between the wave vector of the incident wave K_0 and the reflecting plane is not equal to the Bragg angle θ, then specimen absorption is calculated

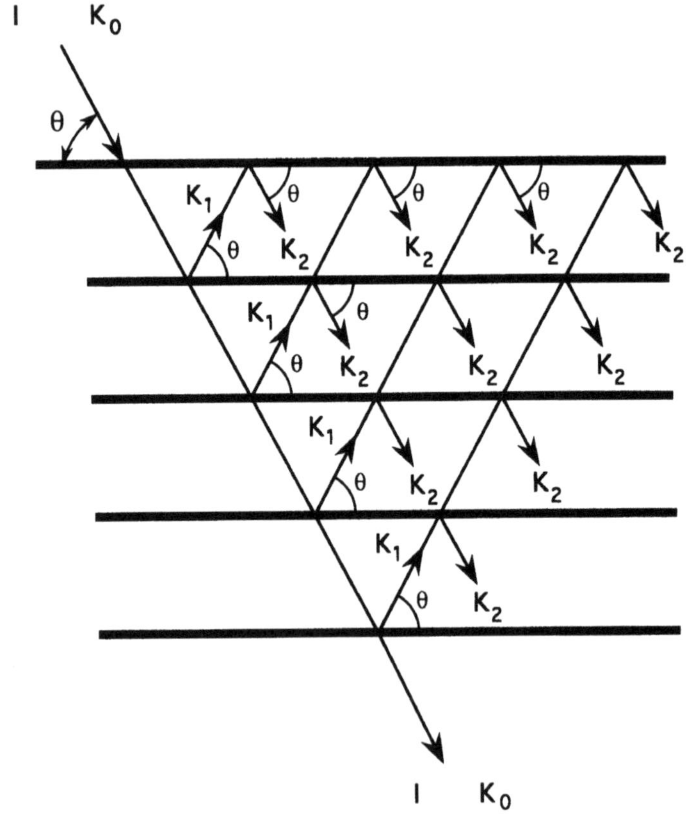

Figure 2.5. Schematic representation of primary extinction of an X-ray beam.

according to the rules discussed in the previous sections. As soon as the incident angle becomes equal to Bragg angle θ, reflection takes place. The wave vector of the reflected wave K_1 then makes the same angle θ with the reflecting plane, and it is therefore reflected once more. Hence, a double-reflected wave with wave vector K_2 is propagated in the ideal crystal in the direction of the incident wave. Each reflection is accompanied by a phase shift of $\pi/2$, and the double-reflected wave and the incident wave have opposite phases. Interference between them decreases the intensity of the incident wave reaching the reflecting planes located inside the crystal and hence decreases the diffracted intensity. The scattering effect is evident for strong reflections, and it increases as the size of the crystal increases. In crystal structure analysis, this phenomenon is termed *primary extinction*. The dynamic diffraction theory takes into account the interaction between the incident and reflected waves by means of a correction term applied to the intensity calculated by the kinematic theory, which ignores the extinction effect:

$$I_{obs} = I_{kin} \cdot E \qquad (2.24)$$

where I_{obs} is the observed diffracted intensity for a real crystal, I_{kin} the intensity calculated by the kinematic approach and E a correcting term. In order to estimate the effect of primary extinction, we will consider the correcting term to be in the form proposed by Darwin (1914):

$$E = \frac{\tanh (nq)}{nq} \qquad (2.25)$$

where n is the number of reflecting planes and q is a dimensionless constant equal to:

$$q = 2d^2 \frac{e^2}{mc^2} \frac{F}{V_c}$$

where d is the interplanar spacing and the other terms are as explained in equation (2.19). The values of the correcting term are given in Table 2.1.

The correcting term is negligible for $nq < 0.2$. For the 200 reflection of rock salt $q = 2.10^{-4}$ (Lonsdale 1949) and for n = 1000 (coherent domain size 1000 x 5.6 = 5600 Å), the correction is negligible, although it starts to be substantial for a coherent domain of 5 μm.

In the majority of QXRD cases the grain size is greater than 5 μm. If each grain is a perfect (one-domain) single crystal, we cannot neglect extinction effects. Indeed, substantial primary extinction has been detected in powders with a grain size in the 10–20 μm range (Cline and Snyder 1983, 1987; Sabine 1988). Similarly, Hubbard et al. (1976) consid

Table 2.1. Correcting term for primary extinction

nq	tanh (nq)/nq
0.2	0.97
0.4	0.95
0.6	0.90
0.8	0.83
1.0	0.76
2.0	0.48
4.0	0.25

ered primary extinction to be the cause of significant discrepancies between observed and calculated reference intensity ratios (see Section 4.7) for well-crystallized powders. A correction factor for primary extinction will be discussed in Chapter 5.

Although primary extinction does not exist in an ideally imperfect crystal consisting of small coherent domains, another extinction phenomenon is manifest in these crystals as a result of the screening of the incident beam by upper coherent domains. In the reflecting position the upper domain removes a considerable part of the primary beam's intensity, in addition to the usual beam attenuation. If in the path of the attenuated primary beam there is another domain with reflecting planes strictly parallel to the reflecting planes of the first domain, then the reflection from the second domain will be reduced. It is assumed that in a powder sample secondary extinction may be completely neglected (Zachariasen 1945).

2.3 Background—angle variation and intensity

2.3.1 Introduction

Discrete Bragg reflections are superimposed on the continuous (angular) distribution of scattered or emitted X-rays. The measured quantity is the total intensity or the sum of the Bragg reflection and the background. The background must be measured in the vicinity of the diffraction peak and subtracted from the total intensity. This procedure can be inaccurate if the total intensity and the background intensity are close, i.e., for weak reflections and a high background level.

Details of the counting strategy used to decrease the background-related experimental error will be considered in Chapter 5. Here, we will discuss the sources of background intensity and feasible ways of reducing it.

In the following discussion, we will assume that radiation entering the detector is scattered or emitted only by the specimen material and not by parts of the specimen holder or other parts of the instrument. The first condition necessary for our discussion is that the primary beam is intercepted by the specimen. The length of the irradiated surface $l_p \approx R\alpha/\sin\theta$ (Section 3.2, Fig. 3.3), where R is the goniometer radius and α is the angular divergence of the primary beam. The second condition is proper adjustment of the collimator defined by two slits, the receiving slit RS and the antiscatter slit AS. The convergent angle of the diffracted beam ω must be equal to the divergent angle α, and the same area of the specimen that is irradiated by the primary beam must be "seen" through the collimator. Usually, the diffractometer manufacturer provides instructions as to how to achieve this objective.

These conditions are, however, not sufficient to prevent scattering by foreign materials in the special case of diffraction by thin films or other low-mass samples. As a result of beam penetration, scattering by the specimen material is superimposed on scattering by the substrate material. The substrate might be either a polycrystalline material, such as a metal in the case of hard coatings, or a single-crystal material, such as a Si wafer, in the case of various functional deposits. These practical cases will be discussed in Chapter 5.

The radiation emitted by common X-ray tubes is polychromatic. In addition to K_α lines, which are of primary interest for powder diffraction, there is a K_β line and a continuous white spectrum. The entire primary spectrum can contribute to the background; we will discuss the contribution of each spectral component.

2.3.2 Scattering of continuous radiation

Continuous radiation is scattered coherently by the crystalline phases of the specimen, and due to the continuous nature of this radiation the diffracted radiation is also distributed over a wide angular range, contributing to the background. For a tube voltage of 45 kV, the peak of the continuous radiation has a photon energy of ~30 keV, namely, a wavelength of ~0.4 Å, which is much less than that of the commonly used characteristic radiation. The diffraction of continuous radiation is concentrated in the low-angle range and might be the main reason for the increase in the background towards $2\theta = 0$.

This effect is somewhat reduced in contemporary diffractometers in which partial monochromatization is achieved by pulse height discrimination in scintillation and proportional counters. For example, for CuK_α radiation (photon energy ~ 8.9 keV) and a scintillation counter with an energy resolution of ~50%, scattering with a wavelength of ~0.4 Å (photon energy ~ 30 keV) will be almost completely eliminated. The fraction of white radiation adjacent to the CuK_α line on the high-energy (lower λ) side is removed by a β filter. Increased low-angle background is more visible for the photographic technique (blackening around the exit hole) than for the diffractometric technique. Of course, scattering by continuous radiation is completely eliminated by monochromatization of either incident or diffracted beams.

2.3.3 *Scattering of characteristic radiation*

2.3.3.1 Incoherent scattering

Incoherent (Compton) scattering is the most obvious source of background, which is significant for organic compounds. We have used data from the International Tables for X-Ray Crystallography to present the ratio of incoherent to coherent scattering I_{in}/I_{coh} for two values of $\sin\theta/\lambda = 0.1$ Å^{-1} ($2\theta = 17.7°$, CuK_α) and $\sin\theta/\lambda = 0.5$ Å^{-1} ($2\theta = 100.8°$, CuK_α) (Table 2.2). The intensity of coherent scattering was taken as the square of the corresponding atomic factor f^2. For light elements, incoherent scattering is overwhelming at high scattering angles and remains appreciable at low angles. For crystalline compounds of these elements, particularly for organic compounds, we can expect that the background will consist entirely of incoherent scattering, because the fraction of fluorescent radiation in the background of these compounds is nil. Nevertheless, for well-crystallized organic compounds, the background in the particularly important low-angle range is rather low and does not constitute a problem in quantitative analysis. The problem is acute, however, in the analysis of poorly crystallized material, for example, in the application of the X-ray technique for the determination of polymer crystallinity (see Section 4.12). In this case, incoherent scattering

Table 2.2. Ratio of incoherent scattering to coherent scattering for a number of elements

	I_{in}/I_{coh}	
	$\sin\theta/\lambda = 0.1$	$\sin\theta/\lambda = 0.5$
3 Li	0.210	1.39
5 B	0.083	1.65
6 C	0.040	1.32
8 O	0.018	0.94
13 Al	0.019	0.25
20 Ca	0.010	0.167
26 Fe	0.004	0.84
32 Ge	0.002	0.051

must be removed physically or computationally from the nondiscrete total scattering of the amorphous component. The wavelength difference [equation (2.7)] may hardly be exploited for this objective, since it is small at low Bragg angles. Relative values for $2\theta = 20°$ are $\Delta\lambda/\lambda \approx 1.10^{-3}$ for CuK_α and ~ 2.10^{-3} for MoK_α. The diffracted beam graphite monochromator, having a wide transmission band, is certainly not good for this purpose. Monochromators with perfect crystals and narrow rocking curves may eliminate incoherent scattering but at the expense of a significant reduction of total scattered intensity.

2.3.3.2 Coherent scattering by imperfect crystals

Any crystal imperfection that violates the identity of the ideal lattice points decreases the intensity of Bragg reflections. The energy balance is distributed more or less evenly between Bragg reflections and thus contributes to the background.

Imperfections that are important here include thermal vibration of atoms, static atom displacements in solid solutions, disordered substitutions in solid solutions, imperfections in stacking, and orientation of layers in layered structures. A significant increase of background may be expected in the presence of amorphous or highly disordered crystalline matter. A typical example is shown in Fig. 2.6, representing the diffraction pattern of fly ash from a coal power station. Diffraction peaks of

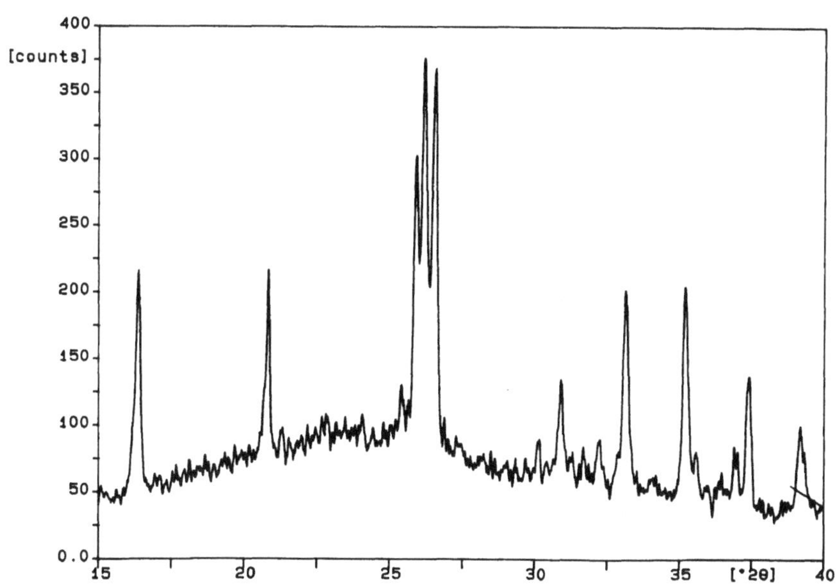

Figure 2.6. Diffraction pattern of coal ash.

crystalline phases are superimposed on the high hump-like background due to diffuse scattering of the glass phase, which is abundant in coal ash. It is clear that the background due to coherent scattering of characteristic radiation cannot be eliminated by any monochromatization techniques.

2.3.3.3 Fluorescent radiation

This source of background is insignificant for elements below Ti due to considerable absorption of their fluorescence in the sample itself and in the air path between the specimen and the detector. For example, CuK_α radiation is reduced by 11% in passing through 10 cm of air at atmospheric pressure. However, for heavy elements, fluorescent radiation commonly constitutes the major component of the background. Fluorescence is excited by characteristic radiation of the tube and by continuous radiation. The first case is especially prominent if the energy of the incident photon is only slightly greater than the energy of the K_α or L_α absorption edges of the major elements in the specimen. This always happens if the atomic number of the anode material is greater than the atomic number of any abundant element in the specimen and the difference ΔZ lies between 2 and 4. A typical example is Fe-fluorescence in CuK_α radiation. The yield of fluorescence is so high that a few percent of Fe may increase the background several times.

Fluorescent radiation is removed from the spectrum by a diffracted beam monochromator. The curved graphite monochromator is excellent for this purpose. Most X-ray detectors, such as a proportional counter, are not usually adequate for discrimination of fluorescent radiation without significant intensity losses.

There are two unwanted side effects connected with selective absorption of the incident characteristic radiation near the absorption edge:

1. Total increase of the absorption coefficient and corresponding reduction of the diffracted intensity.
2. Selective increase of absorption by the phase abundant in fluorescent elements and corresponding emphasis of the microabsorption phenomenon.

These negative effects are not eliminated by diffracted beam monochromatization. The only way to reduce them is careful choice of the type of X-ray tube, even if a diffracted beam monochromator is employed. The same is true, of course, if a diffracted beam monochromator cannot be used, for example, if a position-sensitive detector is used for data acquisition. The simple rule is to avoid such a situation when the atomic number of the anode element is greater than the abundant elements in the specimen by 2 and 4 digits.

Continuous radiation might cause fluorescence of elements with atomic numbers greater than the atomic number of anode material.

Fluorescence due to continuous radiation can be eliminated by monochromatization of either the incident or the reflected beam.

2.4 X-ray diffraction by nonhomogeneous polycrystalline materials

2.4.1 *Introduction*

The basic equation for quantitative X-ray phase analysis derived by Alexander and Klug (1948) is expressed as:

$$I_{ij} = K_{ij}\, c_j / (\rho_j\, \mu^*) \tag{2.26}$$

where I_{ij} is the intensity of the *ith* diffraction peak of the *jth* phase, c_j is the weight fraction of phase j, ρ_j is the density of phase j, and μ^* is the mass absorption coefficient of the specimen. In developing this equation (see Section 3.2.2), the specimen was considered to be completely homogeneous from the standpoint of X-ray absorption, on the assumption, for example, that there was no difference between bulk and powder specimens. In other words, it is assumed that the mean absorption coefficient of the specimen is applicable to its tiny volume element. In fact, this is not true because in a multiphase specimen the absorption coefficient of a reflecting particle (grain) usually differs from the mean absorption coefficient of the specimen. Thus, some sort of microscopic approach to specimen absorption must be developed.

Let us consider a spherical volume V in our specimen. The attenuation of X-rays passing through this volume may be approximated by $\sigma \approx \exp(-\mu_V D)$, where μ_V is the linear absorption coefficient of the sample material confined in the volume V, and D is the diameter of the sphere. If D is large, we can expect that:

$$\mu \approx \mu_V \tag{2.27}$$

where μ is the mean absorption coefficient of the specimen. By decreasing D, we decrease the attenuation. Ultimately, we will come to a volume so small that $\exp(-\mu_V D)$ will be negligible, say, less than 1%. If condition (2.27) is retained for this small volume, we can regard the specimen as homogeneous. According to this definition, a coarse-grained powder with appreciable absorption in a single grain is not homogeneous, even if it exists in a single phase. Indeed, in this case the absorption coefficient of the particle μ_V is certainly different from the mean absorption coefficient of the specimen, the latter being equal to $v_j \mu_j$, where v_j is the volume fraction of the solid grains in the powder specimen. Fineness of the particles is a necessary condition for homogeneity.

Homogeneity of the powder can be characterized by the product $\mu_j D_j$, where D_j is the particle diameter. The condition for homogeneity in the above-described context can then be defined approximately as:

$$\mu_j D_j \leq 0.01 \tag{2.28}$$

for all phases.

A bulk sample is somewhat different in this respect, because it lacks the natural nonhomogeneity of the powder caused by porosity. Of course, condition (2.28) is sufficient to treat a multiphase bulk specimen as homogeneous. However, a multiphase coarse-grained bulk specimen may be considered homogeneous even if it deviates from equation (2.28) but the absorption coefficients of various phases μ_j are close.

In fact, the condition for homogeneity is frequently violated, and appropriate corrections for intensity formulation must be developed to make allowance for the heterogeneity of the powder or multiphase solid specimen.

2.4.2 *Two extreme cases: homogeneous and nonhomogeneous specimens*

In order to obtain good insight into the problem and the possible consequences for QXRD, let us consider a hypothetical layered specimen consisting of two phases, 1 and 2, as shown schematically in Fig. 2.7. The layer thickness d is equal to the mean grain diameter. Let us now consider two extreme cases of grain size.

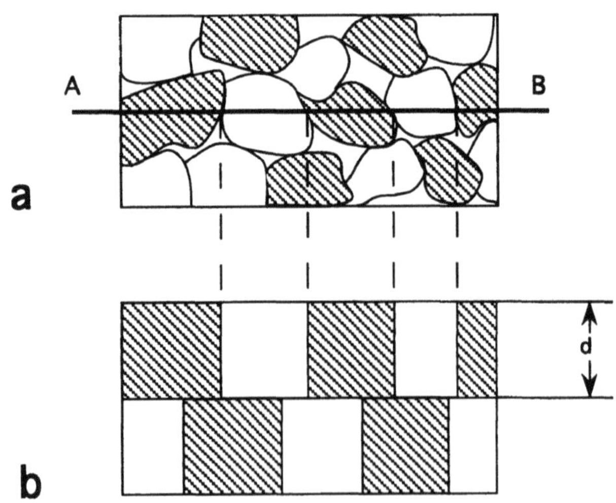

Figure 2.7. Model of a two-phase specimen. (a) specimen surface and (b) specimen cross section through AB.

The first:

$$\mu_1^* \rho_1 D \leq 0.01; \; \mu_2^* \rho_2 D \leq 0.01 \qquad (2.29)$$

where μ_1^* and μ_2^* are the mass absorption coefficients, and ρ_1 and ρ_2 are the densities of phases a and b.

The second:

$$\mu_1^* \rho_1 D >> 1; \quad \mu_2^* \rho_2 D >> 1 \qquad (2.30)$$

The first case is the general case of a homogeneous specimen. According to the Klug-Alexander theory, the diffraction intensities for the two phases are expressed as:

$$I_1' = \frac{K_1 c_1}{\rho_1 \mu^*}; \qquad I_2' = \frac{K_2 c_2}{\rho_2 \mu^*} \qquad (2.31)$$

where K_1 and K_2 are calibration constants, c_1 and c_2 are the weight fractions of the two phases, and $\mu^* = \mu_1^* c_1 + \mu_2^* c_2$ is the mass absorption coefficient of the sample.

In the second case, X-rays will be completely absorbed in the first layer of grains. We also neglect the grain boundary effects when X-rays penetrate a grain of one phase (say "1") through a grain of the second phase. Under these conditions, the presence of phase 1 does not effect the absorption of X-rays diffracted by phase 2 and vice versa. The intensity of diffraction by any phase is proportional to the volume fraction of that phase and in our special case to the fraction of sample surface covered by that particular phase. We thus obtain the following expression for the diffracted intensities:

$$I_1'' = \frac{K_1 c_1}{\rho_1 \mu_1^*}; \qquad I_2'' = \frac{K_2 c_2}{\rho_2 \mu_2^*} \qquad (2.32)$$

It is clear that the intensities are quite different from those for a homogeneous sample. Indeed, the intensity ratio is expressed as:

$$I_1''/I_1' = \mu^*/\mu_1^* = (\mu_1^* c_1 + \mu_2^* c_2) / \mu_1^* \qquad (2.33)$$

In the coarse-grained (nonhomogeneous) sample, the intensity will be enhanced for light, low-absorbing phases ($\mu_1 < \mu$) and diminished for high-absorbing phases ($\mu_1 > \mu$).

The intensity ratios for the two phases will also be different for homogeneous and nonhomogeneous specimens:

$$I_1'/I_2' = (K_1 c_1 \rho_2)/(K_2 c_2 \rho_1); \qquad I_1''/I_2'' = (I_1'/I_2')(\mu_2^*/\mu_1^*) \tag{2.34}$$

One can see that deviation from the case of a homogeneous sample could be extremely significant. Real samples lie between these two extreme cases: homogeneous samples ($\mu D \leq 0.01$) and completely nonhomogeneous samples ($\mu D >> 1$). For such real samples, the effect of nonhomogeneity should be estimated.

2.4.3 Particle absorption factor

The most important contributions to the understanding of the phenomenon of particle absorption were made by Brindley (1945) and de Wolff (1947). Their results were later improved in a series of subsequent works (Wilchinsky 1951; Harrison and Paskin 1964,; Shimazu 1967, 1968; Suortti 1972; Hermann and Ermrich 1987, 1989; Ermrich and Hermann 1990). We will, however, follow the idea of the first two authors who showed clear physical insight into the problem. Unfortunately, as the reader may already sense, all corrections to the homogeneous case formulation require a precise knowledge of the grain size, which is usually not available in QXRD studies. Thus, at least an estimate must be made, and for this purpose the theories of Brindley and de Wolff are satisfactory.

As can be seen from the above-discussed example, absorption by a nonhomogeneous sample differs from that by a homogeneous one. This difference arises from two phenomena:

1. The existence somewhere along the absorbing path of a reflecting grain with absorption differing from the mean absorption by the sample.
2. Statistical fluctuations of the number of absorbing grains in the absorbing path.

Suppose that $l_j = \sum_k l_{kj}$ is the total path traversed by incident and reflected X-ray beams in the grains of phase j (l_{kj} is the path traversed in the grain k of phase j, Fig. 2.8). Then, the intensity reflected by a small grain of phase j, located at depth x, is given by:

$$I_{ij} = I_0 E_{ij} [\exp(-\sum_j \mu_j l_j)] \, \delta V_j \tag{2.35}$$

where I_0 and E_{ij} have the same meaning as in equation (2.20), μ_j is the absorption coefficient of phase j, and δV_j is the volume of the reflecting grain. The factor

$$A = \exp(-\sum_j \mu_j l_j) \tag{2.36}$$

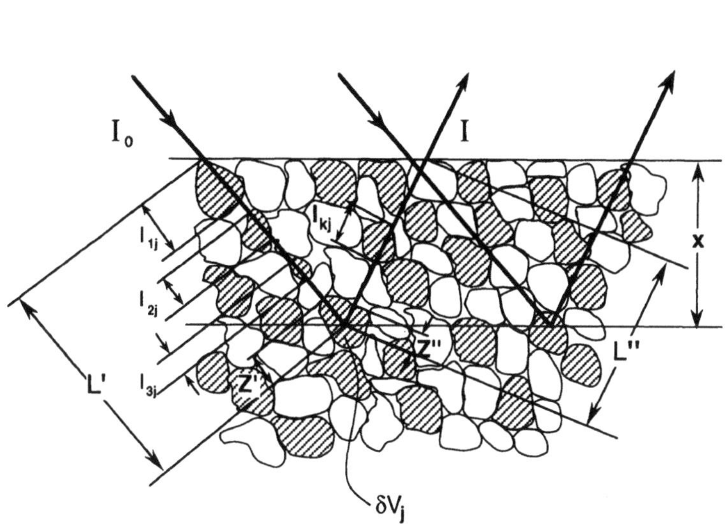

Figure 2.8. Definition of the particle absorption factor. Z = Z' + Z" is the total absorption path in a reflecting crystallite of phase j. L = L' + L" is the total path of the incident and reflected beams in the specimen.

expresses the intensity reduction due to sample absorption. If we assume that the mean grain size of all component phases is much less than L = L'+L", the total X-ray path in the specimen (Fig. 2.8), then:

$$l_j/L = v_j$$

where v_j is the volume fraction of phase j and

$$A = \exp(-\mu L) \tag{2.37}$$

where $\mu = \Sigma\mu_j v_j$ is the linear absorption coefficient of the specimen. The validity of equation (2.37) is assumed tacitly in the development of the basic QXRD equation, which we come to in the next chapter. However, this condition is violated in coarse-grained materials.

Let us consider the significance of the fact that a reflecting particle of a particular phase (say phase j) is located somewhere within the absorbing path L. We assume that along the path L-Z of the X-rays the absorption coefficient is close to the absorption coefficient of the specimen μ and that along the path Z = Z'+Z" the particle absorption coefficient is μ_j (Fig. 2.8). Strictly speaking, this assumption is not correct. By removing a reflecting particle from the absorption path L, we violate the statistical distribution of particles along L. Thus, the absorption coefficient operating along L-Z may differ from μ. However, considering this assumption as a first approximation, we have to reflect it in the calculation of absorption in these cases when μ_j and μ differ significantly and/or the path Z constitutes a significant fraction of L (i.e., a coarse-grained specimen). Thus,

$$A = \exp[-\mu(L - Z)] \exp(-\mu_j Z) = \exp[-Z(\mu_j - \mu)] \exp(-\mu L) \quad (2.38)$$

In calculations of the intensity of the diffraction peak, we have to follow exactly the argument of Section 3.2.2 and express the intensity diffracted by the thin layer Δx at a depth x, as follows:

$$\Delta I_{ij} = I_0 E_i \langle \exp[-Z(\mu_j - \mu)] \rangle \exp(-\mu L) v_j S\Delta x$$

where I_0 and E_i were defined in equation (2.20), S is the irradiated specimen surface, $S\Delta x = v$ is the irradiated volume of our thin layer, v_j is the volume fraction of the phase j, and the total diffraction path $L = 2x/\sin\theta$. The first exponential should be averaged over all possible orientations of the particle in the incident beam. The mean value of this exponential may be expressed as:

$$\tau_j = \frac{1}{v_j}\int_0^{v_j} \exp[-Z(\mu_j - \mu)]dV \quad (2.39)$$

and integration is carried out over the whole particle volume v_j. Integrating the expression for ΔI and assuming the sample to be infinitely thick, we obtain:

$$I_{ij} = \tau_j \frac{K_{ij} v_j}{\mu} \quad (2.40)$$

Brindley (1945) suggested a slightly different form for factor τ_j, which he called the *particle absorption factor*. Suppose that L is the geometric path of X-rays in a powder specimen; then the X-ray path in the solid matter will be $L\rho/\bar{\rho}$, where ρ is the apparent density of the powder specimen and $\bar{\rho}$ is the density of the solid matter. Brindley supposed that along the path $(\rho/\bar{\rho})L$-Z the mean absorption coefficient of solid matter μ operates, while along the path Z in the particle the absorption coefficient μ_j is operative. Thus, the absorption factor may be expressed as:

$$A = \exp\{-\bar{\mu}[(\rho/\bar{\rho}) L - Z] - \mu_j Z\} \quad (2.41)$$

and provided that $\mu/\bar{\mu} = \rho/\bar{\rho}$, the reflected intensity may be written as:

$$\Delta I = I_0 E_i \exp(-\mu L) \langle \exp[-Z(\mu_j - \mu)] \rangle = I_0 E_i \tau_j \exp(-\mu L) \quad (2.42)$$

where $\tau_j = \frac{1}{v_j}\int \exp[-(\mu_j - \bar{\mu})\,dz]$ is another form of the particle absorption factor. This form of the particle absorption factor is approximately the same as the form expressed in equation (2.39). The suggestion that the mean absorption coefficient $\bar{\mu}$ is operative along the path $(\rho/\bar{\rho})L - Z$ is inaccurate, because a particle with a substantial size Z and an absorption coefficient μ_j, different from $\bar{\mu}$, was removed from the path L. Thus, if the absorption coefficient $\bar{\mu}$ is statistically operative along the path $L\rho/\bar{\rho}$, it is not operative along the path $L\rho/\bar{\rho} - Z$. Taking the approximate character of expressions (2.39) and (2.42) into account, we will retain expression (2.39) in further discussions.

Sometimes the effect of absorption by an individual particle expressed by the factor τ_j is called *microabsorption* in contrast to specimen absorption, or macroabsorption. The extent of microabsorption depends on the difference in absorption coefficients μ_j for the phase diffracting X-rays and μ for the specimen. For a low-absorbing phase $\mu_j < \mu$, microabsorption will enhance the intensity (in full agreement with the extreme case of this effect discussed at the beginning of this section).

2.4.4 *Absorption by a nonhomogeneous system*

Although the particle absorption factor accounts for most of the microabsorption effects, certain approximations were made in its derivation. The theory of Brindley (1945) required some corrections, which were introduced by de Wolff (1947). Let us consider again the reflection of X-rays by particles of phase j located at a depth x inside the specimen. In order to account for the reduction of intensity, we have to average the factor A given by equation (2.36) over all reflecting particles of phase j at depth x (Fig. 2.8):

$$\langle A \rangle = \langle \exp(-\sum_j \mu_j l_j) \rangle \tag{2.43}$$

where the angular brackets $\langle\ \rangle$ imply averaging over the reflecting particles at a depth x.

It is usually assumed that:

$$\langle \exp(-\sum_j \mu_j l_j) \rangle \approx \exp(-\sum_j \mu_j \langle l_j \rangle) \tag{2.44}$$

and consequently that $v_j = \langle l_j \rangle / L$, leading to the generally accepted equation (2.37). However, equation (2.44) does not hold for the general case and may be regarded only as a first approximation of the real situation.

By adding and subtracting the same term $\sum_j \mu_j \langle l_j \rangle$ to the power of the exponent in equation (2.43), we come to an expression for the absorption term:

$$A = \exp(-\sum_j \mu_j \langle l_j \rangle) \times \langle \exp[-\sum_j \mu_j (l_j - \langle l_j \rangle)] \rangle \tag{2.45}$$

The second exponent can be expanded in a power series:

$$\langle \exp[-\sum_j \mu_j (l_j - \langle l_j \rangle)] \rangle = 1 - \langle \sum_j \mu_j (l_j - \langle l_j \rangle) \rangle + 1/2 \langle [\sum_j \mu_j (l_j - \langle l_j \rangle)]^2 \rangle$$

The middle term vanishes because the distribution of $\sum_j \mu_j l_j$ over all the reflecting particles in the sample layer is assumed to be symmetric around the mean value $\langle \sum \mu_j l_j \rangle$. Thus,

$$A = [\exp(-\sum \mu_j \langle l_j \rangle)] \; [1 + 1/2 \langle [\sum \mu_j (l_j - \langle l_j \rangle)]^2 \rangle] \tag{2.46}$$

Assuming that the second term in brackets is much smaller than l, we obtain the expression:

$$A = [\exp(-\sum_j \mu_j \langle l_j \rangle)] \; [\exp 1/2 \langle [\sum_j \mu_j (l_j - \langle l_j \rangle)]^2 \rangle] \tag{2.47}$$

The second exponent, in fact, represents the correction to the generally accepted absorption term or, in other words, it expresses the contribution of specimen nonhomogeneity to X-ray absorption. This term is essentially positive and tends to increase the diffracted intensity or, in other words, to decrease the absorption coefficient to any apparent or effective value. de Wolff (1947) further developed the correcting term in equation (2.47), leading to the following form of the absorption factor:

$$A = \exp(-\sum_j \mu_j \langle l_j \rangle + \Delta\mu)L = \exp(-\mu + \Delta\mu) \tag{2.48}$$

where $\Delta\mu = 1/2 \sum_j v_j \langle d_j \rangle (\mu_j - \mu)^2$ is considered as the correction to the specimen absorption coefficient; $\mu = \sum_j \mu_j \langle l_j \rangle / L = \sum_j v_j \mu_j$ is the absorption coefficient of the specimen; and $\langle d_j \rangle$ is the mean grain size for phase j.

Of course, the correction $\Delta\mu$ from equation (2.48) is fully applicable to absorption measurements, i.e., the measured absorption coefficient for a nonhomogeneous specimen will be less than $\mu = \Sigma\mu_j v_j$ by the quantity $\Delta\mu$ defined in equation (2.47).

The next correction to the theory of Brindley is related to the definition of $<l_j>$, the mean beam path through the grains of phase j. Suppose that the reflecting phase is designated phase p. Then, $<l_j>$ is defined as:

$$<l_j> = \int f_{pj}(s)\, ds \qquad (2.49)$$

where f_{pj} is the conditional probability of finding phase j at a distance s from the reflecting particle of phase p located at a distance L from the beam entrance.

For a monophasic porous specimen, Hermann and Ermrich (1987) derived the following expression for f_{pp}:

$$f_{pp} = \alpha^2 \exp(-\ln\alpha) \exp(-s/\beta)] \qquad (2.50)$$

where α is the volume fraction of the solid material and β is related to pore size. de Wolff evaluated $<l_j>$ from equation (2.48) for a multiphase specimen and finally reached the following form for the absorption factor for phase j:

$$A_j = \exp[-(\mu+\Delta\mu)L] \times \{<\exp[-d_j(\mu_j - \mu)]> \exp[-\sum_j v_j <d_j> (\mu_j - \mu)]\} \qquad (2.51)$$

The intensity reflected by phase j in an infinitely thick specimen (see Section 3.2.2) is expressed as:

$$I_{ij} = \frac{K_{ij} v_j}{2(\mu - \Delta\mu)} \{<\exp[-d_j(\mu_j - \mu)]>\exp[-\sum v_j <d_j>(\mu_j - \mu)]\} \qquad (2.52)$$

where $\Delta\mu$ is defined in equation (2.48) and the two exponents in brackets are correcting terms independent of the total absorption path L. The first exponent, which is related only to the reflecting phase, is the particle absorption factor τ_j defined earlier. The second exponent corrects approximations made in deriving the particle absorption factor.

In a multiphase powder sample, the power of the second exponent in equation (2.52) will include negative ($\mu_j > \mu$) and positive ($\mu_j < \mu$) terms, which at least partially compensate for one another. Two major microabsorption corrections remain: the particle absorption factor τ_j and the correction for the absorption coefficient $\Delta\mu$. These two factors are estimated in the next section.

For the monophase porous specimen, $v_j = \alpha$, the volume fraction of solid matter. Approximating $\tau_j \approx \exp[-<d_j> (\mu_j - \mu)]$, we have from equation (2.52):

$$I_{ij} = \frac{K_{ij} v_j}{2(\mu - \Delta\mu)} \exp[-\langle d_j \rangle \mu_j (1 - \alpha)^2] \tag{2.53}$$

or expanding the exponent:

$$I_{ij} = \frac{K_{ij}\, v_j}{2(\mu - \Delta\mu)} [1 - \langle d_j \rangle \mu_j (1 - \alpha)^2] \tag{2.54}$$

The term in brackets gives the correction for the diffracted intensity for a porous specimen in comparison with the bulk material. Cooper (1965) suggested, and proved experimentally, a similar parabolic dependence of intensity reduction on the pore fraction $1 - \alpha$. The theoretical developments of Hermann and Ermrich (1987) also resulted in a parabolic correcting term:

$$1 - 2\mu_j \langle l \rangle (1 - \alpha)^2$$

where $\langle l \rangle$ is the mean chord length of the particles. The numerical coefficient 2 in the latter equation may be easily accounted for by the differences in the definitions of $\langle d_j \rangle$ and $\langle l \rangle$. The magnitude of the intensity reduction can be seen from the data obtained by Suortti (1972) in studying the reduction of CuK_α fluorescence emitted by copper powder specimens with varying granularity and porosity (Fig. 2.9). The letters A, B, C, D, and E represent powders of mean sizes 1, 2.2, 3.7, 4.1, and 6.4 μm, respectively. Grains D and E were very irregular in shape. Numbers in parentheses are the solid matter fraction $\times 10^3$. The drop in intensity toward lower scattering angles is due to the surface roughness effect discussed in Section 2.4.7. Pure porosite produces angle-independence intensity reduction. In QXRD practice, this effect will not interfere with analytical results as long as the specimen porosity is kept constant.

2.4.5 Intensity correction for nonhomogeneous specimens

Our task is to estimate the magnitude of the two effects of nonhomogeneity—the microabsorption effect and the reduction of the apparent absorption coefficient of the sample. Both effects depend on particle size.

Taking into account the intensity diffracted by a homogeneous sample, I_h, we may rewrite expression (2.52) as:

$$I = \tau_j I_h \mu / (\mu - \Delta\mu) \tag{2.55}$$

where the particle absorption factor τ_j and the correction for absorption coefficient $\Delta\mu$ are defined in equations (2.52) and (2.48), respectively. The

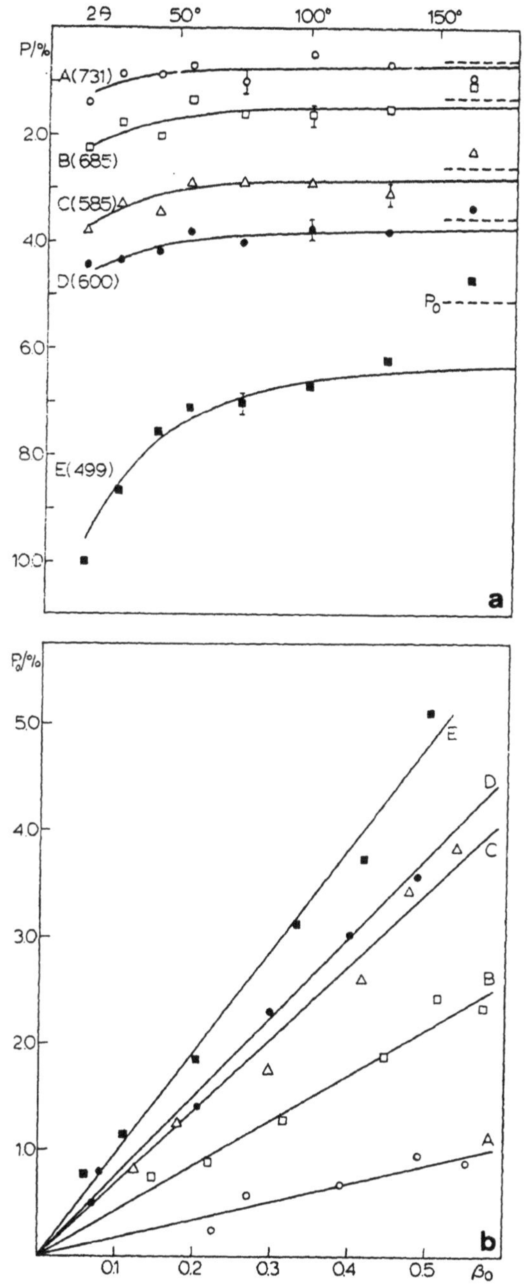

Figure 2.9. (a) Percent granularity reduction of fluorescent intensity and (b) the porosity effect for various Cu specimens. Taken from Suortti (1972).

relative intensity variation due to nonhomogeneity of the sample will be equal to:

$$\frac{I - I_h}{I_h} \approx (1 - \tau_j) + (\Delta\mu/\mu)\tau_j \tag{2.56}$$

The value of τ_j may be calculated by equation (2.39), assuming that the particle has a definite shape. In practice, if the particles are more or less isometric and randomly oriented, a spherical form would be a good approximation. Similar calculations were carried out by Brindley (1945), who tabulated the results as a function of $(\mu_j - \mu)D$ and the Bragg angle. The correction is appreciable even at values of $(\mu_j - \mu)D \approx 0.02$ and reaches the 10% level at $(\mu_j - \mu)D \approx 0.15$. Thus, no correction is necessary if $(\mu_j - \mu)D \leq 0.01$, and this condition will, of course, be met if $\mu_j D_j < 0.01$ for any phase in the multiphase mixture. Brindley classified such powders as fine.

The second correction term of equation (2.56) $(\Delta\mu/\mu)\,\tau_j$ is less significant than the first. Equation (2.48) shows that the value $\Delta\mu/\mu$ does not exceed a few percent if $\mu_j D_j < 0.1$ and the ratio μ_j/μ_k for any two phases constituting the sample does not exceed 5. Powders with $0.01 \leq \mu_j D_j \leq 0.1$ were regarded by Brindley as medium-grade powders. For these powders, the first correction term from equation (2.56)—the particle absorption factor—is significant, but the second correction term is not. The next grade of powders—coarse powders—has $0.1 \leq \mu_j D_j \leq 1$. For coarse powders, both correction terms from equation (2.56) are significant. For very coarse powders, when $\mu_j D_j > 1$, significant absorption takes place in the first layer of particles. The statistical treatment used in deriving the correction terms from equation (2.56) is not appropriate for these powders, and the intensity is expressed by the formulas derived at the beginning of this section. Thus, the division of nonhomogeneous powders into four groups—fine, medium, coarse, and very coarse—as proposed by Brindley, has definite physical meaning, as follows:

1. Almost no correction is needed for fine powders.
2. The particle absorption factor should be applied to medium-grade powders.
3. The particle absorption factor and a decrease of the mean absorption $(\Delta\mu/\mu)$ should be taken into account for coarse powders.
4. "One-layer" or "surface" diffraction is essential for very coarse powders.

The boundaries between these groups are, of course, only approximate, and the probable deviation from homogeneity can be calculated for each particular case, using the formulation given in this section. The ratios of the absorption coefficients of the phases in the sample are, of course, exceptionally important factors.

Hermann and Ermrich (1987, 1989), using the stochastic distribution of Poisson polyhedra to model a multiphase powder specimen, and de Wolff (1947), following a somewhat similar approach to the definition of the absorption path, arrived at the following expression for diffracted intensity:

$$I_{ij} = \frac{K_{ij} V_j}{2\,\mu'_j} \tag{2.57}$$

where μ'_j is the effective specimen absorption coefficient, in which the subscript emphasizes the phase dependence. The value of μ'_j is expressed as:

$$\mu'_j = \frac{\mu}{1 - 2\,(\mu_j - \mu)\, b_j}$$

where

$$b_j = -\langle l_j \rangle \frac{(1 - V_j)}{V_j} \ln(1 - V_j)$$

in which $\langle l_j \rangle$ is the mean chord length of the particles of phase j and other symbols have their usual meanings. In this case, the intensity variation due to nonhomogeneity of the specimen is equal to:

$$\frac{I - I_h}{I_h} = \mu/\mu'_j \tag{2.58}$$

With regard to the practical aspects of QXRD, it is often not the correction term itself—but its variation with specimen composition—that is most important. As is to be expected, the second correction term $\Delta\mu/\mu$ is much smaller than the particle absorption factor for comparable μD values, although it can be quite appreciable for μ_1/μ_2 differing greatly from unity. The curves for high $\mu_1 D_1$ values are strongly composition dependent, and intensity variations in this case can be ignored only over a narrow composition interval.

On the other hand, microabsorption effects in intensity ratios for any two phases in the sample may be much less affected by variations in the composition. Indeed, from the theorem for the mean integral, the expression for the particle absorption factor may be represented as:

$$\tau_j = \exp\,[-(\mu_j - \mu)\overline{D}] \tag{2.59}$$

where $\overline{D}$ is any effective dimension of the particle confined between 0 and the maximal value.

Taking into account expression (2.52) for the intensity and expression (2.59) for the particle absorption factor and ignoring the diffraction peak index, we obtain for the intensity ratios for two phases j and k:

$$I_j/I_k = (K_j/K_k)(v_j/v_k) \exp[-(\mu_j\bar{D}_j - \mu_k\bar{D}_k)] \exp[-\mu(\bar{D}_j - \bar{D}_k)]$$

If the particle dimensions for the two phases are similar, $D_j \approx D_k \approx D$, the second exponential in this expression approximates to unity, and

$$I_j/I_k = (K_j/K_k)(v_j/v_k)\tau_{jk} \tag{2.60}$$

where $\tau_{jk} = \exp[-(\mu_j - \mu_k)\bar{D}]$ expresses the microabsorption effect. This factor will be very significant if μ_j and μ_k are different, but it is constant or almost constant throughout the whole composition interval—a fact that has great importance for certain methods of QXRD.

2.4.6 *Application of the theory*

The question is how to apply the theory discussed in the previous sections to real cases of QXRD. Unfortunately, direct application of the correction terms outlined in equations (2.56) and (2.58) is hardly realistically possible for two main reasons:

1. Correction terms are strongly dependent on particle size, and any error in estimation of particle-size distribution will diminish the benefit of the correction. The two-wavelength approach suggested by Hermann and Ermrich (1989) in order to eliminate grain-size measurements is too tedious to be practical.
2. The theory discussed in this section is itself approximate. It was derived on the basis of a completely random arrangement of the numerous particles in the absorbing path. In fact, the number of particles may not be that great. Moreover, in the reflection case, usually only a thin upper layer of the sample participates in reflection, especially at relatively low Bragg angles. Although the number of particles in the absorbing path may be large, the randomness of their arrangement will inevitably be affected by the nearby sample surface.

A few attempts to apply microabsorption corrections in particle QXRD have been based on the observed deviation of the intensity-concentration curve from the theoretically expected one. This approach will be discussed later, in Section 4.2.2. Thus, the usefulness and significance of the theory discussed in this section lies in establishing a strategy for proper sample preparation and the proper QXRD method in order to avoid unwanted microabsorption effects. Table 2.3 gives a rough estimation of the significance of microabsorption for various

Table 2.3. Homogeneity of various materials (CuK_α radiation)

Compound or element	Absorption coefficient (cm^{-1})	Particle (grain) size, μm				
		1	5	10	29	40
Quartz SiO_2	92	Fine	Medium	Medium	Coarse	Coarse
Calcite $CaCO_3$	205	Medium	Medium	Coarse	Coarse	Coarse
Rutile TiO_2	570	Medium	Coarse	Coarse	Coarse	Coarse*
Barite $BaSO_4$	1000	Medium	Coarse	Coarse	Coarse*	Coarse*
Galena PbS	1600	Coarse	Coarse	Coarse*	Coarse*	Coarse*
Be	2.5	Fine	Fine	Fine	Fine	Fine
Al	131	Fine	Medium	Medium	Coarse	Coarse
Cu	470	Medium	Coarse	Coarse	Coarse	Coarse*
Mo	1670	Coarse	Coarse	Coarse*	Coarse*	Coarse*
W	3300	Coarse	Coarse*	Coarse*	Coarse*	Coarse*

*Very coarse

compounds by arranging them in the four groups proposed by Brindley (1945). In practice, the grain size is seldom less than 5-10 μm, and the microabsorption correction is essential for most compounds. The real magnitude of the effect should be estimated for each particular case. Several examples are given later.

Example 1. Analysis of small amounts of Si in $TaSi_2$

This analysis is essential for establishing the homogeneity range of $TaSi_2$. The absorption coefficients of these two phases are quite different: 141 cm^{-1} for Si versus 1270 cm^{-1} for $TaSi_2$ (CuK_α is assumed). The sample was obtained in the form of a fine powder with silicon particles of about 5 μm. A sample porosity of 25% was assumed, and the sum of the volume fractions of the two phases may be given by:

$$V_1 + V_2 = 0.75$$

The intensity correction due to nonhomogeneity of the specimen was calculated for several weight fractions of silicon by equations (2.56) and (2.58). The particle absorption factor was then taken from the International Tables for X-Ray Crystallography, Vol. III (1968), assuming spherical silicon particles. The results are given in Table 2.4.

Intensity corrections deduced from two approaches to the problem, that of de Wolff (1947) [equation (2.56)] and that of Hermann and Ermrich (1989) [equation (2.58)], show the same trend in composition dependence. The absolute values of the corrections are different,

Table 2.4. Intensity correction for the Si-$TaSi_2$ system due to nonhomogeneity of the sample

Fraction of Si				Intensity correction $\Delta I/I$	
Weight	Volume	τ-1	$(\Delta\mu/\mu)\ \tau$	Equation (2.56)	Equation (2.58)
0	0	0.42	0.14	0.56	0.64
5	17	0.34	0.19	0.53	0.61
10	33	0.27	0.20	0.47	0.35

although the difference is not as large as could be anticipated due to quite different lines of calculation of the two corrections.

The intensity correction is itself significant, as would be expected from the ratios of the absorption coefficients for Si and $TaSi_2$. Thus, the analysis of Si over the whole 0%-100% range under these conditions is hardly possible. However, Si is present only as a minor phase, with a weight fraction not exceeding 5%. In this concentration range, the intensity correction, although significant, remains almost constant. Thus, by using standard mixtures with the same granularity as the mixtures being analyzed, an analysis is certainly possible.

Example 2. Analysis of iron oxides in soil

In this example, we will compare two X-ray wavelengths, CuK_α and CoK_α. Although Fe fluorescence with CuK_α will increase the background, nowadays this obstacle is largely overcome by using diffracted beam monochromators. However, Fe-bearing compounds absorb CuK_α radiation strongly and thus create the condition for high microabsorption. This example gives a clear illustration of this phenomenon. The main mineral concentrations of soil (i.e., the matrix) are quartz and clay minerals. We will characterize the matrix absorption in terms of the mean absorption coefficients of quartz and kaolinite. As an iron-bearing mineral, we will choose hematite. The absorption coefficients are listed in Table 2.5.

The intensity correction was calculated by equations (2.48) and (2.56) for various Fe oxide abundances. The value of the particle absorption factor was estimated from Table 2.5, assuming spherical particles. We also assumed a total volume fraction of solid material of ~ 0.75 in the powder sample. The results of calculations for the two types of radiation are given in Table 2.6. The Fe oxide particle size was assumed to be 5 μm.

The intensity correction is much greater for CuK_α and also varies considerably with concentration. For CoK_α, the correction is almost stable, and analysis could be performed in the whole range of concentrations.

Table 2.5. Absorption coefficients for soil components

	CuK_α		CoK_α	
	μ^* cm^2/g	μ cm^{-1}	μ^* cm^2/g	μ cm^{-1}
Matrix	33	85	50.5	130
Fe oxides	230	1170	51.0	260

Table 2.6. Intensity correction for analysis of Fe oxides in soils

Weight fraction (%)	Volume fraction (%)	CuK_α			CoK_α		
		$\tau - 1$	$(\Delta\mu/\mu)\tau$	$\Delta I/I$	$\tau - 1$	$(\Delta\mu/\mu)\tau$	$\Delta I/I$
0	0	0.335	0	0.33	0.11	0	0.11
5	2	0.330	0.07	0.40	0.11	0	0.11
20	8.6	0.310	0.146	0.47	0.10	0	0.10
50	25.6	0.260	0.140	0.40	0.09	0.1	0.10
100	75	0.10	0.015	0.12	0.04	0	0.04

Example 3. Analysis of fluorite in a matrix composed of calcite, gypsum, anorthite, and albite

In order to simplify the calculation, we will regard this system as a two-phase system, in which the first phase is fluorite (300 cm^{-1}, CuK_α) and the second is a hypothetical matrix with an absorption coefficient of 145 cm^{-1}, an average for the four minerals given above, for CuK_α.

The calculations were performed along the same lines as those in the two previous examples, using the same equations (2.48) and (2.56) and assuming a 25% porosity of the powder sample. In this case, we calculated the intensity correction $\Delta I/I$ for three different mean grain sizes of fluorite: 10 μm, 20 μm, and 40 μm. The results are given in Table 2.7.

The advantages of fine grinding are evident. For a 10-μm powder, the analysis may be performed freely over the whole 0%-100% range. Even for coarse-grained powders, the correction is stable over a wide concentration range, making QXRD reasonable. However, in the case of coarse-grained powders, it is extremely important to maintain the granularity of the standard mixtures used for calibration as close as possible to the granularity of the samples investigated.

2.4.7 *Surface roughness*

The intensity corrections for a nonhomogeneous specimen discussed in the previous sections are θ independent. However, it was found that there was an appreciable reduction of diffracted intensity towards low

Table 2.7. Intensity corrections for the analysis of fluorite

Weight fraction of fluorite (%)	ΔI/I for various grain sizes		
	10 μm	20 μm	40 μm
0	0.12	0.28	0.46
10	0.13	0.28	0.45
20	0.13	0.27	0.45
50	0.12	0.23	0.41
80	0.08	0.16	0.30
100	0.06	0.12	0.24

scattering angles (de Wolff 1956; Paakkari and Sourtti 1968). This effect was related to surface roughness. Indeed, a thin surface layer with inevitable irregularities in particle packing contributes more to intensities of front reflections than of back reflections. Hermann and Ermrich (1987) described surface irregularities as an exponential decrease of the volume fraction of a solid material:

$$\alpha = \begin{cases} \alpha_0\,[1 - \exp(-z)] & \text{for } z \geq 0 \\ 0 & \text{for } z < 0 \end{cases} \quad (2.61)$$

where α_0 is the mean volume fraction of the solid material, $z = t/t_0$, in which t is the coordinate perpendicular to the specimen surface, and t_0 is a parameter of surface irregularity. The final expression for the intensity correction due to weak surface roughness is given by:

$$I/I_h = 1 - 2\,\mu\beta\alpha_0\,(\tau/\sin\theta)\,(1 - \tau/\sin\theta) \quad (2.62)$$

where I_h is the intensity diffracted by a homogeneous specimen without surface irregularities, and $\beta = -\langle l\rangle \ln \alpha_0$, in which $\langle l\rangle$ is the mean chord length of the particles. $\tau = t_0/\beta$, in fact, establishes the ratio of the thickness of the irregular surface layer to the grain size.

The magnitude of the effect for weak surface roughness and $\mu\langle l\rangle$ values smaller than 0.2 can be estimated from Fig. 2.9, which shows intensity reduction due to specimen porosity and surface roughness. The latter is responsible for the θ-dependent intensity drop especially pronounced at $2\theta < 30°$. The magnitude of the effect more or less corresponds to the theory of Hermann and Ermrich (1987). The role of μd in the systematic change of intensities vs. 2θ was given by Schreiner and Kimmel (1987) and Kimmel and Schreiner (1989), who showed a systematic reduction of XRD intensities at lower 2θ values for powders with high μd values.

For very coarse particles with $\mu d >> 1$, the intensity reduction can be much greater and extended to higher scattering angles. Vasudevan et al. (1992) reported an almost sixfold intensity drop at $2\theta \approx 40°$ for lead filings of a median size of 47 μm and CuK_α radiation ($\mu d \approx 13$). However, the same authors did not detect any angle-dependent intensity variation at $2\theta > 40°$ for a specimen of nickel filings of the same median size, but $\mu d >> 2$.

2.5 Orientation of reflecting particles

Although the discussion in this section is restricted to the Bragg-Brentano scheme, it can easily be extended to other geometries. A brief analysis of the diffraction geometry of the conventional Bragg-Brentano focusing scheme (Fig. 3.2) gives the impression that only crystallites with the hkl crystallographic plane parallel (or almost parallel) to the sample plane (diffraction vector parallel to the sample normal) will reflect X-rays. However, a detailed examination reveals that due to the finite size of all of the geometric elements (i.e., the focal spot, the specimen, and the detector slit), significant spreading of normals is allowed but to a lesser extent in the focusing ("horizontal") plane than in the axial ("vertical") plane. The situation in the focusing plane is illustrated in Fig. 2.10. Suppose that the angle between the central rays F_0O and OC_0 is

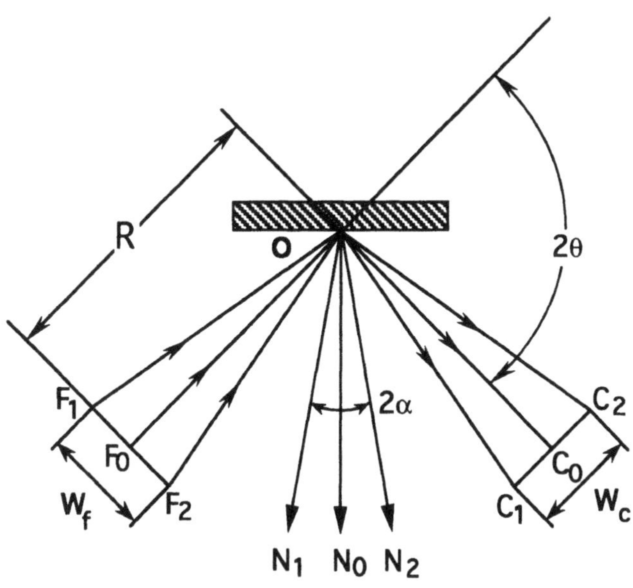

Figure 2.10. Spread of crystallite orientation in the equatorial plane.

equal to the angle $2\theta_0$, where θ_0 is the Bragg angle for a certain hkl plane. Then N_0, the normal to the reflecting plane, is parallel to the sample normal. ON_1 is the bisector of the angle F_1OC_1, where F_1O and OC_1 are rays coming from and going to the edge points of the optical focus F_1 and the reflecting slit C_1. Although the angle between F_1O and OC_1 is not equal to $2\theta_0$ at a particular moment, the condition will be realized sometime during the scan of the diffraction peak. At that moment, ON_1 will coincide with the normal to the reflecting crystallite (diffraction vector). Thus, the orientation of the reflecting crystallites in the focusing plane spreads in the range of $+\alpha$ to $-\alpha$, where α is easily defined from Fig. 2.10:

$$\alpha = (w_f + w_c) / 4R \tag{2.63}$$

where R is the radius of the goniometer.

This spread of normals is independent of the angle θ_0 and is rather small. For the conventional $w_f \approx w_c \approx 0.2$ mm and R = 173 mm, $\alpha \approx 0.07°$ The spread of normals in the vertical plane is much greater.

Figure 2.11 illustrates a three-dimensional view of diffractometer geometry. Let us consider X-rays diffracted by crystallites near the central point of the sample O. The rays $\overline{F_0O}$ and $\overline{OC_0}$ define the equatorial plane of the diffractometer, with ON_0 coinciding with the sample normal. The ray emerging from the edge point F_2 of the tube focus and detected at the edge point C_2 of the receiving slit must be reflected by the crystallite with the normal to the reflecting plane striking the plane $F_1F_2C_1C_2$ at the point N_2 with inclination φ to the horizontal (focusing) plane. The value of φ is easily defined from Fig. 2.11:

$$\varphi = \arcsin \left[(H_f + H_c) / 4R\sin\theta \right]$$

where H_f and H_c are the heights of the focal spot and receiving slit, respectively. The maximal inclination can be expected for a crystallite at the edge point of the sample P_1 (or P_2) reflecting the ray from F_2 to the edge detector point C_2. Suppose that the heights of the sample $H_p = H_f = H_c = H$, then the maximal inclination angle will be given by:

$$\varphi_1 = \arcsin \left[(H/(R\sin\theta) \right] \tag{2.64}$$

For a diffractometer with two sets of Soller slits with divergence $\delta = s/l$, where l is the slit length and s is the distance between plates (Zevin 1990):

$$\varphi_m = \arcsin (\delta/\sin\theta) \tag{2.65}$$

This angle might be quite substantial at low Bragg angles. For example, for the conventional $\delta = 2.3°$ and $\theta = 5°$, the value of φ_m is equal

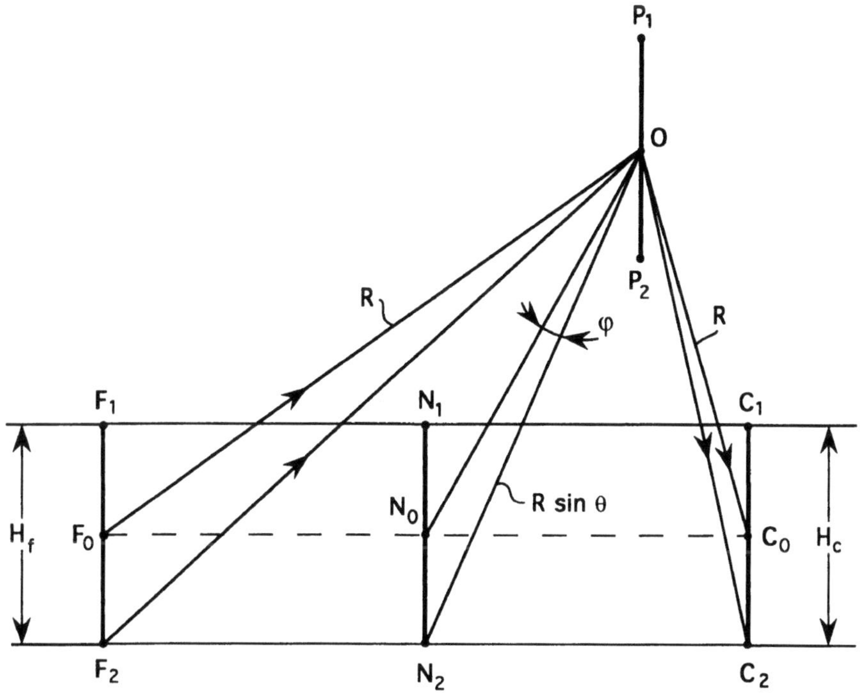

Figure 2.11. Three-dimensional view of diffractometer geometry. $F_1\ F_2$ = focus line, $P_1\ P_2$ = specimen, $C_1\ C_2$ = receiving slit.

to an astonishing 27°. The number of crystallites with various normal orientations is, however, not equal. It is maximal for poles with $\varphi = 0$ (i.e., parallel to the focusing plane) and minimal (zero) for poles with $\varphi = \varphi_m$. The frequency function for the pole orientations emphasizing the density of states with a given angle φ may be represented as a first approximation (Zevin 1990) as follows:

$$f(\varphi) = \begin{cases} 1 - 2\,(\varphi/\varphi_m)^2 & \text{if } \varphi_m/2 \le \varphi \le \varphi_m \\ 2\,(1 - \varphi/\varphi_m)^2 & \text{if } 0 \le \varphi \le \varphi_m/2 \end{cases} \tag{2.66}$$

The fraction of reflecting crystallites for a specimen with randomly oriented crystallites is defined as:

$$f = \frac{1}{4\pi} \int_{-\alpha}^{\alpha} \int_{-\varphi_m}^{\varphi_m} f(\varphi)\, d\alpha\, d\varphi \approx h/\sin\theta \tag{2.67}$$

where h is a constant defined by the instrumental parameters, such as Soller slit divergence, and the widths of the focusing and receiving slits.

Thus, the conventional "powder" supplement to the Lorentz factor equal to $(\sin\theta)^{-1}$ is obtained. For a sample with preferred orientation and a particular distribution of the pole density $P(\alpha, \varphi)$, the fraction of reflecting crystallites is:

$$f = \frac{1}{4\pi}\int_{-\alpha}^{\alpha}\int_{-\varphi_m}^{\varphi_m} P(\alpha,\varphi)\, f(\varphi)\; d\varphi\, d\alpha \tag{2.68}$$

which reduces to

$$f = \frac{2\alpha}{4\pi}\int_{-\varphi_m}^{\varphi_m} P(\alpha,\varphi)\, f(\varphi) d\varphi \tag{2.69}$$

because $P(\alpha, \varphi)$ is usually regarded as a constant within very small angle α.

For a sharp texture and low Bragg angles, equation (2.69) might reveal an angle-dependent term that is different from the conventional $(\sin\theta)^{-1}$. This fact must be taken into account in the analysis of oriented clay mounts (Zevin and Viaene 1990).

3 Geometric aspects of X-ray diffractometry

3.1 Geometric schemes in X-ray diffractometry

Powder diffractometers were initially designed for the consequent or scanning mode of operation. In these instruments, the detector scans the desirable range of scattering angles, and thus a diffraction pattern is produced. There is a limited time (usually seconds) in which to measure the intensity of the narrow diffraction peak in comparison with minutes or even hours of exposure time in a simultaneously operated unit—a Debye-Sherrer camera. When diffracted intensity is gathered from an extended specimen surface and converged to a point or line in front of a detector, the diffractometer requires a geometric scheme with an enhanced intensity output, namely, a focusing scheme. Recently, as a result of the development of new detectors and X-ray sources, interest has revived in parallel-beam geometries.

The following geometric schemes have practical value in X-ray diffractometry:

1. Focusing-reflection mode
2. Focusing-transmission mode
3. Parallel-beam mode.

The fundamental geometric difference between reflection and transmission modes lies in the characteristics of the convergent-divergent beam. In the reflection mode, the specimen changes the divergent beam into a convergent beam and vice versa (Fig. 3.1a,b). In the former case, the reflected beam converges to a point, thus utilizing the focusing geometry. In contrast, in the transmission mode the specimen does not change the divergent characteristics of the beam, i.e., the diffracted beam remains divergent for a divergent primary beam and convergent for a convergent primary beam (Fig. 3.1c,d). In order to fulfill the focusing condition in the transmission mode an additional device is needed to convert the divergent beam from the X-ray tube or the divergent beam reflected by the transmission specimen into a convergent diffracted beam striking the detector. This task is accomplished by a reflection crystal monochromator (Fig. 3.2c; see Section 3.2).

There are two types of focusing geometry for both reflection and transmission modes:

1. Geometry with a fixed specimen-detector distance and a variable focusing circle.
2. Geometry with fixed focusing circle and variable specimen-detector distance.

Most currently available diffractometers have the first—reflection mode—type of scheme, which is known as *Bragg-Brentano parafocusing geometry* (Fig. 3.2b; see Section 3.2). The first type of transmission mode scheme (Fig. 3.12; see Section 3.4) is used in the diffractometer manufactured by STOE. The second type of reflection mode scheme is known as *Seeman-Bohlin focusing geometry* (Fig. 3.9; see Section 3.3). The corresponding transmission mode scheme (Fig. 3.15; see Section 3.4) was first proposed by Guinier for the focusing camera, and it usually bears Guinier's name in diffractometric applications as well.

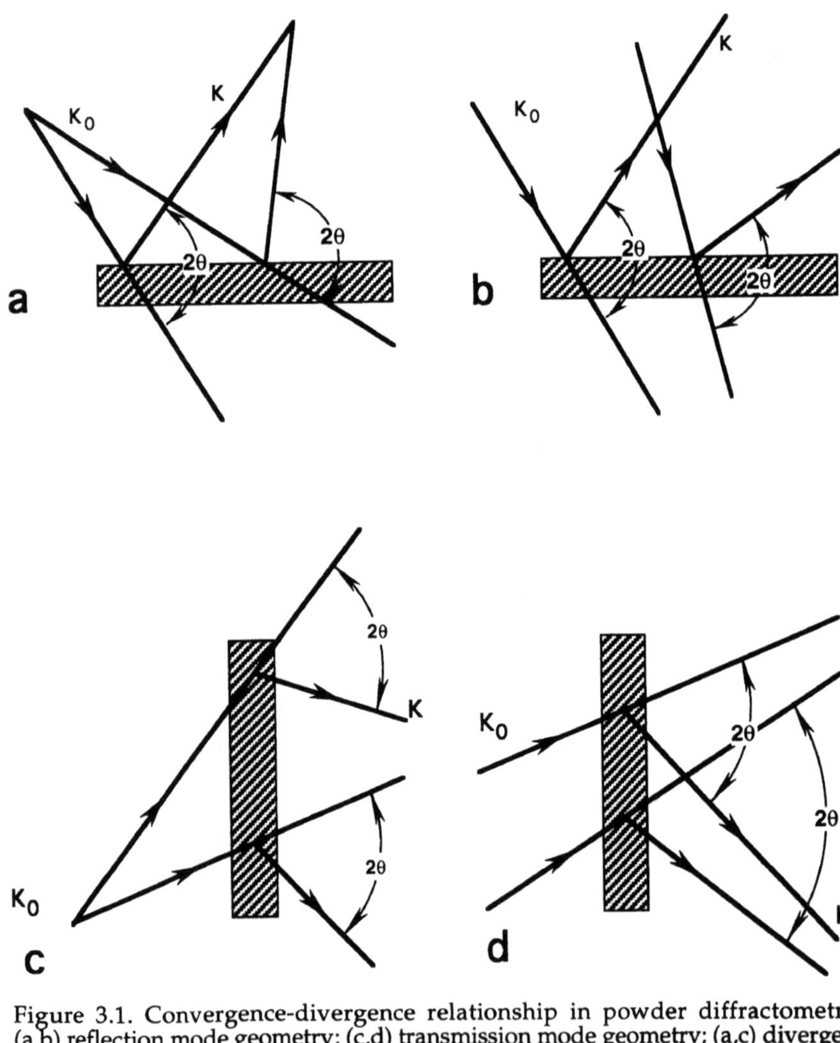

Figure 3.1. Convergence-divergence relationship in powder diffractometry. (a,b) reflection mode geometry; (c,d) transmission mode geometry; (a,c) divergent primary beam; (b,d) convergent primary beam.

In order to fulfill the focusing condition in the equatorial (diffraction) plane containing the incident and diffracted rays, the specimen must be placed adjacent to the focusing circle. The best solution is to bend the specimen to the curvature of the focusing circle. This solution is, however, not practical for the first type of focusing geometry with a variable focusing circle. In that case, a good approximation to focusing is obtained by arranging the flat specimen at a tangent to the focusing circle. In order to achieve this condition, the angle between the sample and the incident beam must be modified during the course of the run by sample rotation and/or movement of the source. [There are three possibilities: source fixed, sample rotates, and detector moves ($\theta/2\theta$); sample fixed, source and detector move (θ/θ); detector fixed, source moves, and sample rotates ($2\theta/\theta$).] In contrast, for the second type of focusing geometry, both the specimen (bent or flat) and the source remain stationary relative to the primary beam during the diffractometric run. Thus, the first type of focusing geometry is intrinsically suitable only for consequent mode diffraction measurements, which include continuous or step-wise specimen rotation. The second type of geometry is suitable for both consequent and simultaneous mode measurements.

Parallel-beam nonfocusing diffractometric geometry has gained practical significance in the past five to seven years. We are currently witnessing the development of two quite different types of parallel-beam geometry in terms of time (age), beam shape, and availability—the initial fundamental Debye-Sherrer geometry and a novel synchrotron radiation diffractometry.

The parallel or almost parallel incident beam in Debye-Sherrer geometry is obtained by means of a narrow, usually circular, diaphragm. This geometry was rejected when X-ray diffractometry first became an analytical tool due to the relatively low diffracted intensity, which is not adequate for the sequential mode measurements used in X-ray diffractometry. However, the recent development of position-sensitive detectors (PSDs) and simultaneous-mode measurements has revived interest in Debye-Sherrer geometry as a result of its suitability for simultaneous measurements and the normal incidence of the diffracted beam (Fig. 3.19; see Section 3.6).

In synchrotron radiation diffractometry, synchrotron storage rings naturally provide a parallel X-ray beam with a cross section large enough to illuminate a specimen of a few square centimeters, i.e., a measurement common in focusing X-ray diffractometry. The parallel diffracted beam is formed by a set of narrow Soller slits (Fig. 3.23; see Section 3.7). The consequent and simultaneous operation modes are equally suitable for synchrotron radiation, due to the absence of focusing restrictions.

All the geometric schemes described so far will be thoroughly discussed in later sections. Although our main aim is an examination of the impact of the diffractometer geometry on diffracted intensity, we will also discuss briefly the instrumental aberrations intrinsic to the particular scheme. Naturally, the geometric schemes discussed here must be compared. We have chosen Bragg-Brentano reflection focusing geometry as the basic unit on which to carry out the comparative study. We do not intend to compare various diffractometer designs but rather different focusing and nonfocusing geometric schemes. Each scheme has to be characterized by the level of diffracted intensity and the level of instrumental distortion of the peak profile, such as peak shift or peak broadening. These two characteristics are not independent. If the intensity of the primary beam is assumed to be constant [I_0 in equation (2.19)], then increasing the diffracted intensity can usually be achieved only at the expense of peak distortion. Thus, the best way to evaluate the merits of particular diffractometer schemes would be to equalize instrumental distortions and compare the diffraction intensities or vice versa. We have illustrated the applicability of such an approach in several cases (Zevin et al. 1963; Zevin and Umanski 1964; Zevin 1981). This approach has been modified somewhat by contemporary developments in diffractometer instrumentation (digitalized patterns) and data processing (profile fitting). In many cases, it is essential to keep instrumental distortion as low as possible, even at the expense of a substantial decrease in diffracted intensity. However, statistical errors can be maintained at the desirable low level simply by increasing counting time (overnight runs, etc.), thus compensating for the decrease of diffracted intensity flux. The general expression for diffracted intensity (2.20) provides a basis for comparison of the various geometric schemes. If we assume the incident beam intensity I_0 and the relative length of the measured Debye ring chord H/R to be constant, then the intensity is influenced by the product absorption factor A and the illuminated specimen volume V. These factors will be evaluated for each geometric scheme discussed here.

3.2 Diffractometers with Bragg-Brentano reflection focusing

3.2.1 *General geometric features*

A detailed description of Bragg-Brentano geometry can be found elsewhere (Wilson 1963; Kheiker and Zevin 1963), but we feel that the basic features should be described here in order to ensure a comprehensive understanding of the material.

Bragg-Brentano geometry is shown schematically in Fig. 3.2 (a three-dimensional view in Fig. 3.2a and a projection in the focusing plane in Fig. 3.2b). The divergence angle of the primary beam from the line focus of the X-ray tube F is defined by the divergence slit DS in the focusing plane and the Soller slits S1 in the axial plane, normal to the focusing plane. The collimation effect of the Soller slits is defined by the axial divergence $\delta = s/l$, where s is the distance between the parallel plates in the Soller slit and l is the length of the plate. In order to ensure convergence of the diffracted beam in front of the receiving slit RS (focusing condition), the specimen SP must be bent by the radius of the focusing circle, passing through the focal point, the receiving slit, and the center point of the specimen (broken line circle in Fig. 3.2b). Note that the radius of the focusing circle is not constant but is angle dependent:

$$r_f = R/(2 \sin \theta) \tag{3.1}$$

where R is the radius of the diffractometer.

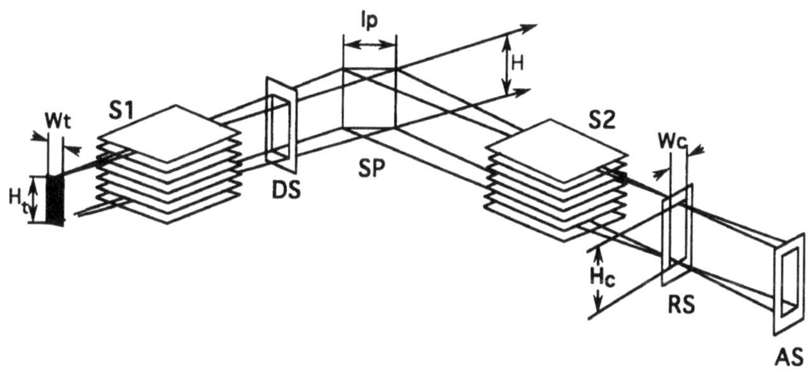

Figure 3.2a. Bragg-Brentano focusing reflection geometry. Three-dimensional view.

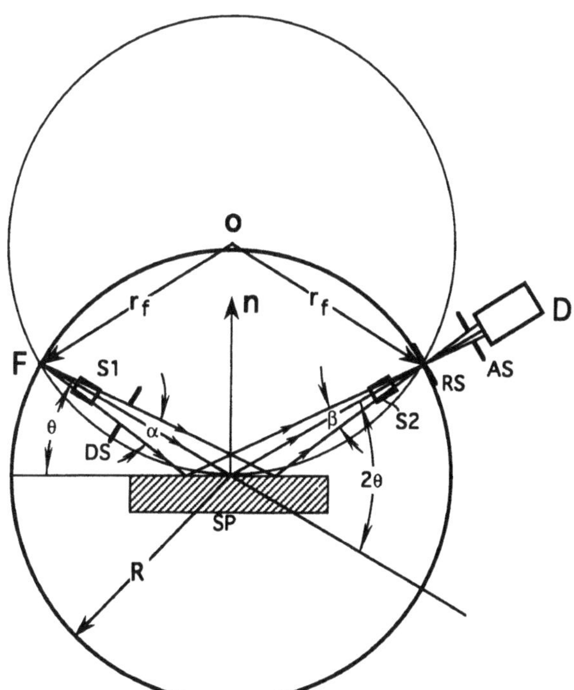

Figure 3.2b. Bragg-Brentano focusing reflection geometry.

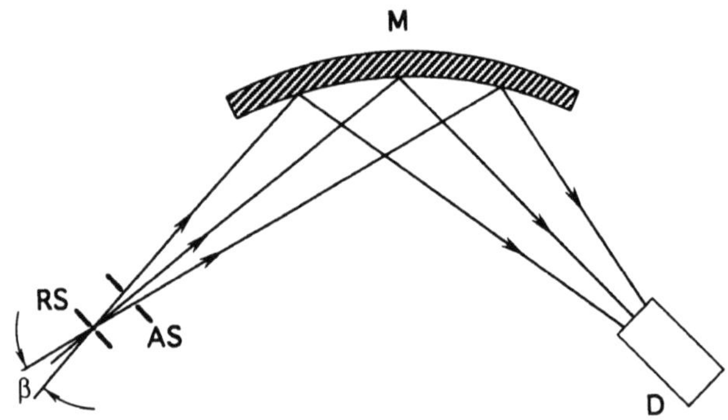

Figure 3.2c. Bragg-Brentano focusing reflection geometry. Reflection beam monochromator (M).

Continuous specimen bending is apparently not practical, and in the Bragg-Brentano scheme flat-surface specimens are used. In order to ensure minimal departure from the ideal focusing condition, the surface of the specimen must lie in a plane midway between the incident and diffracted beams and thus constitute an angle θ with the incident beam. If this condition is met, the specimen surface remains at a tangent to the focusing circle.

Note that the focusing we are discussing here is quite different from reflection focusing in the optical sense. In our case, the curvature of the specimen surface has a second-order effect on focusing, and beam convergence (with certain aberrations) is achieved even for a flat specimen. This is, of course, impossible in optics. This is the reason why the term parafocusing is sometimes used for the beam-convergent phenomenon in powder diffractometry.

The cross section B of the primary beam near the specimen may be described by:

$$B = 2RH \tan(\alpha/2) \approx RH\,\alpha \tag{3.2}$$

where R is the radius of the goniometer, H is the width of the beam,[1] α is the divergence of the primary beam, and tan $\alpha/2 \approx \alpha/2$ because α does not exceed a few degrees. If we assume that the specimen surface intercepts the primary beam, then the irradiated area of the specimen is given by:

$$S = B/\sin\theta = HR\,\alpha/\sin\theta \tag{3.3}$$

In constant divergence geometry (α = const), the irradiated area decreases continuously with increasing Bragg angle. In order to keep the irradiated area constant and to enhance the intensity diffracted at higher 2θ angles, Jenkins and Paolini (1974) proposed a special device ("θ-compensating slit") that continuously increases the divergence

---

[1] H should be no greater than the sample width. As a rule, we assume that the sample size exceeds the beam size.

angle α in proportion to sin θ. In this case, we have a fixed value of S, and the divergence α can be derived from equation (3.3).

Two slits, the receiving slit RS and the antiscatter slit AS, comprise a collimator that "sees" the illuminated specimen surface (Fig. 3.2b). These two slits define the convergent angle of the diffracted beam β, which is very close to the divergence angle α. The rays that have passed through the receiving and antiscatter slits strike detector D.

The illuminated volume of the specimen increases linearly with the beam divergence α, and the integrated intensity of the diffraction peaks is proportional to α. If the illuminated width of the specimen and the receiving slit are kept constant, then variations in the divergence of the Soller slits δ have little effect on the illuminated area, but they do affect the intensity of the primary or diffracted beam by obscuring oblique rays. As a first approximation valid for low or medium axial divergence ($\delta R \leq H/2$), we can assume that the intensity of the X-ray beam passing through the Soller slits is proportional to δ (Zevin 1981).

An important constituent of contemporary Bragg-Brentano geometry is the focusing diffracted-beam monochromator placed between the receiving and antiscatter slits and the detector D (Fig. 3.2c). The monochromator crystal M (pyrolytic graphite) combines high reflecting power and high mosaic spread, ensuring effective reflectance of the diffracted beam. The main purpose of the monochromator is to remove nonmonochromatic (fluorescent and white) radiation from the diffracted spectrum. Although it cannot resolve the two components of the K_α doublet, the mosaic spread is low enough to control the diffracted beam divergence in the axial direction. Thus, the second set of Soller slits S2 (Fig. 3.2a) can be removed if a monochromator is inserted. If we take into account the fact that the β filter must also be removed, it will be not surprising to discover that the graphite monochromator does not reduce diffracted intensity.

3.2.2 *Absorption correction*

We will derive the absorption correction for an ideal specimen satisfying the following conditions:

1. The specimen is considered to be infinitely thick, i.e., the irradiated volume is thick enough to reduce the intensity of the primary beam to a negligible value, say, to less than 1×10^{-3} of its original value on the sample surface.
2. The material under study is homogeneous in the sense defined in Section 2.4.

Let us now express equation (2.21) in a reduced form:

$$I_{ij} = I_0 E_{ij} A_i v_j V \qquad (3.4)$$

where v_j is the volume fraction of phase j in this specimen, A_i is the absorption correction term, and E_{ij} includes all the remaining terms of the right-hand side of equation (2.21).

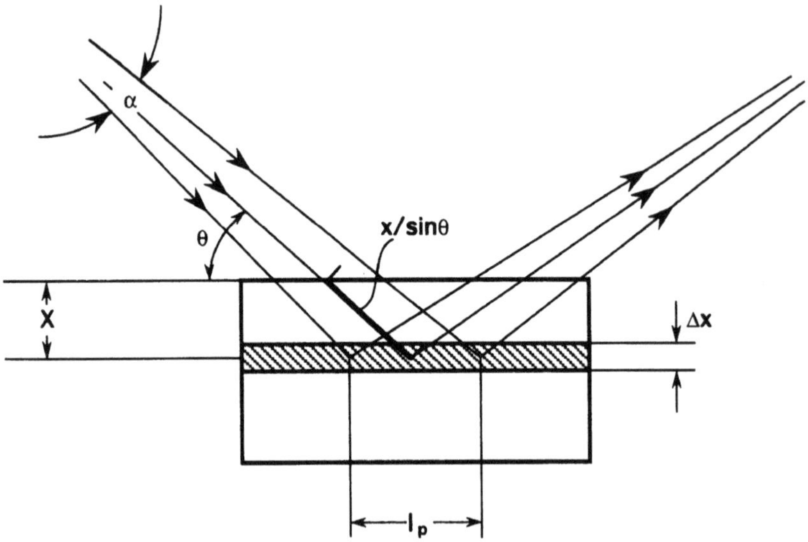

Figure 3.3. Derivation of the absorption factor in the Bragg-Brentano reflection scheme.

Let us consider the intensity diffracted by a thin layer of material of thickness Δx located at a depth x (Fig. 3.3). The cross section of the incident beam B is defined by equation (3.2). The irradiated volume V is given by:

$$V = l_p H \Delta x = B\Delta x/\sin\theta$$

where l_p and H are the length and the width of the illuminated specimen surface, respectively.

The intensity of the primary beam at depth x is attenuated by a factor of $\exp(-\mu x/\sin\theta_i)$, where $x/\sin\theta_i$ is the absorption path of the primary beam inside the specimen. The diffracted beam is subject to the same reduction (due to symmetry of this geometry). Thus, the intensity diffracted by the layer Δx is given by:

$$dI_{ij} = I_0 E_{ij}\,[\exp(-2\mu x/\sin\theta_i)](B/\sin\theta_i)v_j\,\Delta x \tag{3.5}$$

and the intensity diffracted by the whole sample is given by:

$$I_{ij} = \int_0^{\infty} dI_{ij} = \frac{I_0 E_{ij} B}{2\mu} v_j \tag{3.6}$$

Thus, the term $A_i V$ from equation (2.21) is equal to $B/2\mu$.

If the irradiated area S is kept constant by a θ-compensating slit, then $V = S\Delta x = \text{const}$, and $B/\sin\theta$ should be replaced by S in equation (3.5). Integration leads to the following expression:

$$\frac{I_0 E_{ij} S \sin\theta_i}{2\mu} v_j = \frac{I_0 E_{ij} B_i}{2\mu} v_j \tag{3.7}$$

where $B_i = S \sin \theta_i$ is the cross section of the primary beam near the specimen and subscript i indicates that B_i is variable in this case. In a constant-divergence scheme, the primary beam cross section B = const, and the absorption correction is independent of the Bragg angle θ and is equal to $1/2\,\mu$. If the irradiated surface S is kept constant by means of a θ-compensating slit, then the absorption correction is equal to $\sin \theta/2\mu$. We must keep this difference in mind when we compare experimental results obtained with different instruments. For example, powder patterns in the Powder Diffraction File of the International Centre for Diffraction Data are presented in a "constant-divergence" version if obtained with a reflection Bragg-Brentano diffractometer.

Now, let us derive the final equation for the intensity diffracted by an ideal specimen in a Bragg-Brentano diffractometer. We replace E_{ij} from equation (3.6) with its value from equation (2.21). After rearranging the terms, we obtain:

$$I_{ij} = \left[\frac{r_e^2}{32\pi}\right] \left[\frac{I_0 B H_c \lambda^3}{R}\right] \left[\frac{1 + \cos^2 2\theta_i}{\sin^2\theta_i \cos \theta_i}\right] \left[\frac{|F_{ij}|^2 D_{ij} M_{ij}}{\vartheta_j^2}\right] \left[\frac{v_j}{2\mu}\right] \tag{3.8}$$

The first term includes fundamental physics constants. The second term is defined by the experimental conditions and remains constant if these conditions are constant. The third term is a function of the scattering angle, which depends on neither the structure of the phase nor the composition of the specimen. The fourth term is defined by the crystal structure of the analyzed phase (analyte) and does not depend on the composition of the specimen. The fifth term, which is the most important for this book, represents the dependence of the intensity on the composition of the specimen.

We will thus combine the first four composition-independent terms into a single coefficient, K_{ij}:

$$I_{ij} = K_{ij}\, v_j/\mu \tag{3.9}$$

K_{ij} is known as the calibration coefficient or calibration constant. Usually, weight fractions and not volume fractions are used in analytical practice. The weight fraction of any phase j may be expressed as:

$$c_j = v_j\, \rho_j/\rho \tag{3.10}$$

where ρ_j and ρ are the densities of phase j and the sample, respectively. Taking into account that $\mu/\rho = \mu^*$ is the mass-absorption coefficient of the sample, we obtain the final equation:

$$I_{ij} = \frac{K_{ij}\, c_j}{\rho_j\, \mu^*} \tag{3.11}$$

This is the famous equation derived by Klug and Alexander in 1948. This equation constitutes the starting point for the development of various QXRD methods discussed in the next chapter.

3.2.3 *Real irradiated volume in the Bragg-Brentano scheme*

Equation (3.11) was developed on the assumption that the irradiated volume of the specimen is infinitely thick. In fact, this is not strictly true, not only because the real volume of sample that occupies the sample holder is finite, but also because of the geometry of reflection. The volume of material reflecting X-rays must be a common volume for the incident and diffracted beams. If we assume that the cross sections of the incident and diffracted beams are equal, or, in other words, that the divergent angle α is equal to the convergent angle β, then this common volume is a triangular prism with the cross section shown in Fig. 3.4 and a height equal to the width of the illuminated surface H in the axial direction. The lower point D of the prism ACD is located at a depth:

$$T = (l_p/2) \tan\theta = R\,\alpha/2\cos\theta$$

where $AC = l_p = (R\alpha/\sin\theta)$, the length of the irradiated area of the sample (Fig. 3.4). Thus, the layer of the sample at depth x, which diffracts X-rays within the convergent angle α, has the volume:

$$dV = (l_p - 2\,x\cot\theta)H\,\Delta x$$

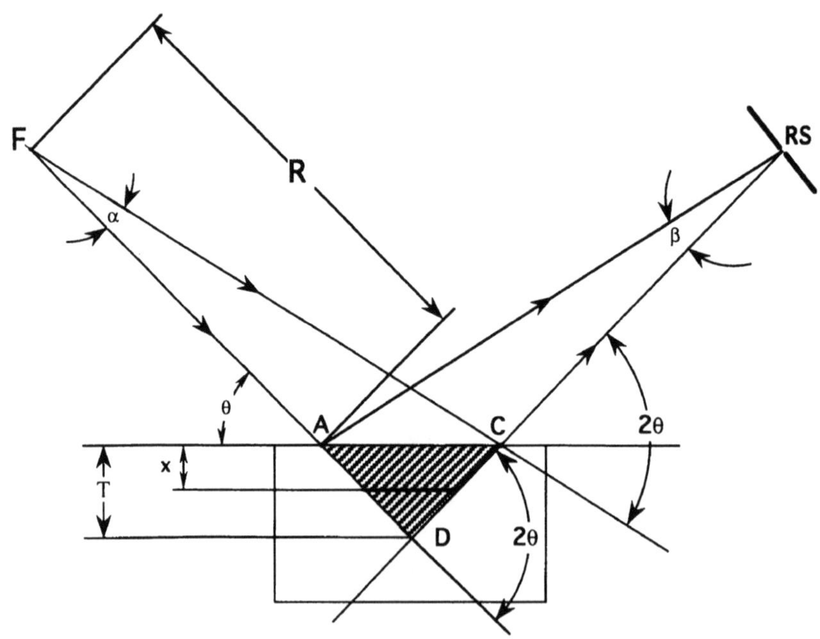

Figure 3.4. Real irradiated volume in the Bragg-Brentano reflection scheme.

Table 3.1. Ratio of the actual intensity diffracted by the powder sample (I'_{ij}) to the assumed intensity in the "infinite thickness" approach (I_{ij})

	I'_{ij}/I_{ij} x 100	
μ, cm^{-1}	$2\theta = 15°$	$2\theta = 30°$
1000	1.00	1.00
500	1.00	1.00
100	99.6	99.2
50	99.2	98.4
10	95.8	91.7
5	91.4	83.5

We may now define the term dI_{ij} from equation (3.6) as:

$$dI_{ij} = I_0E_{ij} \exp[-2\mu x/\sin \theta_i]\, H\,(l_p - 2x \cot \theta_i)\, v_j\, \Delta x$$

and integration should be performed in the limits of x between 0 and T:

$$I'_{ij} = \int_0^T dI_{ij} = \frac{K_{ij}c_j}{\rho_j\, \mu^*} \left\{ 1 - \frac{\sin 2\theta_i}{2R\alpha\mu} [1 - \exp(-2R\,\alpha\mu/\sin 2\theta_i)] \right\} \tag{3.12}$$

where the coefficient K_{ij} has the same meaning as in equation (3.11).

We can expect a tangible difference between the precise expression (3.12) and the "infinite thickness" approximation (3.11) only for low-absorbing materials and low beam divergencies. A comparison of these two equations for the conventional experimental setting $\alpha = 1°$; $R = 173$ mm; $R\alpha \cong 3$ mm, and $2\theta = 15°$ and 30° is shown in Table 3.1.

Thus, only for very low-absorbing materials, such as organics, are deviations from the "infinitely thick" approximation significant, although even in this case the intensity losses do not exceed 10%.

A more detailed discussion of this problem, including cases of unequal divergence and convergence angles and various positions of the sample surface, may be found in the work of Wilson (1972).

3.2.4 *Instrumental aberrations*

The geometric theory of X-ray diffractometry was developed in the 1950s and 1960s and is fully documented in the works of Klug and Alexander (1954), Kheiker and Zevin (1963), and Wilson (1963). We do not intend to repeat this theory here, but we will set out to illustrate the basic results. Each source of instrumental aberrations may be represented by a corresponding instrumental function $G(\varepsilon)$. The argument ε is the deviation from the scattering angle $2\theta_0$, corresponding to Bragg's law. The combined effect of the instrumental aberrations is calculated as their convolution:

$$G_{1,2}(\varepsilon) = \int_{-\infty}^{\infty} G_1(u)\, G_2(\varepsilon-u)\, du \tag{3.13}$$

and the final instrumental function G(ε) is obtained by consequent convolution of a series of individual functions. The profile of the diffraction peak is considered to be a convolution of the pure diffraction profile F(ε) and the instrumental function G(ε).

$$H(\varepsilon) = \int_{-\infty}^{\infty} G(u)\, F(\varepsilon-u)\, du \tag{3.14}$$

The profile F(ε) is defined by chromatic dispersion and various phase-related effects, such as strains, finite coherent length, and various lattice defects.

As a starting point in the estimation of peak width, we will take an instrumental function defined by the finite width of the receiving slits w_s. The instrumental function is a θ-independent rectangular function with width $\omega_t = w_s/R$, where R is the goniometer radius. For typical values of $w_s \cong 0.2$ mm and R = 173 mm, $\omega_t \cong 0.07°$ and the width of the diffraction peak will be no less than this value. The instrumental functions defined by the beam divergence (horizontal and vertical) are asymmetric and strongly dependent on θ. As a first approximation, the variance of these functions may be taken to decay as cot θ, emphasizing the rapid decrease of instrumental broadening with increasing scattering angle. Another instrumental function is determined by the penetration of X-rays, and the extent of its action depends, of course, on specimen absorption. The variance of this function is dependent on θ and varies with sin 2θ. Peak shift and peak broadening due to specimen transparency are almost negligible for materials with moderate absorption coefficients ($\mu \geq 100$ cm^{-1}). However, for organic compounds it might be a major source of peak distortion.

Consider now the factors defining the width of the diffraction peaks for an ideal powder, which do not reveal intrinsic broadening effects. In two scattering angle ranges—front reflection (2θ < 20°-30°) and back reflection (2θ > 100°-120°)—the situation is clear. In the former range, instrumental factors are overwhelming, producing asymmetric diffraction peaks (Fig. 3.5). The broken-line profiles in this figure are obtained by transformation of the spectral distribution of the $K\alpha$ doublet I(λ) according to Bragg's law:

$$I(2\theta) = I[2 \arcsin (\lambda/2d)]$$

The full width at half maximum (FWHM) of the diffraction peak in the low-angle range depends on geometric parameters, such as slit width and beam divergence, but in the best case it is close to 0.10°-0.12°.

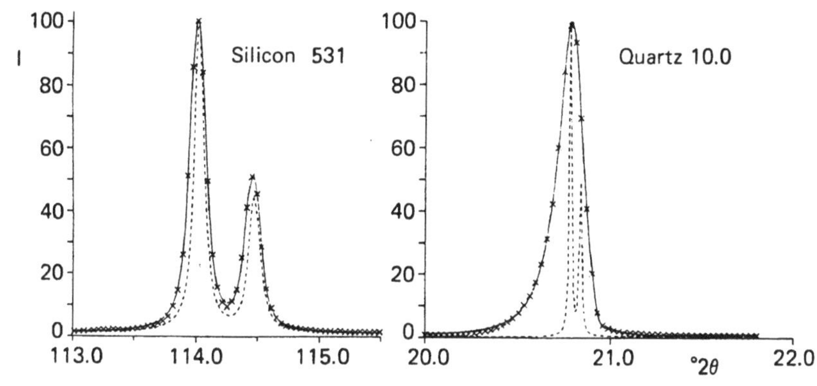

Figure 3.5. Powder diffractometer experimental data (x), fitted profiles (——), and spectral distribution (- - - - - - -). The differences result from the aberrations. Reprinted with permission from the American Crystallographic Association, Inc. (Parrish et al. 1976).

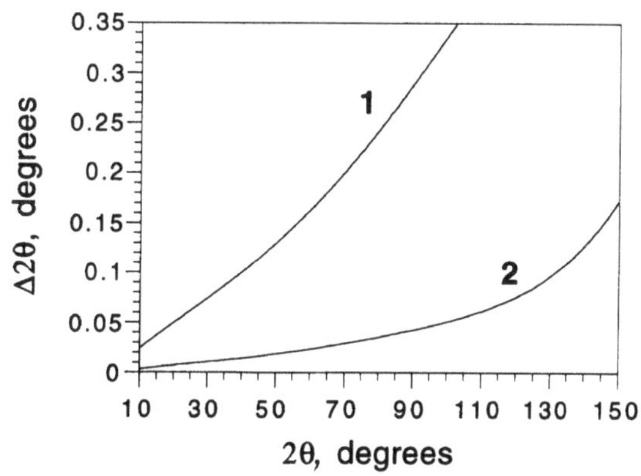

Figure 3.6. Chromatic dispersion in X-ray diffractometry. 1 - Angular separation of the $CuK_{\alpha1} - K_{\alpha2}$ doublet; 2 - Full width at half maximum of the α_2 component.

In the back-reflection range, instrumental aberrations diminish, and peak shape is defined almost completely by chromatic dispersion (Fig. 3.5) and curve 2 in Fig. 3.6. In the medium range of scattering angles, both instrumental and chromatic factors are important. Especially significant is the chromatic aberration caused by an unresolved K_α doublet. The angular separation of the two components is given by:

$$\Delta 2\theta = 2\frac{\Delta\lambda}{\lambda}\tan\theta \qquad (3.15)$$

where $\Delta\lambda = \lambda_2 - \lambda_1$; λ_1, λ_2 are the wavelengths of the doublet components; and λ is the mean wavelength. $\Delta 2\theta$ is plotted against the scattering angle for the CuK_α doublet in Fig. 3.6. Although instrumen-

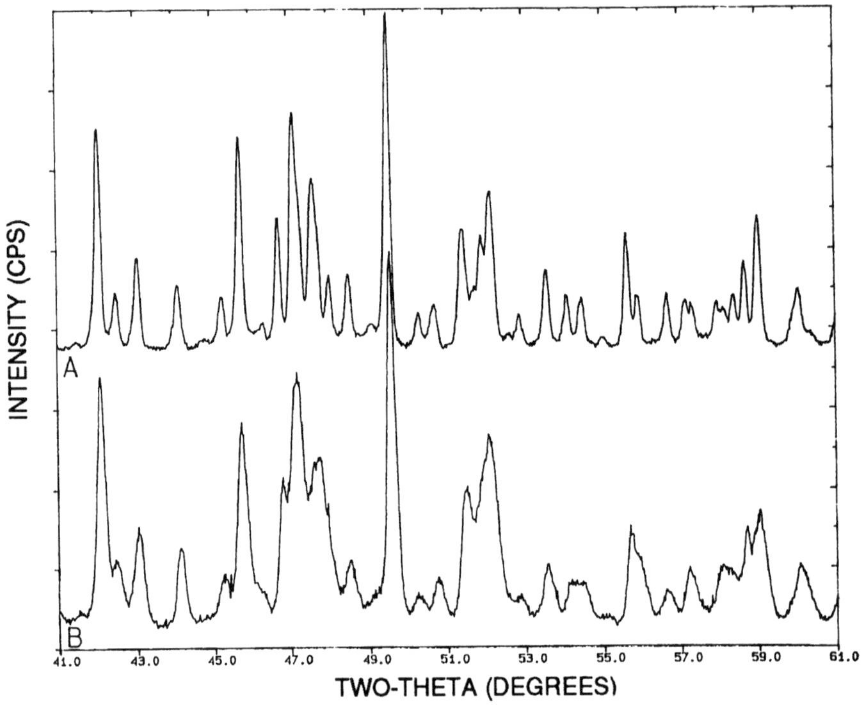

Figure 3.7. Comparison of X-ray diffraction patterns of $Nd(OH)_2NO_3.H_2O$. A - $CuK_{\alpha1}$, B - $CuK_{\alpha1} - K_{\alpha2}$. Reprinted with permission from The Royal Swedish Academy of Sciences (Louër 1988).

tal aberrations decrease with increasing scattering angle, chromatic aberration increases, resulting in an almost constant peak width for scattering angles of up to 50°-60°. Chromatic aberration may be reduced if the α_2 component is removed from the primary beam by a focusing crystal monochromator, usually quartz or germanium, having a narrow spectral band. The intensity of the primary beam reflected by the monochromator is also reduced significantly, in many cases by as much as tenfold. However, two intensity-reducing components are not needed if a monochromator is used:

1. β filter (since the K_β component is removed by the monochromator), and
2. Primary beam Soller slits (since the axial divergence is suppressed by the monochromator).

Removal of these two components gives an approximately fourfold intensity gain, which compensates for intensity losses due to reflection by the monochromator. Louër and Langford (1988) reported a peak intensity of α_1 reflections with a primary beam monochromator of as high as 80% of the corresponding peak in conventional diffractometry.

Another popular technique of monochromatization proposed by Rachinger (1948), and further developed in different ways by Delhez and Mittemeijer (1975) and Ladell et al. (1975), is analytical and not physical (with a monochromator). The elimination ("stripping") of the α_2 component is carried out through data processing, assuming a certain model for the $K_{\alpha 1}$–$K_{\alpha 2}$ doublet. There is no intensity loss, but systematic errors may result from errors in the doublet modeling. An impressive comparison of two patterns, one run on a K_α doublet and the other on monochromatized $K_{\alpha 1}$ radiation, is given in Fig. 3.7. Removal of α_2 drastically reduces peak width and peak overlapping.

3.2.5 *Diffraction by a thin-layer specimen*

Although the infinite thickness in the meaning used in Section 2.2 is often achieved with a depth of less than 1 mm, the amount of sample available may not be sufficient to meet this condition. In order to obtain a suitable expression for the intensity, we will replace the upper limit of integration in equation (3.6) by t, the thickness of the sample. The actual volume of the material diffracting X-rays into the detector's receiving slit depends on the relationship between the convergent angle of the diffracted beam β and the divergent angle of the incident beam α (Fig. 3.2b). Let us consider the two possible cases.

Case 1. Convergent angle β > divergent angle α

When the convergent angle β is greater than the divergent angle α, the width S_r of the reflected beam is greater than the width S_i of the incident beam (Fig. 3.8a). Then, the reflecting volume is a rectangular parallelepiped ACKM with a cross section equal to that shown in Fig. 3.8a and a height equal to the width of irradiated area in the axial direction. Then, by integrating equation (3.6) in the limits 0 and t, instead of an infinite upper limit, we obtain:

$$I_{ij} = I_0\, E_{ij}\, (B/2\mu)\, v_j\, [1 - \exp(-2\mu t/\sin\theta_i)] \qquad (3.16)$$

After performing rearrangements and substitutions identical with those made in the derivation of equation (3.11), we obtain:

$$I_{ij} = \frac{K_{ij}\, c_j}{\rho_j\, \mu^*}\,[1 - \exp(-2\mu t/\sin\theta)]$$

$$= \frac{K_{ij} c_j}{\rho_j\, \mu^*}\,[1 - \exp(-2\mu^* A/\sin\theta)]$$

$$= (I_{ij})'\,[1 - \exp(-2\mu t/\sin\theta)] \qquad (3.17)$$

where K_{ij} is the calibration constant, ρ_j is the density of phase j, μ^* is the mass absorption coefficient of the sample, $A = \rho t$ [cm^2/g] is the mass of the sample per 1 cm^2 ("surface density"), and $(I_{ij})'$ is the intensity diffracted by an infinitely thick specimen.

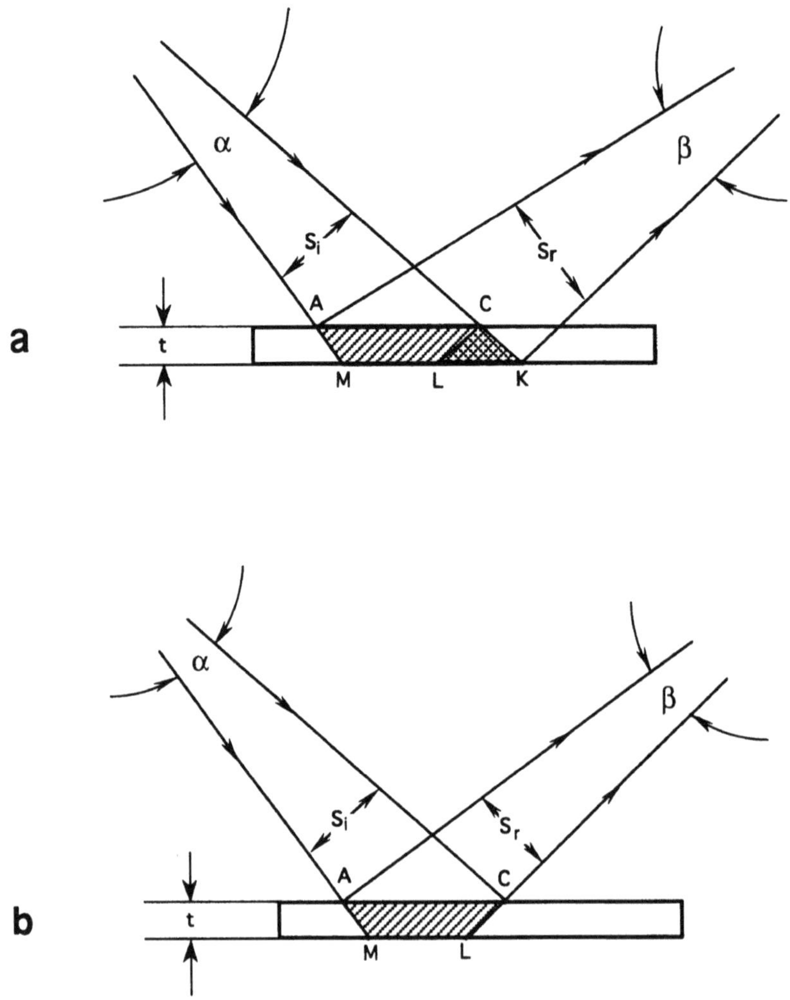

Figure 3.8. Irradiated volume in thin-layer diffraction. a - Convergent angle β is greater than divergent angle α. b - Convergent angle β is equal to divergent angle α.

Case 2. Convergent angle β = divergent angle α

When the convergent angle β is equal to the divergent angle α, the reflecting volume is a truncated prism ACLM (Fig. 3.8b), which has a smaller volume than the parallelepiped ACKM. The difference in volumes is equal to the volume of the triangular prism CLK, cross-hatched in Fig. 3.8a. The ratio of volumes:

$$\frac{\text{volume CLK}}{\text{volume ACKM}} = \frac{t\cos\theta}{S_i}$$

where $S_i = R\alpha$, the width of the incident beam near the specimen is easily deduced from Fig. 3.8a. Depending on the divergence α, S is equal to 1-3 mm, whereas t is less than 0.1 mm. Thus, the difference between the volumes ACKM and ACLM is insignificant, and in the forthcoming discussion we will refer only to equation (3.16).

Case of a thin layer

For extremely thin layers, the exponent in the right-hand side of equation (3.16) may be expanded in a power series with only the first two terms retained:

$$\exp[-2\mu t/\sin\theta] \approx 1 - 2\,\mu t/\sin\theta$$

and equation (3.16) transforms into:

$$I_{ij} = I_0\, E_{ij}\, (B/\sin\theta_i)t\, v_j = I_0\, E_{ij}\, V\cdot v_j \tag{3.18}$$

where $V = (B/\sin\theta_i)t$, the volume of material illuminated by the X-ray beam, i.e., for a low-absorbing specimen, the intensity increases linearly with the specimen volume—the result expected in accordance with the basic equation (3.4).

From equations (3.17), we obtain an expression that is practical for analytical purposes:

$$I_{ij} = \frac{2\,K_{ij}\;b_j\,A}{\rho_j \sin\theta_i} = \frac{K'_{ij}\;M_j}{\sin\theta_i} \tag{3.19}$$

where $K'_{ij} = 2\,K_{ij}/\rho_j$ and $M_j = c_jA$ is the mass of phase j on 1 cm^2 of specimen surface.

If the irradiated area of the specimen is kept constant by a "θ-compensating" divergence slit, then the term (Kimmel 1987):

$$V = \frac{B_i}{\sin\theta_i}\cdot t \;=\; S\,\frac{\sin\theta_i}{\sin\theta_i}\cdot t \;=\; S\cdot t$$

appears in equation (3.18), giving it a simpler form that is, however, valid only for a thin-layer specimen:

$$I_{ij} \;=\; K'_{ij}\, M_j$$

Equation (3.18) is very important in the analysis of low-mass deposits and thin films, and a detailed discussion is presented in Section 4.13.

The virtues of the Bragg-Brentano reflection focusing scheme are as follows:

1. Angle-independent (for constant divergence mode) absorption correction;
2. Constant specimen-detector distance;
3. Normal incidence of the diffracted beam relative to the focusing circle;

4. Almost nonrestricted scattering angle range; and
5. Moderate or high level of diffracted intensities ensuring good detection of minor phases and good counting statistics.

The disadvantages of this scheme are:

1. High level of instrumental aberrations in the low-angle range ($2\theta < 60°$);
2. Sensitivity to surface-related effects, such as surface roughness and surface-layer preferred orientation; and
3. Intrinsic sequence measurement mode.

An additional important feature of the Bragg-Brentano diffractometer is that the diffracted planes are in the same position relative to the sample surface at all 2θ angles. If the surface is in a mirror position, then the focus condition is fulfilled and there is no sensitivity to residual stresses.

3.3 Diffractometers with Seeman-Bohlin reflection focusing

3.3.1 *Absorption correction*

The focusing plane of the Seeman-Bohlin arrangement is shown schematically in Fig. 3.9. Soller slits S1 and S2, divergence slit DS, receiving slit RS, and antiscatter slit AS define the divergence and the dimensions of the primary and diffracted beams. The primary beam with divergence α strikes the specimen SP at a constant angle γ. The diffraction beam converges at the point of the focusing circle with radius r_f. The source-specimen distance FP is constant and is equal to:

$$r_1 = 2\, r_f \sin \gamma \tag{3.20}$$

The specimen-receiving slit distance P–RS varies with the angle 2θ:

$$r_2 = 2\, r_f \sin (2\theta - \gamma) \tag{3.21}$$

The irradiated surface of the specimen is constant and is equal to:

$$S = B/\sin \gamma = r_1 H \alpha/\sin \gamma = 2r_f H \alpha$$

where H is the effective beam width along the axial direction. Note that in this scheme the diffraction vector $\bar{S}$ is not parallel to the specimen normal $\bar{n}$ but is inclined by an angle $\theta - \gamma$ (Fig. 3.9). In other words, the reflecting planes are not parallel to the specimen surface, as in the Bragg-Brentano arrangement, but are oblique to the specimen surface by the same angle. This fact may be very important in calculation corrections for preferred orientation of crystallites, as discussed in Chapter 5.

Although in the Seeman-Bohlin scheme the actual reflecting volume is a finite prism, as in the Bragg-Brentano arrangement, we will

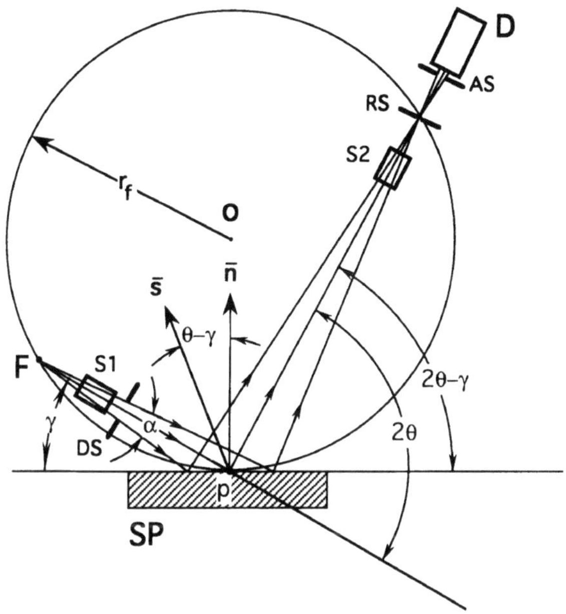

Figure 3.9. Focusing plane of the Seemen-Bohlin reflection focusing scheme.

consider it infinite, taking into account that the difference between these two approaches is significant only for very low absorbing materials. Thus, the volume of the thin layer at a depth x (Fig. 3.10) may be expressed as:

$$V = (B/\sin\gamma)\,\Delta x = H(r_1\alpha/\sin\gamma)\,\Delta x = 2\,H\,r_f\,\alpha\,\Delta x$$

and the intensity diffracted by this layer as:

$$dI_{ij} = I_0\,E_{ij}\exp[-\mu x/\sin\gamma\,]\exp[-\mu x/\sin(2\theta-\gamma)]\,(B/\sin\gamma)\,v_j\,\Delta x$$

where E_{ij} is defined in equations (2.19) and (2.20).

Integrating for semi-infinite media, we obtain:

$$I_{ij} = \int_0^\infty dI_{ij} = \frac{I_0\,E_{ij}\,(B/\sin\gamma)}{\mu[1/\sin\gamma + 1/\sin(2\theta - \gamma)]}\,v_j \qquad (3.22)$$

The final intensity equation is obtained from equation (3.22) by substituting E_{ij} with its expression from equation (2.21) and taking into account the fact that the specimen-detector distance is not constant but

$$I_{ij} = \left[\frac{r_e^2}{32\,\pi}\right]\left[\frac{I_0\,B\,H_c\,\lambda^3}{2r_f}\right]\left[\frac{1+\cos^2 2\theta_i}{\sin^2\theta_i\cos\theta_i}\right]\left[\frac{1}{\sin(2\theta_i-\gamma)+\sin\gamma}\right]\left[\frac{|F_{ij}|^2\,D_{ij}\,M_{ij}}{\vartheta_j^{\,2}}\right]\left[\frac{v_j}{\mu}\right] \qquad (3.23)$$

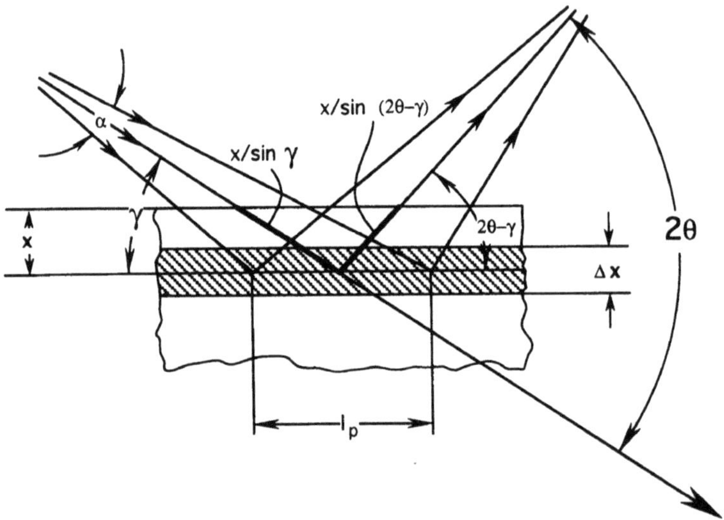

Figure 3.10. Derivation of the absorption factor in Seeman-Bohlin reflection focusing geometry.

Combining the constant terms of equation (3.23) into one calibration constant K_{ij} and replacing the volume fraction v_j by the weight fraction c_j, as was done in Section 3.2, we reach the basic equation for quantitative analysis for a diffractometer with Seeman-Bohlin reflection focusing:

$$I_{ij} = \left[\sin\,(2\theta_i - \gamma) + \sin\,\gamma\right]^{-1} \frac{K_{ij}}{\rho_j} \frac{c_j}{\mu^*} \tag{3.24}$$

Note that in equations (3.23) and (3.24),there is an additional 2θ-dependent term, $1/[\sin\,(2\theta_i\text{-}\gamma) + \sin\gamma]$, in comparison with the corresponding equations (3.8) and (3.11) for the symmetrical Bragg-Brentano case.

If the thickness of specimen t is finite, then by integration of expression (3.22) from 0 to t, we will obtain the following expression for the diffracted intensity:

$$I_{ij} = \frac{K_{ij}\, c_j}{\rho_j\, \mu^*} \left[\sin(2\theta_i - \gamma) + \sin\,\gamma\right]^{-1} \left\{1 - \exp\left[-\mu t \left((\sin\,\gamma)^{-1} + (\sin\,2\theta_i - \gamma)^{-1}\right)\right]\right\} \tag{3.25}$$

For a very thin specimen, the exponent in expression (3.25) may be expanded in a power series with only the first two members. Following the same computation as we carried out in Section 3.2, we will then obtain:

$$I_{ij} = \frac{K_{ij}\, M_j}{\rho_j} \left[\sin\,(2\theta_i - \gamma)\,\sin\,\gamma\right]^{-1} \tag{3.26}$$

where M_j is the mass of the phase j on 1 cm^2 of the specimen surface.

3.3.2 *Instrumental aberrations*

The major sources of instrumental aberration are the same as those in the symmetrical Bragg-Brentano case, namely, X-ray beam divergence, finite width of the tube focus and receiving slit, specimen transparency, and specimen shift in the normal direction. The expressions for corresponding instrumental functions can be found elsewhere (Segmüller 1957; Zevin and Kheiker 1967; Parrish and Mack 1967; Mack and Parrish 1967). As in the symmetrical case, the range of low scattering angles is most severely affected by instrumental aberrations, which fall to a minimum at $\theta = \gamma$, i.e., for the symmetrical case.

We will illustrate this by a peak centroid shift ε due to a flat-specimen aberration given by:

$$\varepsilon_{S\text{-}B} = -\frac{\alpha^2 \sin 2\theta}{12\sin(2\theta - \gamma)\sin\gamma} \tag{3.27}$$

for the Seeman-Bohlin scheme, and

$$\varepsilon_{B\text{-}B} = -\frac{\alpha^2 \cot\theta}{6} \tag{3.27'}$$

for the Bragg-Brentano scheme.

The ratio of the peak shift in Seeman-Bohlin geometry ($\gamma \neq \theta$) to that in symmetrical Bragg-Brentano geometry ($\gamma = \theta$) $\varepsilon_{S\text{-}B}/\varepsilon_{B\text{-}B}$ shows a sharp increase with increasing scattering angle (Fig. 3.11). Moreover, the reduction of angle-dependent aberrations with 2θ is much sharper for the symmetrical scheme. As a result, in Seeman-Bohlin focusing we can expect a greater contribution of instrumental aberrations to the overall peak broadening in the intermediate scattering range of 30°-60°.

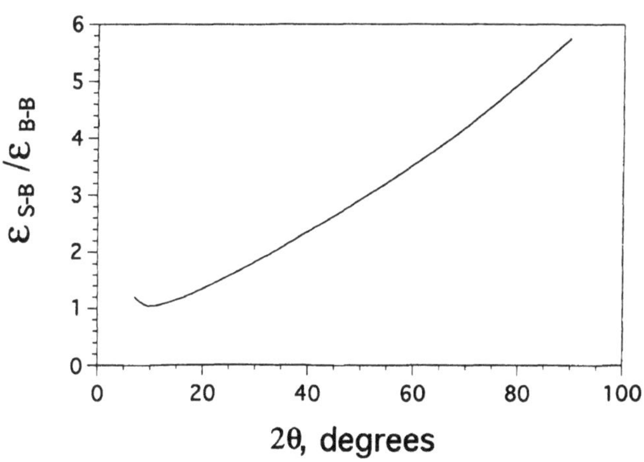

Figure 3.11. Ratio of flat specimen aberrations in Seeman-Bohlin and Bragg-Brentano focusing schemes.

3.3.3 *Comparison with the Bragg-Brentano scheme*

As usual, we will consider three terms in the general intensity equation (2.21): the intensity of the primary beam I_0, the absorption factor A, and the illuminated specimen volume V. The basic symmetrical scheme (for purposes of comparison) is obtained from the regular Seeman-Bohlin arrangement by shifting the specimen along the focusing circle to a position $\gamma = \theta$. Thus, the expressions for the absorption correction, the integrated intensity, etc., can be obtained from the corresponding Seeman-Bohlin expressions by simple substitution of θ for γ. In other words, we suppose that primary beam divergence in the focusing plane α and irradiated specimen surface and radius of focusing circle r_f are equal for both schemes.

If a point source of X-rays is used, then the intensity of the X-rays at a distance R from the source will be inversely proportional to R^2. In practical X-ray diffractometry, line sources are utilized, and the axial divergence of the primary beam is controlled by Soller slits.

As a first approximation, the intensity of the primary beam at the specimen surface is proportional to the angle of divergence of the element controlling the divergence of the X-rays in the axial direction. If this angle is kept constant and independent of the source-specimen distance, then our scheme is transformed into a two-dimensional scheme with the primary beam divergent in the focusing (diffraction) plane. Thus, the intensity of the primary beam is proportional to R^{-1} and not to R^{-2}.

The source-specimen distance is equal to $r_1 = 2r_f \sin\gamma$ for Seeman-Bohlin focusing and $R = 2r_f \sin\theta$ for the symmetrical case. The ratio of the primary beam intensities is equal to:

$$\frac{(I_0)_{S\text{-}B}}{(I_0)_{B\text{-}B}} = \frac{R}{r_1} = \frac{\sin\theta}{\sin\gamma} \tag{3.28}$$

which emphasizes an intensity gain for Seeman-Bohlin geometry if $\gamma < \theta$.

We will now treat the absorption correction and the volume of the illuminated specimen material. It follows from equations (3.21) and (3.22) that the product of these two factors for the Seeman-Bohlin case is equal to:

$$(AV)_{S\text{-}B} = \frac{B/\sin\gamma}{\mu[1/\sin\gamma + 1/\sin(2\theta - \gamma)]} = \frac{2\, r_f H \alpha}{\mu[1/\sin\gamma + 1/\sin(2\theta - \gamma)]}$$

where B is the cross section of the primary beam near the specimen. The ratio:

$$\frac{(AV)_{S\text{-}B}}{(AV)_{B\text{-}B}} = 2\frac{\sin\gamma}{\sin\theta}\,\frac{1}{1 + \sin\gamma/\sin(2\theta - \gamma)} \tag{3.29}$$

is advantageous for the symmetrical case.

Combining two intensity-related factors expressed in equations (3.28) and (3.29) into one factor and supposing that the fraction of the Debye ring intercepted by the receiving slit is constant in both schemes, we obtain the ratio of diffracted intensities for two geometrical schemes:

$$\frac{I_{S\text{-}B}}{I_{B\text{-}B}} = \frac{(I_0)_{S\text{-}B}}{(I_0)_{B\text{-}B}} \cdot \frac{(AV)_{S\text{-}B}}{(AV)_{B\text{-}B}} = \frac{2}{1 + \sin\gamma / \sin(2\theta - \gamma)} \tag{3.30}$$

The ratio is higher for the Seeman-Bohlin scheme and increases from an obvious 1 for $\gamma = \theta$ up to ~2 for $\gamma << 2\theta$.

However, as was emphasized earlier, the Seeman-Bohlin scheme is subject to a greater level of instrumental aberrations. If we increase the intensity gain for the latter up to equalizing the instrumental distortion in both geometries, the benefits of Seeman-Bohlin geometry are reduced (Zevin 1988). In a previous comparison (Mack and Parrish 1967; Feder and Berry 1970; Parrish 1974) of a number of specific instruments, which showed the greater intensity advantages of the Seeman-Bohlin scheme, intensity rather than peak distortion was of primary interest, and not all the geometric factors were kept constant.

Seeman-Bohlin focusing geometry has a real advantage in the analysis of the composition and structure defects of thin layers and films, especially on polycrystalline substrates. Since the incidence angle is usually low, penetration depth of the X-ray beam is small, sometimes no more than a few hundred angstroms for heavy atom materials. Thus, diffraction by the substrate material, which might be the major constituent of the background, is reduced. It must be emphasized that the effect is caused by eliminating the substrate and not by enhancing the film diffraction.

Seeman-Bohlin geometry provides a means to study preferred orientation or even integration (averaging) over crystallite orientations. Indeed, if the specimen can be moved continuously or stepwise along the focusing circle, then the angle between the diffraction vector and the specimen normal (Fig. 3.9) will vary. The diffracted intensity at each specimen position corresponds to the reflection by grains with a certain orientation. By averaging these intensities (of course, after the relevant absorption correction), one can partially compensate for the effect of preferred orientation. If the same operation is performed with a Bragg-Brentano diffractometer, significant defocusing and peak broadening result.

The virtues of the Seeman-Bohlin reflection scheme are:

1. Availability of both sequence and simultaneous measurement modes;
2. A high level of diffracted intensity, ensuring detection of minor phases; and
3. Availability of glancing incident angle geometry and ability to study the surface layer composition.

The disadvantages of this scheme are:

1. High level of instrumental distortion in the low scattering angle range ($2\theta<60°$);

2. Angle dependence of the absorption correction;
3. Restriction of the scattering-angle range on the low-angle side;
4. Sensitivity to the surface-related effects, like surface roughness, preferred orientation, and residual stress; and
5. Oblique incidence of the diffracted beam, making it difficult to maintain high angular resolution in PSD.

3.4 Transmission technique with constant specimen-detector distance (Bragg-Brentano transmission analog)

Here we will consider two transmission geometries, both employing an incident beam monochromator to converge the beam emitted by the X-ray tube. Although other arrangements with diffracted beam monochromators are possible (Parrish and Roberts 1962; de Wolff 1968; Lang 1956), the results are similar. Moreover, since commercially manufactured transmission mode diffractometers include an incident beam monochromator, it is preferable to discuss this arrangement.

3.4.1 *Geometric scheme*

The optical principle of focusing is shown in Fig. 3.12. The primary beam from the tube focus F_0 is reflected by a curved crystal monochromator M. The monochromatic beam passes through the center of the diffractometer circle C and converges at the point F on this circle. The specimen is placed in the center of this circle. The normal $\bar{n}$ to the specimen surface is the median of the scattering angle 2θ, giving this scheme a symmetrical appearance. In order to maintain this arrange

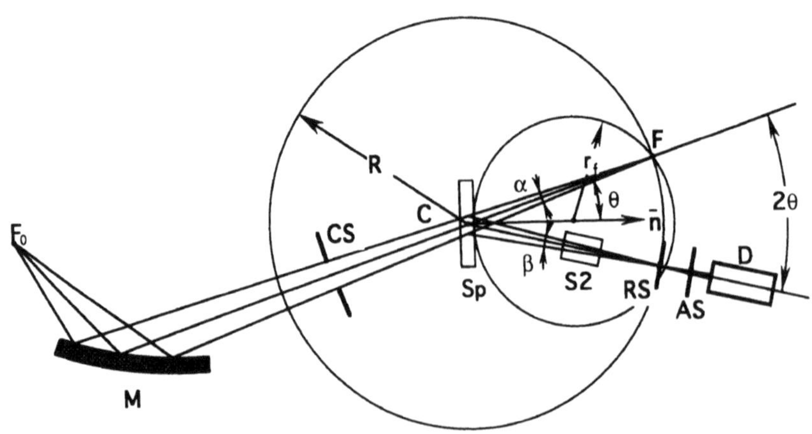

Figure 3.12. Focusing plane of the transmission scheme with constant specimen-detector distance (Bragg-Brentano transmission analog).

ment during a diffractometric run, the specimen is rotated continuously at half the speed of rotation of the detector. The diffracted beam converges in front of the receiving slit RS. Note the normal incidence of the diffracted beam on the diffractometer circle. The specimen-receiving slit distance is constant and equal to the diffractometer radius R. The slit CS defines the convergent angle α of the primary beam, and the two slits RS (receiving slit) and AS (antiscatter slit) define the convergent angle β of the diffracted beam. Divergence of the incident beam in the axial direction is controlled by the monochromator, and there is no need for primary beam Soller slits. However, one set of Soller slits S2 is needed to control the axial divergence of the diffracted beam.

If we consider the monochromator focus point F as an imaginary source, then the entire arrangement is very similar to the back-reflection Bragg-Brentano case. The focusing circle with radius $r_f = R/2 \sin \theta$ is tangent to the specimen surface and passes through the focal point F and the center of the receiving slit RS.

3.4.2 *Absorption correction*

The irradiated surface of the specimen is equal to $RH\alpha/\cos \theta$, and the volume of a thin layer at a depth x inside the specimen irradiated by the X-rays is equal to (Fig. 3.13):

$$V = (HR\alpha/\cos \theta)\, \Delta x = (B/\cos \theta)\, \Delta x \tag{3.31}$$

where H is the width of the irradiated surface in the axial direction and B is the cross section of the incident beam near the specimen. The distance traversed by the incident beam inside the sample to reach the volume V is equal to $x/\cos \theta$ (Fig. 3.13), and the absorption pass of the diffracted beam is equal to $(t - x)/\cos \theta$. The intensity diffracted by the volume V, according to equation (2.20), is:

$$dI_{ij} = \frac{I_0\, E_{ij}\, HR\alpha}{\cos \theta_i} \exp[-\mu x/\cos \theta_i]\, \exp[-\mu(t - x)/\cos \theta_i]\; v_j\, \Delta x$$

Integration over the whole sample thickness leads to:

$$I_{ij} = \int_0^t dI_{ij} = \frac{I_0\, E_{ij}\, Bt}{\cos \theta_i} \exp[-\mu t/\cos \theta_i]\; v_j \tag{3.32}$$

Before substitution of E_{ij} by its expression from equation (2.19), we must take into account the fact that our scheme uses an incident-beam monochromator, which changes the polarization of the incident beam. The new polarization factor is defined by equation (2.22). Keeping this in mind, we obtain the following expression for the diffracted intensity:

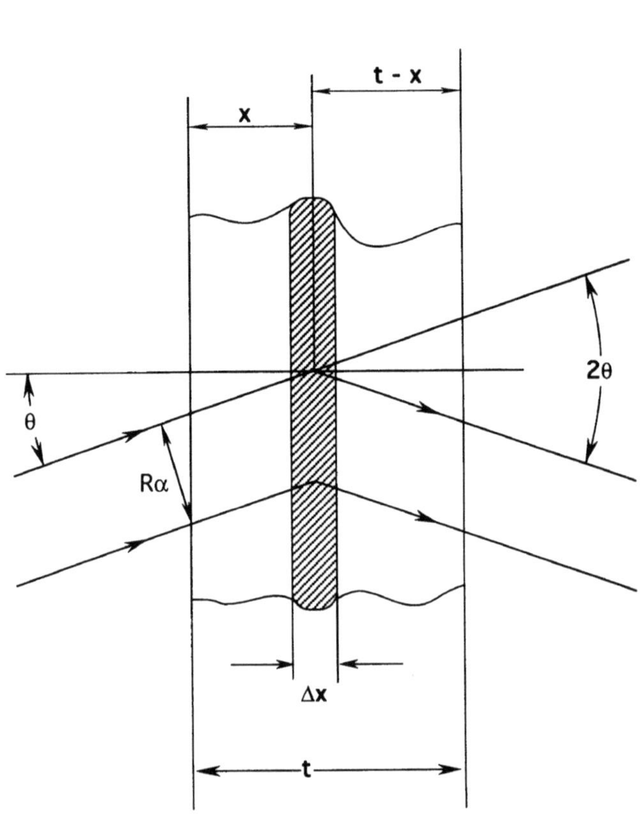

Figure 3.13. Derivation of the absorption factor for the symmetrical transmission scheme.

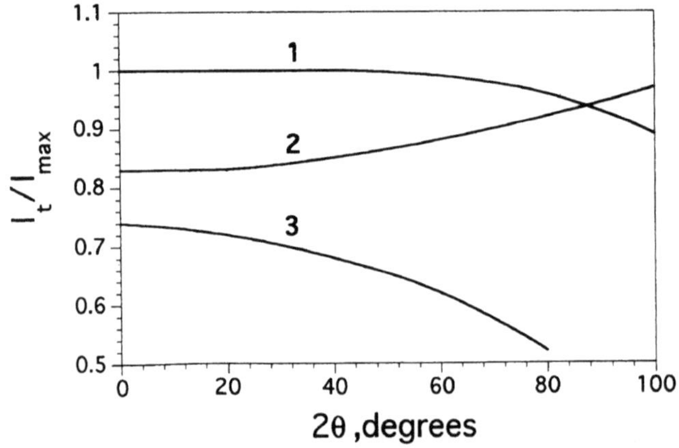

Figure 3.14. Relative intensity in the transmission mode for various specimen thicknesses t. $t = \mu^{-1}$(1); $t = 0.5\,\mu^{-1}$ (2); and $t = 2\,\mu^{-1}$ (3).

$$I_{ij} = \left[\frac{r_e^2}{32\pi}\right]\left[\frac{I_0 H_c \lambda^3}{R}\right]\left[\frac{1 + K\cos^2 2\theta_i}{2\sin^2\theta_i \cos\theta_i}\right]\left[\frac{|F_{ij}|^2 D_{ij} M_{ij}}{\vartheta_j^2}\right][\exp(-\mu^* A/\cos\theta_i]\left[\frac{Bt}{\cos\theta_i} v_j\right] \tag{3.33}$$

where A (=ρt) is the surface density or the mass of material per 1 cm^2 of the specimen. The term $V = Bt/\cos\theta_i$ is the irradiated volume of the specimen. It is more convenient to convert the product Vv_j into the product of the illuminated mass of the specimen G and the weight fraction c_j:

$$Vv_j = G\,c_j/\rho_j$$

Finally, assigning the first four terms of equation (3.33) to be a calibration constant K_{ij}, we arrive at the equation for QXRD in transition geometry with a fixed diffractometer circle:

$$\begin{aligned} I_{ij} &= K_{ij}\exp[-\mu t/\cos\theta]\frac{Bt}{\cos\theta} v_j \\ &= K_{ij}\exp[-\mu^* A/\cos\theta]Vv_j \\ &= \frac{K_{ij}}{\rho_j}\exp[-\mu^* A/\cos\theta]c_j \end{aligned} \tag{3.34}$$

where t is the specimen thickness, B is the cross section of the incident beam near the specimen, A is the surface density of the specimen, V is the volume, and G is the mass of the illuminated part of the specimen.

There is an optimal specimen thickness t_o (illuminated specimen volume V_o or mass G_o) that produces the maximal diffracted intensity. This is easily found by taking t as a variable and differentiating the left-hand side of equation (3.32). Thus,

$$t_o = \cos\theta/\mu \tag{3.35}$$

and under these conditions, assuming $v_j = 1$:

$$(I_{ij})_{max} = I_0\,E_{ij}\,(B/\mu)e^{-1} \tag{3.36}$$

The optimal specimen thickness is dependent on θ and can be satisfied only for one scattering angle. However, an approximation

$$t_o \approx \mu^{-1} \tag{3.37}$$

is quite acceptable for a front reflection range of $2\theta < 60°$. Deviation of the specimen thickness from $t \approx \mu^{-1}$ will reduce the peak intensity, although the function I(t) is rather shallow in the vicinity of its maximum. The series of curves I_t/I_{max} [where I_{max} is defined in equation (3.36) and I_t corresponds to the specimen of constant thickness t] is given in Fig. 3.14. Deviations of specimen thickness from $0.5\mu^{-1}$ up to $2\mu^{-1}$ lead

to a reduction of intensity by no more than 10%-30% for $2\theta \leq 60°$, i.e., in the range where the transmission technique is most effective.
Another very important feature of the I_t/I_{max} curves shown in Fig. 3.14 is their relatively slow variation with the scattering angle in the front reflection range. In other words, if the specimen thickness $t \approx 1/\mu$, the absorption correction is almost constant for $2\theta < 60°$. This phenomenon resembles the constant-divergence Bragg-Brentano reflection geometry for an infinitely thick specimen, where this correction is also angle-independent. The important consequence for QXRD is that peak ratios are not dependent on experimental conditions, such as specimen thickness or radiation used. Analytical techniques based on that assumption are discussed in Chapter 4. Some experimental results may be found in the paper of Davis and Spilde (1990).

3.4.3 Instrumental aberrations

Looking at the schematic presentation of transmission geometry in Fig. 3.12, we realize that there is a striking similarity with the back-reflection Bragg-Brentano case (Fig. 3.2). We can imagine the focus of the monochromator F to be the X-ray source, and then the diffraction peaks appear to be reflected by the planes parallel to the specimen surface. Of course, this analysis is only imaginary, since real reflecting planes are normal to the specimen surface and not parallel to it. Nevertheless, the entire theory of instrumental aberration developed for Bragg-Brentano reflection focusing is applicable here, with the simple substitution of the angle θ with the complementary angle $90° - \theta$. This means that the level of instrumental aberration for the scheme discussed here is equal to that for the back-reflection range in reflection Bragg-Brentano geometry, i.e., it is rather low; for example, the specimen displacement S in the direction normal to the specimen surface causes a diffraction peak shift by:

$$\Delta 2\theta = (2\,S/R)\sin\theta \qquad (3.38)$$

i.e., much less than the corresponding shift in the reflection Bragg-Brentano scheme, which is equal to $(2S/R)\cos\theta$ provided that $\theta < 45°$. The same is true for peak shift and peak broadening due to X-ray divergence in the focusing and axial (vertical) planes. The sample transparency effect, however, cannot be derived from the reflection analogy due to fundamental differences in absorption characteristics. In the transmission case discussed here, the contributions of each specimen layer to the total diffracted intensity are equal. The instrumental function defined by finite specimen thickness is, therefore, rectangular, with the width being determined by equation (3.38), where the specimen shift S is replaced by the specimen thickness t:

$$\Delta 2\theta = (2t/R)\sin\theta \qquad (3.39)$$

For $t \sim \mu^{-1}$ and $R \approx 100$ mm this contribution to the overall peak broadening will be negligible in the front reflection range.

Symmetric transmission focusing, like the asymmetric Guinier scheme, has a low level of chromatic aberrations, as discussed in detail in Section 5.3.

3.4.4 Comparison between reflection and transmission Bragg-Brentano geometries

The perfect or nearly perfect crystal monochromator required to converge the primary (or reflected) beam in transmission focusing reduces the intensity flux I_0 by at least 5-10 times in comparison with reflection geometry, where the common diffracted beam graphite monochromator does not cause any intensity losses.

Let us now compare the absorption correction, both on an "equal specimen mass" and on an "equal beam divergence" basis. Suppose first that only a small amount of material is available and that in the transmission case it is spread over particular areas to produce a specimen with an optimal thickness of $\cos \theta/\mu$. The diffracted intensity is expressed by equation (3.36). If the same amount of material is used in the reflection scheme, then in order to obtain the maximal intensity, it must be spread over a large area to make the absorption negligible. In this case, equation (3.18) is valid. By converting equation (3.36) to the comparable form, we obtain:

$$(I_{ij})_T = (I_0)_T \, E_{ij} \, V \, e^{-1} \tag{3.40}$$

Assuming equality of the E_{ij} terms, including equality of the goniometer radius and the receiving slit width for the two geometric schemes, we come to the ratio of intensities:

$$I_R/I_T = [(I_0)_R/(I_0)_T] \, e \tag{3.41}$$

and the additional gain for the reflection case is equal to e. This benefit, however, might only be theoretical, because in order to achieve it, we have to prepare (in the reflection case) a very thin uniform layer of material spread over a very large surface, which might prove difficult.

Let us consider another case, that involving an abundant amount of material and a constant beam divergence α (beam cross section B) for both schemes. Comparing the intensities of the reflection and transmission cases given in equations (3.7) and (3.36), and assuming $v_j = 1$, we obtain:

$$I_R/I_T = [(I_0)_R/(I_0)_T] \, e/2 \tag{3.42}$$

A modest benefit is achieved in reflection focusing due to a more favorable absorption factor.

Instrumental aberrations for the two schemes have already been discussed. Transmission geometry exhibits a much lower level of instrumental and chromatic aberrations in the front reflection range than its reflection counterpart. Of course, chromatic aberration can also be reduced in the reflection case by using a perfect crystal monochromator. However, in this case, the intensity benefit of the reflection scheme will be lost.

Instrumental distortion in the reflection case may be reduced in principle by decreasing beam divergence at the expense of intensity loss. Although a comparative study has not been carried out, it is unlikely that the intrinsic intensity gain of the reflection mode is sufficient to confer the same "transmission" level of instrumental distortion by a mere decrease of beam divergence. Thus, for the two diffractometric modes discussed here, we have different levels of potential diffracted intensities and of diffraction peak broadening:

1. The transmission scheme, which has a low degree of instrumental and chromatic aberrations and a relatively low diffracted intensity; and
2. The reflection scheme, which has a higher degree of instrumental and chromatic aberrations and a higher diffracted intensity.

The virtues of the symmetrical transmission focusing scheme (Bragg-Brentano transmission analog) are:

1. Very low level of instrumental and chromatic aberrations;
2. Almost constant (θ-independent) absorption correction;
3. Normal incidence of the diffracted beam;
4. Ease of estimation of the absorption correction; and
5. Relative independence of surface-related sample effects.

The disadvantages of this scheme are:

1. Low level of diffracted intensities;
2. Only sequential mode measurements are viable; and
3. Scattering angle range restricted to front reflections.

3.5 Transmission technique with an invariant focusing circle (Guinier diffractometer or the Seeman-Bohlin transmission analog)

3.5.1 General description

This scheme was first suggested by Guinier for the X-ray camera and was adapted to the diffractometer a few years ago. The focusing plane of this geometric arrangement is shown schematically in Fig. 3.15. The X-ray beam emerging from the tube focus F_0 and reflected by the monochromator M converges at point F on the focusing circle. The angle γ between the specimen surface and the central line of the incident beam is less than 90° and is usually ≈ 65°-70°. The diffracted beam converges in front of the receiving slit RS on the focusing circle. The convergent angle of the primary beam α is controlled by the slit CS, and the convergent angle of the diffracted beam β is defined by the pair of slits RS and AS. If B is the cross section of the primary beam, then the irradiated specimen surface can be expressed by:

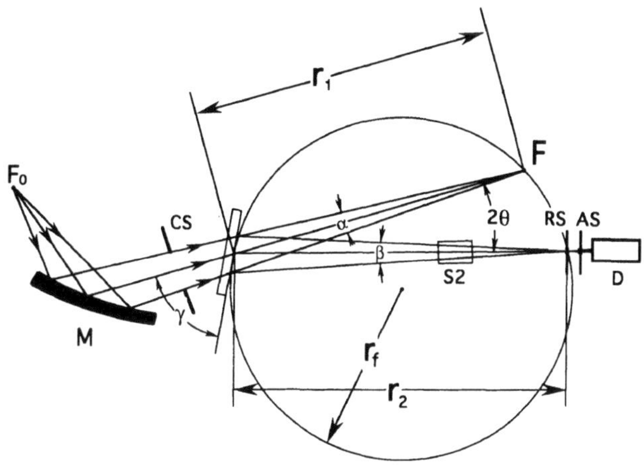

Figure 3.15. Focusing plane of the transmission scheme with constant focusing circle (Seeman-Bohlin transmission analog).

$$S = B/\sin\gamma = H\, r_1\, \alpha/\sin\gamma = 2H\, r_f\, \alpha \tag{3.43}$$

where $r_1 = 2\, r_f \sin\gamma$ is the distance between the specimen and the focal point F, r_f is the radius of the focusing circle, and H is the width of irradiated area in the axial (normal to the focusing plane) direction. The distance between the specimen and the receiving slit varies with the Bragg angle and is equal to:

$$r_2 = 2r_f \sin(2\theta+\gamma) \tag{3.44}$$

As in the symmetric transmission case, the axial divergence of the primary beam is controlled by a monochromator, and only one set of Soller slits S2 is needed to control the axial divergence of the diffracted beam.

If we take point F on the focusing circle as an imaginary source of X-rays, we immediately recognize the characteristic features of the Seeman-Bohlin (back-reflection) arrangement. Although this similarity is only imagined, it enables us to apply the theory of instrumental aberrations developed for the Seeman-Bohlin scheme to the Guinier diffractometer, by simple substitution of the angle θ with the complementary angle 90°– θ. In reflection focusing geometry, the back-reflection range is characteristic for reduced instrumental aberrations. The corresponding range in transmission Guinier geometry is front-reflection, as was noted for the symmetrical transmission scheme. In the Guinier diffractometer, one can expect a low level of geometric aberrations caused by beam divergence, specimen displacement, and transparency.

3.5.2 Absorption correction

The volume of a thin layer at a depth x irradiated by X-rays (Fig. 3.16) is equal to:

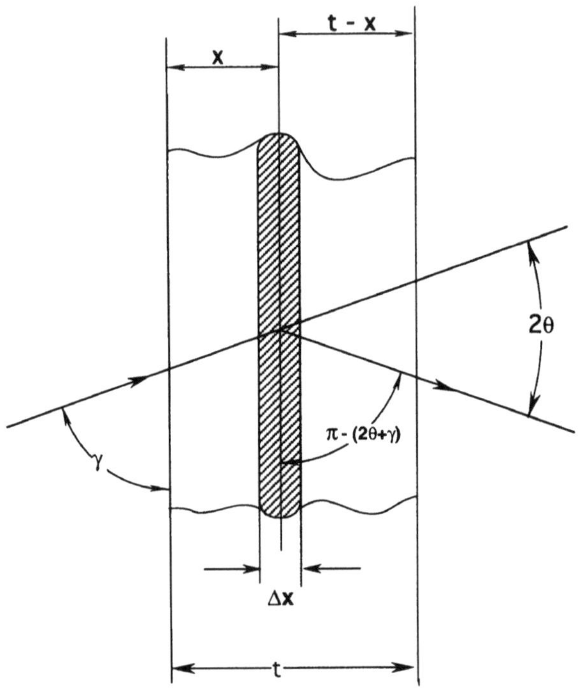

Figure 3.16. Derivation of the absorption factor for the asymmetrical transmission scheme.

$$V = S\,\Delta x = 2r_f\,\alpha\,\Delta x \tag{3.45}$$

where S is the irradiated specimen surface expressed in equation (3.43). The absorption path of the incident beam is given by $x/\sin\gamma$ and that of the diffracted beam by $x/\sin(2\theta + \gamma)$. The intensity diffracted by the layer Δx is expressed as:

$$dI_{ij} = I_0\,E_{ij}\,(B/\sin\gamma)\,\exp[-\mu x/\sin\gamma]\quad \exp[-\mu(t-x)/\sin(2\theta_i+\gamma)]\,\Delta x \tag{3.46}$$

The term E_{ij} has its usual meaning except that the proper polarization factor should be used for an incident beam polarized by a crystal monochromator.

We should also take into account the fact that in this case the specimen-detector distance is angle-dependent [equation (3.44)]. Thus, we retain the constant focusing circle radius r_f in the instrument-dependent second term. The angle-dependent multiplier $\sin(2\theta + \gamma)$ together with the angle-dependent multipliers of the absorption factor constitute a specific term in the expression for the intensity [equation 3.47)].

Integration of equation (3.46) in the usual limits of 0 and the specimen thickness t leads to the following expression for the intensity of diffraction:

$$I_{ij} = \left[\frac{r_e^2}{32\pi}\right]\left[\frac{I_0\,BH_c\,\lambda^3}{2\,r_f}\right]\left[\frac{1+K\cos^2 2\theta_i}{2\sin^2\theta_i\cos\theta_i}\right]\left[\frac{|F_{ij}|^2}{\vartheta_j^{\,2}}\right]\left[\frac{1}{\sin(2\theta_i+\gamma)-\sin\gamma}\right]$$
$$\times\left[\exp\left(-\mu t/\sin(2\theta+\gamma)\right)-\exp(-\mu t/\sin\gamma)\right]\left[v_j/\mu\right] \tag{3.47}$$

Finally, assigning the first five terms from equation (3.47) to be the calibration constant K_{ij}, we arrive at the quantification equation:

$$I_{ij} = \frac{K'_{ij}}{\mu}\left[\exp\left(-\mu t/\sin(2\theta+\gamma)\right)-\exp(-\mu t/\sin\gamma)\right]v_j$$
$$= \frac{K_{ij}}{\mu^*}\left[\exp\left(-\mu^* A/\sin(2\theta+\gamma)\right)-\exp(-\mu t/\sin\gamma)\right]c_j \tag{3.48}$$

where $c_j = v_j\,\rho_j/\rho$ is the weight fraction of phase j, ρ is the specimen density, and A is the specimen mass per cm^2 ("surface density").

As in the symmetric transmission case discussed earlier, the specimen thickness has an optimal value, corresponding to the maximal intensity gain. The optimal specimen thickness is obtained by differentiating the expression for the intensity, and in this case:

$$t_o = (1/\mu)\ln[\sin\gamma/\sin(2\theta+\gamma)]/[1/\sin(2\theta+\gamma)-1/\sin\gamma] \tag{3.49}$$

The optimal thickness is θ-dependent, but $t \sim 1/\mu$ is a rather good approximation.

If the specimen thickness is close to the optimal value, then the angle-absorption correction is almost constant in the front-reflection range. This in an important fact for quantification, as was discussed in Section 3.4.

3.5.3 *Chromatic aberration*

There are two chromatic dispersion elements in the transmission focusing scheme—the monochromator crystal and the specimen itself. With the proper geometric arrangement, these elements can act in opposite directions, resulting in a reduction of the separation (even up to almost coincidence) of α_1 and α_2 components. The monochromator, in fact, produces two focal points, F_1 (α_1 component) and F_2 (α_2 component) (Fig. 3.17), with linear separation:

$$\Delta S_1 = \psi L/\sin\gamma$$

where ψ is the angle between the α_1 and α_2 beams, L is the distance between monochromator crystal M and focus points F_1 or F_2, and γ is the angle of incidence. If the FWHM of the monochromator rocking curve is much less than the angular separation of the K_α doublet, then the α_1 and α_2 beams must be emitted from different points 1 and 2 of the tube focal spot (Fig. 3.17). In this case:

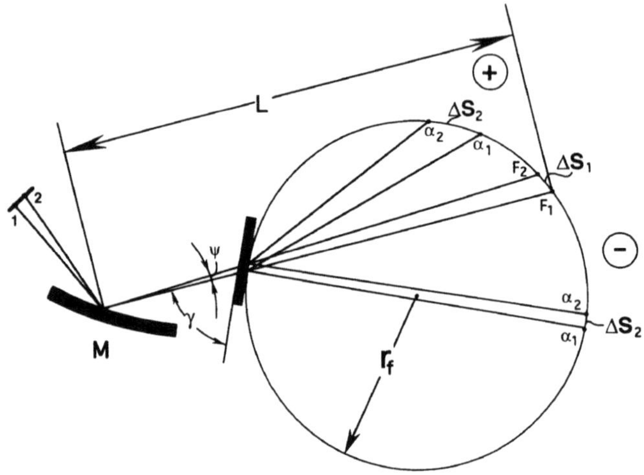

Figure 3.17. Chromatic dispersion in transmission focusing geometry.

$$\psi = \Delta\theta = \frac{\Delta\lambda}{\lambda} \tan \theta_m \qquad (3.50)$$

where $\Delta\lambda = \lambda\alpha_2 - \lambda\alpha_1$, and θ_m is the Bragg angle of the monochromator. For a crystal with a greater mosaic spread covering the angular separation of the doublet, the angle ψ may be larger, up to twice the value expressed in equation (3.50). The chromatic dispersion of the specimen is defined by equation (3.15). Linear separation of the K_α doublet in the diffraction pattern will be equal to:

$$\Delta S_2 = L\,(\Delta\lambda/\lambda)\tan \theta_m/(\sin \gamma) \pm 4r_f\,(\Delta\lambda/\lambda)\tan \theta \qquad (3.51)$$

where θ is the Bragg angle of the reflection under consideration, and the two terms on the right-hand side represent the chromatic dispersion of the monochromator and the specimen. The plus minus sign in equation (3.51) relates to the addition and subtraction portions of the diffraction pattern (upper and lower parts of the focusing circle in Fig. 3.17). In the subtraction parts, the α_1 and α_2 components will overlap, and at any specific angle θ_0 they will coincide completely. This angle depends on the divergence angle ψ, but in the case of a perfect crystal monochromator it is easily defined from equation (3.51):

$$\tan \theta_0 = (\tan \theta_m)\, L/4r_f \sin \gamma \qquad (3.52)$$

For L = 200 mm, r_f = 50 mm, and $\sin \gamma \approx 1$, the angle $2\theta_0 \approx 2\theta_m$, i.e., ~ 26.6° for quartz $(10\bar{1}1)$ and ~28.4° for Si (111) monochromators. This effect is appreciable up to scattering angles of 40°-50°, causing a significant reduction of peak width in comparison with chromatic schemes.

3.5.4 Comparison of symmetric and asymmetric transmission geometries

The comparative characteristics of transmission focusing techniques were discussed in Section 3.4.4. Here, we note only the special features characteristic of the asymmetric (Guinier) scheme. The stationary specimen distinguishes this scheme from the symmetrical one. It is also very important to remember that the diffracted beam is almost normal to the focusing circle, at least in the front reflection range. Indeed, if the primary beam meets the specimen at an angle $\gamma \approx 65°$, then in the range of $2\theta < 50°$ the diffracted beam will meet the focusing circle at any angle between 90° and 65°. This feature is advantageous in comparison with the reflection Seemen-Bohlin scheme, where an oblique incidence of the diffracted beam is characteristic for front-reflection. The stationary specimen and the almost normal incidence of the diffracted beam facilitate the use of a PSD in conjunction with the Guinier scheme. These detectors may compensate for the inherent weakness of the transmission scheme focusing—the low intensity level.

The virtues of the Guinier diffractometric scheme are:

1. A very low level of instrumental and chromatic aberrations;
2. An almost normal incidence of the diffracted beam, facilitating the use of PSDs;
3. An almost constant (θ-independent) absorption correction;
4. The possibility of both sequential and simultaneous measurement modes; and
5. Relative independence of surface-related sample effects.

The disadvantages of this scheme are:

1. Low level of integrated intensities; and
2. Restriction of the scattering angle range to the front-reflection side.

3.6 Debye-Sherrer geometry

3.6.1 General view

In the late 1950s and 1960s, Debye-Sherrer cameras were almost completely ousted from diffraction laboratories by X-ray diffractometers as a result of the better angular resolution of the new instruments, coupled with the reduced time needed for diffractometric runs and the ease of measuring intensity. As a result, Debye-Sherrer cameras were used only in student laboratories and by some amateurs who specialized in the suitability of the technique for analyzing extremely small amounts of material. However, Debye-Sherrer geometry has indisputable virtues (listed at the end of this section), which have contributed to its revitalization in diffractometric schemes with PSDs. In this application, the prowess of Debye-Sherrer geometry for simultaneous-mode measurements and the normal incidence of the diffracted beam are of special importance. The good quantum efficiency

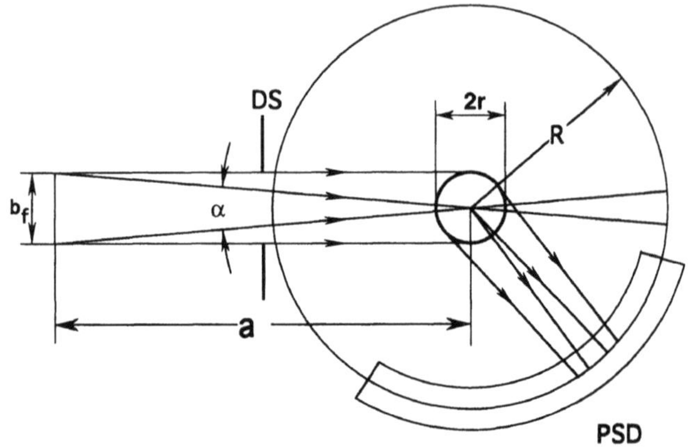

Figure 3.18. Debye-Sherrer geometry.

of PSD aids in overcoming the inherent weakness of the Debye-Sherrer scheme, i.e., the unfortunate absorption coefficient and the low level of diffracted intensity.

Debye-Sherrer diffractometric geometry with a PSD is shown schematically in Fig. 3.18. A cylindrical specimen of radius r is placed in the center of the diffractometer circle of radius R. A parallel or almost parallel incident beam is defined by the tube focus F and the narrow diaphragm DS. The complete theory of Debye-Sherrer geometry is to be found in Buerger (1942), Klug and Alexander (1954), Nelson (1960), and Nuffield (1966). Here we will give only a brief account of the major instrumental functions, which include:

a. *Finite width of the focal spot b_f.* The convergent angle of the primary beam $\alpha \approx b_f/a$, where a is the focus-to-specimen distance (Fig. 3.18). The width of the corresponding instrumental function is given by:

$$W_1 = b_f/R \tag{3.53}$$

b. *Finite specimen size.* If the specimen is bathed in a parallel X-ray beam, then the width of the instrumental function for a low-absorbing specimen is:

$$W_2' = 2r/R \tag{3.53a}$$

and that for a high-absorbing specimen (Buerger 1942) is:

$$W_2'' = (2r/R)\sin^2\theta \tag{3.53b}$$

c. *Beam divergence in the axial direction.* In the case of a point tube focus and a pinhole diaphragm DS, the axial divergence of the diffracted beam is significant due to the finite opening of the PSD in that direction (H_c = 8-10 mm). In this case, the general expression for peak broadening due to axial divergence (Kheiker and Zevin 1963) is reduced to:

$$W_3 \approx (H_c/R)^2 \cot\theta \qquad (3.54)$$

Detailed calculations of the peak profile and the peak displacement are given by Buerger (1942) and Nelson (1960). Here, we will only estimate the expected peak width for the following experimental conditions: $b_f = 0.5$ mm; $2r = 0.3$ mm; $H_c = 10$ mm; $R = 200$ mm; $2\theta = 30°$; a = 250 mm. From equations (3.53) and (3.54), we obtain $W_1 = 0.11°$, $W_2 = 0.08°$, and $W_3 = 0.03°$, i.e., the expected width of the diffraction peak might lie between 0.10° and 0.20°.

3.6.2 *Absorption correction*

Now suppose that a cylindrical sample of radius r is bathed in a uniform beam of X-rays normal to its axis. The absorption factor is calculated as:

$$A = \frac{1}{S}\int_s [\exp(-\mu\,(P + q))]ds \qquad (3.55)$$

where P and q are the lengths of the paths of the incident and reflected beams to and from the reflecting element ds in the cylindrical specimen (Fig. 3.19). Integration in equation (3.55) is performed over the whole specimen cross section S.

Intensity of diffraction in the Debye-Sherrer case is defined as:

$$I_{ij} = I_0 E_{ij} AV v_j \qquad (3.56)$$

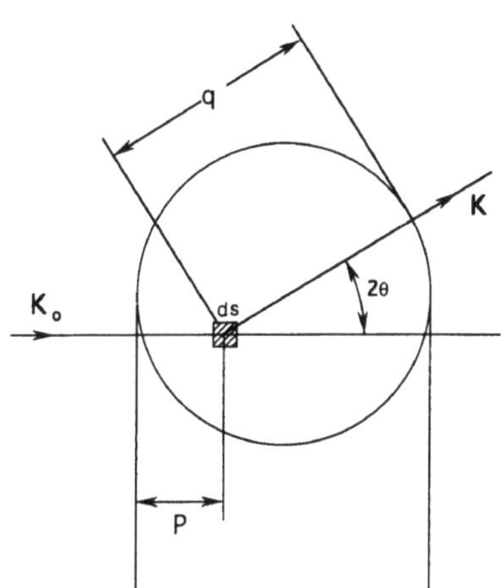

Figure 3.19. Definition of the absorption factor in Debye-Sherrer geometry.

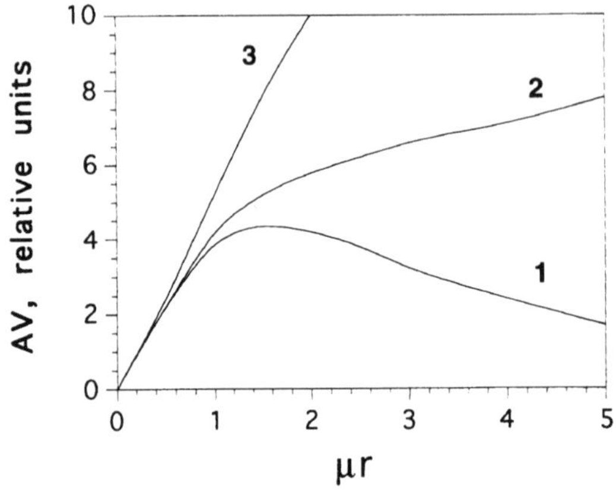

Figure 3.20. Variation of the term AV (where A is the absorption factor and V is the specimen volume) with μr for a cylindrical specimen in Debye-Sherrer geometry. 1) $2\theta = 10°$; 2) $2\theta = 30°$; 3) $2\theta = 60°$.

where E_{ij} is the product of the terms constant for the given phase j and diffraction peak i, V is the specimen volume, and v_j is the volume fraction of phase j. Values of A have been tabulated by Weber (1967) for $0 < \mu r < 31.5$ and by Rouse et al. (1970) for $0 < \mu r < 1$, but in small μr steps of 0.01. The specimen volume V is equal to $\pi r^2 H$, where H is the length of the irradiated part of the capillary tube.

The variation of the term AV with μr for selected scattering angles is illustrated in Fig. 3.20. This term exhibits non-monotonous behavior at low scattering angles (Fig. 3.20), displaying some optimal μr value. Even at moderate scattering angles, after an initial sharp rise with μr, AV slows down at $\mu r > 2$. Thus, $\mu r = 1.5\text{-}2$ may be considered as the optimal value for a cylindrical specimen for achieving the maximal diffracted intensity. The apparent density of a loosely packed powder in a capillary tube is no more than 40%-60% of the bulk density. For silicate materials with a bulk density of ~3g/cm^3 and a mass-absorption coefficient of ~40-50 cm^2/g, the linear absorption coefficient of a capillary specimen $\mu \approx 40 \times 0.5 \times 3 \approx 60$ cm^{-1}. Then, the optimal diameter of the capillary specimen is $d \approx 2\ (1.5/60) = 0.05$ cm. The optimal diameter for highly absorbing materials is smaller than this value.

3.6.3 *Comparison with former methods*

We assume that the intensity of the primary beam and the fraction of the Debye-Sherrer ring scanned by the detector are equal for the Debye-Sherrer and Bragg-Brentano geometric schemes. Thus, from equations (3.16) and (3.56) the ratio of the intensities may be expressed as:

$$\frac{I_{B\text{-}B}}{I_{D\text{-}Sh}} = \frac{B[1 - \exp(-2\mu t/\sin\theta)]}{2\,\mu\, AV} \tag{3.57}$$

where V is the specimen volume in the Debye-Sherrer case, A is defined in equation (3.55), B is the cross section of the primary beam, and t is the specimen thickness in Bragg-Brentano geometry. We assume that in the latter case the primary beam is intercepted by the specimen and that the specimen volume is equal to $Bt/\sin\theta$. First, let us suppose that the specimen volumes are equal in both cases, i.e., $V = Bt/\sin\theta$. In order to calculate the intensity ratio expressed in equation (3.57), we will assume some common experimental conditions: diameter of the capillary tube 0.04 cm, length of the capillary tube H = 1 cm, volume of the specimen ~ 1.25×10^{-3} cm^3, $\mu = 100$ cm^{-1} ($\mu r = 2$), divergence of the incident beam $\alpha = 1°$, and radius of the diffractometer 173 mm ($B \approx RH\alpha \approx 0.3$ cm^2).

The ratio $I_{B\text{-}B}/I_{D\text{-}Sh}$ is shown in Fig. 3.21 as a function of the scattering angle, emphasizing a great intensity benefit achieved with focusing Bragg-Brentano geometry, especially in the front-reflection range. The intensity diffracted in the Bragg-Brentano scheme can be further increased by adding material up to an amount satisfying the condition for infinite thickness. In this case, the exponential term in equation (3.57) vanishes, and the intensity gain in Bragg-Brentano geometry increases even further (curve 2 in Fig. 3.21). This great disadvantage of Debye-Sherrer geometry may be improved by means of a PSD, which facilitates a simultaneous rather than a consequent acquisition pattern.

The virtues of Debye-Sherrer geometry are:

1. Possibility of both consequent and simultaneous mode measurements;
2. Normal incidence of the diffracted beam, facilitating the

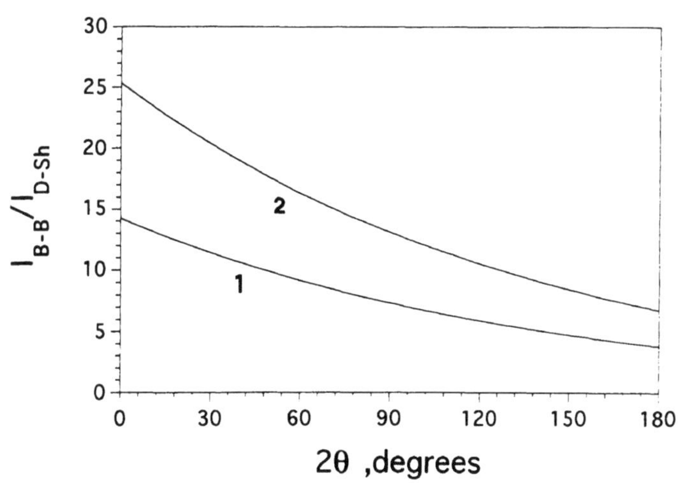

Figure 3.21. Comparison of diffracted intensity in Bragg-Brentano reflection and Debye-Sherrer schemes. 1 - Equal specimen volumes in the two schemes; 2 - "Infinitely" thick specimen in the Bragg-Brentano scheme.

requisite geometry for a PSD; and
3. Reduced preferred orientation effects.

The disadvantages are:
1. A high level of instrumental aberration;
2. A low level of diffracted intensity; and
3. An angle-dependent absorption correction.

3.7 Powder diffractometry with synchrotron radiation

3.7.1 *General description*

The source of the primary beam in this novel technique is a storage ring produced by a synchrotron accelerator, which provides a strictly parallel beam of polarized and very powerful white radiation in the spectral range useful in X-ray diffraction work, namely, 0.5-2 Å. The dimensions of the beam are large enough to illuminate a specimen of a few square centimeters, the conventional size in focusing X-ray diffractometry. The diffraction plane, containing the incident and diffracted rays, is normal to the electric vector of the incident beam. Thus, the polarization factor is equal to one, ensuring an additional intensity gain for synchrotron radiation.

A monochromatic beam is separated from the primary white radiation by means of a Si(111) or Ge(111) double-crystal monochromator CM in the parallel (n,–n) position (Fig. 3.22) (Tanner 1990). The wavelength selected, which is determined by the monochromator Bragg angle θ_m, may be changed by simple rotation of the monochromator block. The important feature of this device is that the position of the monochromatic beam is almost constant with rotation of the monochromator, enabling diffraction measurements at different wavelengths without realignment of the whole instrument. The spectral band cut by the monochromator depends on the crystal perfection and is of the order of $\Delta\lambda/\lambda \approx 3 \cdot 10^{-4}$ for $\lambda = 1.54$ Å (Cox et al.

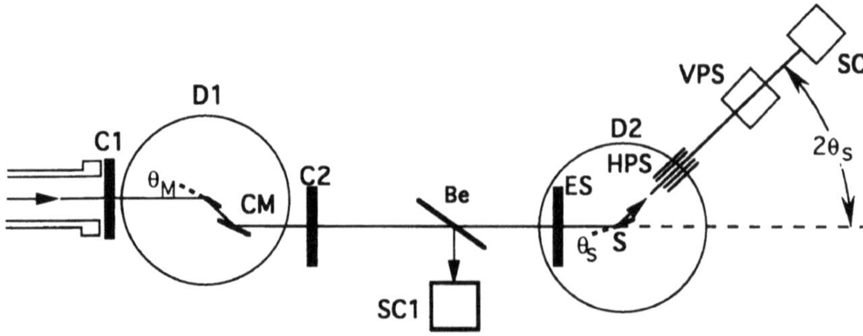

Figure 3.22. Diffractometer with synchrotron radiation. Reprinted with permission from the Australian Journal of Physics (Parrish et al. 1986b).

1983), i.e., approximately ten times lower than the separation of the CuK_{α} doublet. The overall flux of monochromatic radiation at the position of the specimen may vary between 10^9 and 10^{11} photons sec^{-1}, depending on the electron current in the storage ring, the perfection of the monochromator crystals, and the experimental conditions.

The slits C2 and ES define the size of the monochromatic beam in the diffraction plane containing the incident and diffracted beams. The width of ES is a few millimeters. The beam size in the axial direction is approximately 10-15 mm. The cross section of the monochromatic beam is approximately 1-3 x 10-15 mm^2, i.e., a value very close to the beam cross section in conventional focusing diffractometry. However, in this case the beam is not divergent but strictly parallel. The width of the diffracted beam (in the diffraction plane) depends on the scanning mode. If a symmetrical θ-2θ (θ for the specimen, 2θ for the detector) scan is carried out, then the widths of the incident and diffracted beams are both equal to W_{ES}, the width of the entrance slit ES. If, as in the Seeman-Bohlin scheme, a constant incident angle γ is maintained during the scan, then the width of the diffracted beam is equal to

$$W_{ES}\,[(\sin(2\theta-\gamma)/\sin\gamma)]$$

and might be much greater than W_{ES}. Thus, the diffracted beam is rather wide, of the order of several millimeters. Conventional diffractometric geometry with a narrow receiving slit in front of a detector is unacceptable in this case, because during scanning this device will produce broad diffraction peaks of low intensity. A solution to this problem has been found in the form of a perfect crystal analyzer, which acts as a narrow receiving slit but accepts a wide beam (Cox et al. 1983; Hastings et al. 1984). An alternative solution comprises the insertion of a set of horizontal Soller slits HPS (Fig. 3.22) with a very low divergence between the specimen and the detector. In the latter case, the peak width will be defined by the divergence δ of the HPS set, and δ must be of the order of 0.05° for the full potential of synchrotron radiation to be realized. Although such a device seems difficult to manufacture, it has been successfully produced and used (Parrish et al. 1986a,b; Parrish and Hart 1987). Due to the natural collimation of the primary beam, its divergence in the axial direction is almost nil, and no device is needed to control it. However, the diffracted beam is divergent in the axial direction, and its divergence must be controlled by a set of Soller slits (VPS, Fig. 3.22), as is usually the case in conventional diffractometers. Radiation produced by a storage ring is not stable with time, and it must therefore be monitored. In the scheme illustrated in Fig. 3.22, the radiation is monitored by measuring the scattering from a thin beryllium foil with an SC1 detector.

3.7.2 Instrumental aberrations

In parallel-beam geometry concepts such as the focusing circle and the point of beam convergence have lost their meaning. The concepts of specimen displacement, deviation of a flat specimen from the focusing circle, and specimen transparency are important in the definition of the so-called *displacement aberrations* characteristic of reflection focusing. In all these cases, the diffracted material is displaced from the focusing circle, and the point of convergence C' is displaced from the "true" convergence point C (Fig. 3.23a). There is no analogous situation in parallel-beam geometry, in which the effective width of the receiving slit is equal to the beam width in the diffraction plane (Fig. 3.23b). Thus, some of the very important aberrations acting in the focusing mode simply disappear in the parallel-beam mode. The remaining instrumental factors include axial divergence of the diffracted beam, controlled by the set of Soller slits VPS (Fig. 3.22), and finite opening of the Soller slits HPS, which plays the same role as the receiving slit width in focusing geometry. Axial divergence might cause asymmetry of low-angle diffraction peaks.

Due to the low level of instrumental and chromatic aberrations, synchrotron radiation diffractometry is characterized by very high angular resolution, especially if a crystal analyzer is used as the

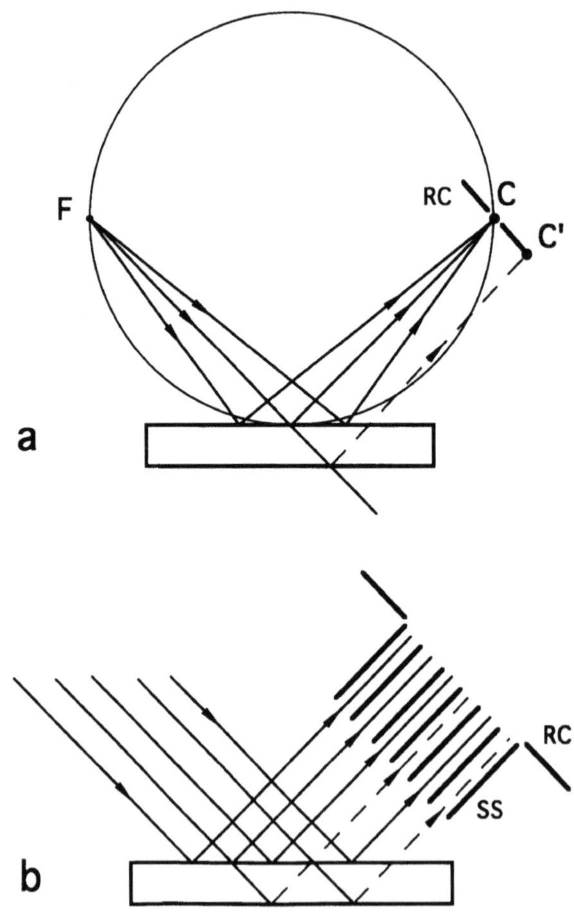

Figure 3.23. Displacement aberrations in focusing (a) and parallel-beam (b) geometry.

effective receiving slit. Cox et al. (1988) reported a diffraction peak with a FWHM of less than 0.02° when Ge(220) was used as the analyzer. The FWHM increases towards higher angles due to chromatic dispersion, which is controlled by the tan θ law. More modest, but still extremely low, values of the FWHM of 0.05° were obtained with a narrow Soller slit collimator (Cernik et al. 1990). Diffraction peaks are symmetric and may be well described by a pseudo-Voigt function, i.e., an algebraic sum of Gaussian and Lorentzian functions. The intensity gain obtained with synchrotron radiation depends on a number of instrumental parameters, including the electron current of the storage ring, and it is therefore somewhat uncertain. Some results published in the literature claim that the intensity gain is greater than that obtained in conventional focusing diffractometers, while others report it to be slightly lower.

3.7.3 Diffractometric modes with synchrotron radiation

Due to the naturally collimated parallel incident beam, the difference in convergent-divergent behavior for reflection and transmission modes discussed in Section 3.1 disappears in synchrotron radiation. Thus, both reflection and transmission schemes may be equally realized within the same instrument by simply rotating the specimen from the reflection to the transmission position. The absence of displacement aberrations in synchrotron diffractometry ensures facile transition from the symmetric scheme, with 2:1 coupling of the detector and the specimen to the asymmetric mode with constant incidence angle and stationary specimen. In the latter case, the detector spans the same diffractometer circle as in the former, i.e., the specimen-to-detector distance remains constant.

Due to the continuous character of the spectra emitted by storage ring electrons, it is possible to perform high-resolution energy diffraction (EDD) (Parrish and Hart 1985). The detector and the specimen are set in a fixed 2:1 position. The monochromator is step-scanned, producing at each step radiation with a particular wavelength in the range of 0.5 Å to 2.0 Å. The detector response is, in fact, an energy-dispersive diffraction pattern. A different range of d-spacings may be spanned by different 2θ settings of the detector and the coupled specimen. In comparison with conventional solid-state detector energy dispersive diffractometry, the techniques discussed in this section have a better energy resolution by at least one order of magnitude.

The following features of synchrotron radiation diffractometry are especially attractive for quantitative diffractometry:

1. Narrow and almost symmetrical diffraction peaks;
2. Ease of switching to various diffractometric modes, ensuring that preferred orientation effects can be taken into account, enabling thin film study, etc.; and
3. Low background level due to elimination of specimen fluorescence by proper selection of the wavelength.

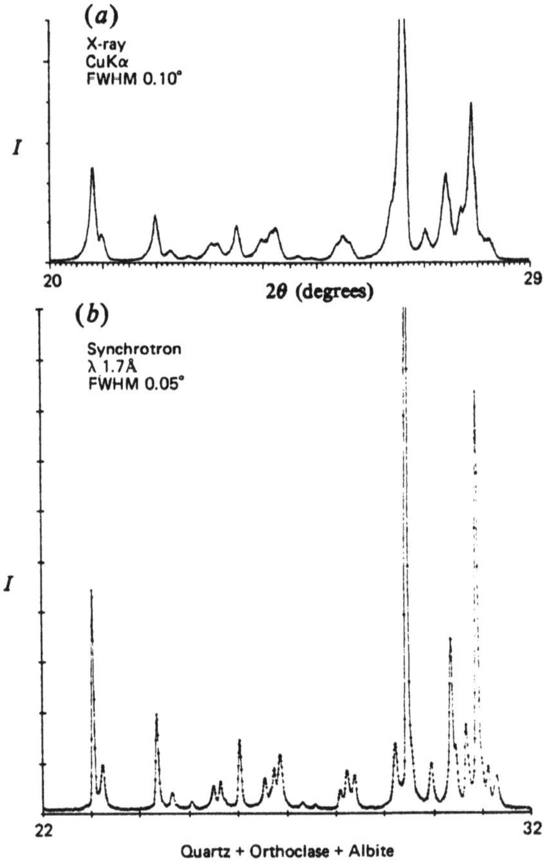

Figure 3.24. Comparison of the complex pattern of a mixture of quartz, orthoclase, and albite obtained with (a) the X-ray tube focusing method and (b) synchrotron radiation with parallel-beam mode. Reprinted with permission (Parrish 1988).

The benefits of synchrotron radiation are illustrated in Fig. 3.24. The diffraction pattern of a three-phase sample consisting of quartz and the feldspars orthoclase and albite was obtained by a conventional diffractometer technique and by synchrotron diffractometry (Fig. 3.24). The tremendous improvement afforded by angular resolution is obvious. Moreover, there is also an improvement of the peak-to-background ratio for the latter case.

Numerous applications of synchrotron radiation for structure refinement, the determination of precision cell parameters, and thin film studies are to be found in the papers cited in this section.

The virtues of synchrotron diffractometry are:

1. Strictly monochromatic radiation with a narrow spectral band;
2. Low level of instrumental aberrations;
3. Very high angular resolution of the diffraction pattern; and
4. Versatility of the diffractometric modes.

The disadvantage of this scheme is its exclusivity.

3.8 Position-sensitive detectors in powder diffractometry

The PSD is not new in X-ray diffraction. The photographic plate used in 1912 by Von Laue and his co-workers is a typical two-dimensional PSD. In powder diffractometry, this technique has been replaced by the more quantum-sensitive photon-counting detectors, which are in fact nondimensional point detectors (extension in the axial direction is not informative in the study of scattering). Point detectors are suitable for the consequent, or scanning, mode of data acquisition, which, of course, means a tremendous loss of time. However, the greater quantum efficiency of contemporary detectors makes the diffractometric run no longer (or even shorter) than the film exposure time in the photographic technique. However, the idea of simultaneous-mode operation is very attractive, and in the 1960s the multichannel diffractometer was designed (Zevin and Kheiker 1967; Evgrafov et al. 1972). In this diffractometer, which is based on Seeman-Bohlin reflection focusing, five scintillation detectors can be positioned to collect scattering by a number of crystalline phases and by the background. Scanning is not required, and the analysis time is reduced to a few minutes. This instrument has been used to control the phase composition of cement in the cement industry.

Further progress over the past 10-15 years in the simultaneous acquisition mode has been linked to the development of the proportional PSD, which combines the efficiency of contemporary particle detectors with the dimensional properties of photographic film. PSDs are suitable for simultaneous acquisition of the whole diffraction pattern and can potentially reduce the time of a diffractometric run by at least one order of magnitude. A detailed description of the construction and the electronic performance of PSDs can be found elsewhere (Göbel 1981; Wölfel 1983). Here, we are interested in the basic geometric scheme of the application of PSDs.

The diffraction pattern of a random powder has a one-dimensional property because absolute values and not directions of reciprocal vectors are of importance. Thus, linear PSDs are suitable for data acquisition. Two characteristics of PSDs are important here:

1. In order to obtain the high precision of contemporary scanning goniometers (0.01° or better), the PSD must have good linear resolution. If, for example, the specimen-to-detector distance is 200 mm, then an angular resolution of 0.01° corresponds to a linear resolution of only 35 μm.
2. A PSD has a finite thickness, and oblique diffracted beams cause gas ionization in a large volume along the position-sensitive element, producing a parallax effect and reducing the angular resolution. This effect is formally equivalent to the additional peak broadening produced by an oblique beam in a two-sided film. Normal incidence of the diffracted beam is thus important in diffractometric schemes involving PSDs.

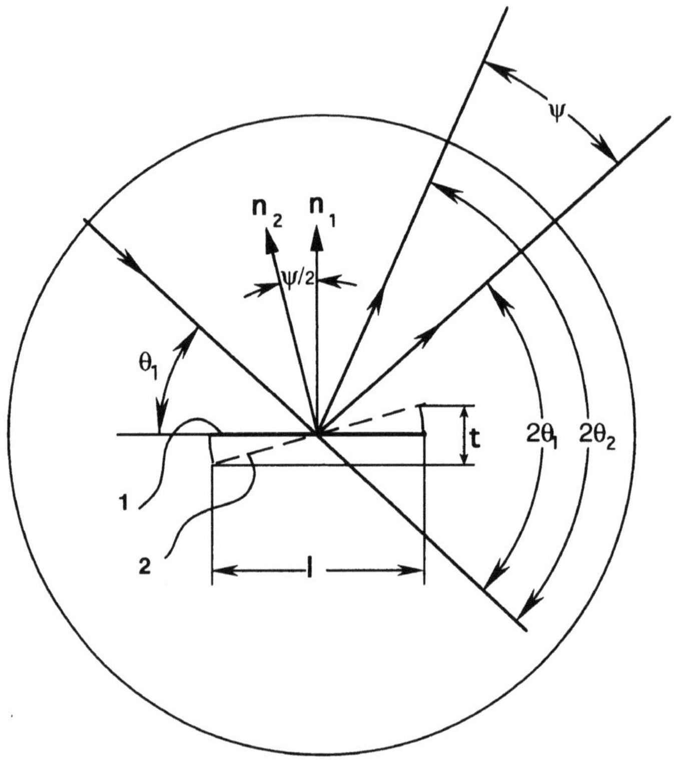

Figure 3.25. Defocusing effect in a PSD with a Bragg-Brentano reflecting diffractometer.

Reviewing the geometric schemes of X-ray diffractometry discussed in previous sections, we immediately come to the conclusion that the Debye-Sherrer scheme, generally used in the photographic technique, is also compatible with the principles of PSDs. Indeed, the first applications of PSDs were based on Debye-Sherrer geometry (Byram et al. 1977). The inherent low-intensity level of the Debye-Sherrer scheme can be compensated for by simultaneous-mode acquisition with PSDs. Of course, a curved PSD is needed in this case, and contemporary commercial PSDs offer a wide range of curved detectors with angular apertures of up to 120°.

The application of PSDs with focusing geometries is problematic. The anode wire (or blade) of the PSD must follow the diffractometer circle in Bragg-Brentano geometry or the focusing circle in the Seeman-Bohlin scheme. In Bragg-Brentano reflection focusing (and its transmission analog), the normal incidence of the diffracted beam is an intrinsic feature. However, for a given specimen position θ, the focusing conditions are maintained only for a scattering angle 2θ, i.e., this geometry is typically suitable for scanning-mode data acquisition. If the specimen remains stationary in any intermediate position, significant defocusing and peak broadening can be expected. Suppose, for example, that the specimen is left in position when the incident angle = θ_1 (Fig. 3.25). Then, in order to focus the reflection at an angle $2\theta_2$, the specimen must be rotated to position 2, $\psi/2$ away from $\theta_1(\psi = 2\theta_2 - 2\theta_1)$. If the specimen nevertheless remains stationary at position 1, a specific instrumental aberration appears, which in the first approximation causes symmetrical broadening of the reflection at the scattering angle $2\theta_2$. The same effect will be produced by a hypothetical nonabsorbing specimen with thickness $t \approx \sin(\psi/2)$ (Fig. 3.25) in position 1 (Kheiker and Zevin 1963). Here, l, the length of the illuminated specimen, is given by $l \approx R\alpha/\sin\theta$, where R is the goniometer radius and α is the incident beam divergence in the focusing plane. Taking into account that displacement of the specimen in the direction normal to its plane s causes a diffraction peak shift by $2s\cos\theta/R$, we can define the width of the instrumental function responsible for out-of-focus specimen rotation as:

$$\Delta 2\theta = 2\,(R\alpha \sin\psi/2)\cos\theta/R\sin\theta \approx 2\alpha\,(\cot\theta)\,(\sin\psi/2) \qquad (3.58)$$

For typical $\alpha = 1°$, instrumental broadening increases to ~0.15° for a PSD aperture of $2\psi \approx 3°$ if $2\theta_1 = 20°$ and 6° if $2\theta_1 = 40°$. This aperture may be twice as large if the primary beam divergence is reduced to $\alpha = 0.5°$. Thus, in the reflection Bragg-Brentano mode only a limited range of scattering angles can be covered by PSD. In order to cover a wider scattering range, while retaining high angular resolution, Göbel (1981) proposed the use of a small aperture linear PSD in the scanning mode. The scanning speed may be increased in proportion to the arc covered by the PSD (~5°), as it can by a narrow slit in the conventional point detector technique (~0.1°). Due to the small aperture of the PSD, inevitable peak broadening is tolerable and at high values of 2θ even negligible. It is interesting to note that in this technique a contribution to the given diffraction peak comes from the specimen at different positions within the range of $\pm\psi/2$, i.e., around the position with exact focusing. This situation is equivalent to specimen oscillation with an amplitude of $\psi/2$, which is an efficient means of increasing the number of reflecting crystallites and improving particle statistics (see Chapter 5). This effect is especially important for minor phases, as has been demonstrated for quartz-alumina mixtures (Göbel 1981).

Transmission focusing geometry, which generally exhibits a smaller degree of instrumental aberrations, is more suitable for wide-aperture PSDs. Indeed, considering the transmission analog of the Bragg-Brentano scheme and estimating the peak broadening by equation (3.58), we have to substitute angle θ by $90° - \theta$ to obtain the new equation:

$$\Delta 2\theta = 2\alpha \tan\theta \sin\psi/2 \qquad (3.59)$$

The same estimations for scattering angles $2\theta = 20°$ and $2\theta = 40°$ will give an acceptable PSD aperture of $2\psi = 50°$ and 25°, respectively. Thus, a stationary PSD with an aperture of ~40°-50° may be used together with a stationary specimen in the transmission focusing technique with, however, modest angular resolution of the diffraction pattern.

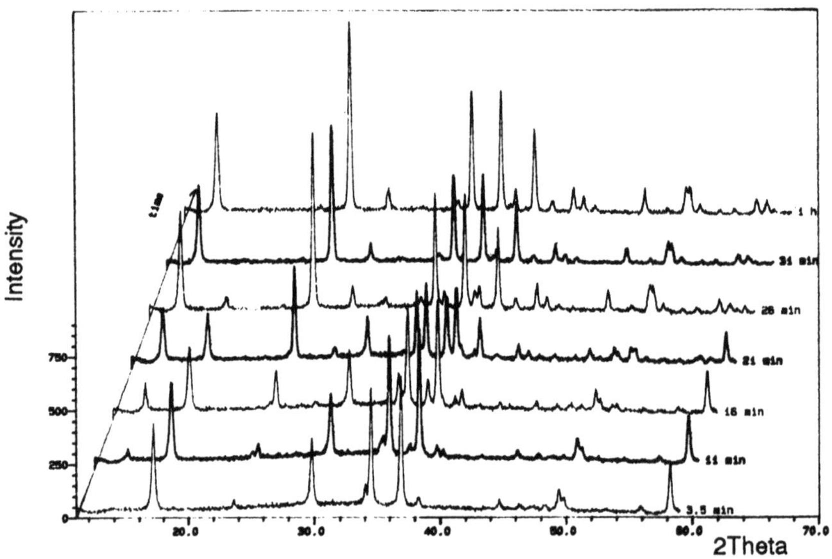

Figure 3.26. Three-dimensional diagram of hydration of calcium sulfate hemihydrate. Reprinted with permission from STOE (1988).

Seeman-Bohlin reflection focusing seems to be no more promising than the Bragg-Brentano scheme. Although the specimen is stationary, the incidence of the diffracted beam is far from normal, being equal to $2\theta-\gamma$ (Fig. 3.9). Oblique incidence at low scattering angles causes a severe parallax effect and corresponding peak broadening. On the contrary, the transmission analog (Guinier diffractometer) is more suitable for PSD. If the incident angle of the primary beam is $90° - \gamma \approx 65°$-$70°$, then this beam diffracted at $2\theta \approx \gamma$ will strike the focusing circle at exactly 90°. For any other scattering angle 2θ, the striking angle will be $90° - \gamma + 2\theta$, for $2\theta \leq \gamma$ and $90° + \gamma - 2\theta$ for $2\theta \geq \gamma$, i.e., close to 90° in rather wide 2θ range.

The weak point of the PSD is the maximal overall counting rate. Although it might exceed 100,000 cps, one must bear in mind that it includes not only diffraction peaks but also background counting through the open aperture of the PSD. The latter may be overwhelming if, for example, sample fluorescence is essential. Selective absorbing foils in front of the PSD may sometimes be a help, but, as a rule, in running a PSD diffractometer one must make a very careful selection of the X-ray tube in order to prevent fluorescence of the major components.

The performance of PSDs is especially impressive in the study of a fast-changing phase composition in the course of phase-transition or chemical reactions. Numerous examples are to be found in the literature, and here we illustrate this type of PSD application with a three-dimensional diagram of the kinetics of gypsum formation from a mixture of plaster of Paris and water (Fig. 3.26). The chemical reaction

$$2CaSO_4.1/2H_2O + 3\,H_2O \rightarrow 2CaSO_4.\,2H_2O$$

is rather fast, and traces of the new phase appear after a few minutes of hydration. The exposure time for each pattern was 1 min. The first diffraction pattern is almost pure calcium sulfate hemihydrate and the last pattern, after 1 h of hydration, is pure calcium sulfate dihydrate. A diffractometer with a PSD will undoubtedly be very useful for fast and sensitive quantitative analysis, especially for control of fast industrial processes.

4 Methodology of quantitative phase analysis

4.1 Introduction

The basic equations of quantitative phase analysis, which relate the X-ray intensity to the abundance of a phase in a specimen, were derived in Chapter 2 for various diffractometric schemes. In the forthcoming discussion, we will concentrate on Bragg-Brentano reflection focusing, since this is the most widely used diffractometric arrangement. This approach has general application, because the results and conclusions obtained can be applied to other schemes using corresponding intensity-weight fraction relationships. The relationship for the Bragg-Brentano scheme, as was first derived by Alexander and Klug (1948) [see the derivation of equation (3.11)], takes the form:

$$I_{ij} = \frac{K_{ij}' c_j}{\rho_j \mu^*} = \frac{K_{ij}' c_j}{\rho_j \sum_{k=1}^{n} \mu_k^* c_k} \tag{4.1}$$

where c_j is the weight fraction of phase j in the specimen, ρ_j is the density of phase j, μ^* is the mass absorption coefficient of the specimen, and c_k, μ_k^* are the weight fractions and absorption coefficients, respectively, of the phases abundant in the specimen. K_{ij}' is a calibration constant, which is a product of several phase-related, instrument-related, and scattering-angle-related terms discussed in detail in Section 3.2. The subscripts i and j in K_{ij}' refer to phase j and diffraction peak i. The density of phase j, i.e., ρ_j—(as a constant for given phase)—may be included in the calibration constant; equation (4.1) is then simplified to:

$$I_{ij} = \frac{K_{ij} c_j}{\mu^*} = \frac{K_{ij} c_j}{\sum_{k=1}^{n} \mu_k^* c_k} \tag{4.2}$$

where $K_{ij} = K_{ij}'/\rho_j$.

In the framework of a given crystal structure of the phase, the density can vary considerably with variations of the chemical composition of the phase. Thus, we have to keep in mind the relationship between K_{ij}' and K_{ij} in calculations of calibration constants (Section 4.9).

In order to determine c_j from equation (4.2), which is our main objective, the following three steps should be followed:

1. Measurement of the intensity I_{ij} of a diffraction peak belonging to phase j;
2. Determination of the absorption coefficient of the specimen, μ^*;
3. Determination of the calibration constant K_{ij}.

Calibration is usually performed by preparation and measurement of analytical standards, i.e., known mixtures of pure phases. Sometimes this tedious work may be replaced by calculation of a calibration constant from known crystal structure data or simply by "borrowing" the constant from previously published work. In some special cases, the calibration constant may be deduced from the intensity interdependence observed in the diffraction patterns of a specially created sample set ("standardless" techniques). In all of these procedures, it is assumed that the structure and (chemical) composition of the analytical standards are identical to the phases of the samples to be analyzed. This requirement is automatically fulfilled in the above-described standardless method and in a novel technique that combines structure refinement with quantification of phase abundances. Calibration problems will be discussed in detail later in this chapter.

Equation (4.2) emphasizes the dependence of the intensity on the absorption characteristics of the sample. We will now illustrate this point. Let us consider our sample as a quasi-binary system, comprising the phase to be analyzed, j, with weight fraction, c_j, and the rest of the sample, or matrix, m, with weight fraction $(1 - c_j)$. Since mass absorption coefficients are additive, we may write the following expression:

$$\mu^* = \mu_j^* c_j + \mu_m^* (1 - c_j) = c_j (\mu_j^* - \mu_m^*) + \mu_m^*$$

Thus, equation (4.1) becomes:

$$I_{ij} = \frac{K_{ij}\, c_j}{c_j(\mu_j^* - \mu_m^*) + \mu_m^*} \qquad (4.3)$$

and the relative intensity of the *ith* diffraction peak may be expressed as:

$$I_{ij}/(I_{ij})_0 = \frac{c_j}{c_j(1 - \mu_m^*/\mu_j^*) + \mu_m^*/\mu_j^*} \qquad (4.4)$$

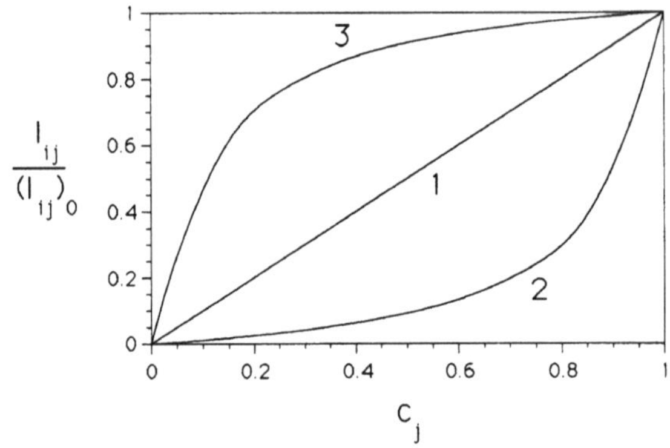

Figure 4.1. Relative intensity versus phase abundance in a binary system. 1) $\mu_j^* = \mu_m^*$; 2) $\mu_j^* = 0.1\, \mu_m^*$; 3) $\mu_j^* = 10\, \mu_m^*$.

where $(I_{ij})_0 = K_{ij}/\mu_j^*$ is the intensity of the *ith* peak in the pure phase j. The sensitivity of I_{ij} to c_j in a binary system is given by the derivation:

$$\frac{\partial I_{ij}}{\partial c_j} = \frac{K_{ij}\, \mu_m^*}{[c_j(\mu_j^* - \mu_m) + \mu_m^*]^2} \tag{4.5}$$

Equation (4.4) is illustrated in Fig. 4.1 for the following three relationships of μ_j^* to μ_m^*.

Case 1. $\mu_j^* = \mu_m^*$

In this case, the sample absorption does not depend on the abundance c_j of phase j. The intensity is linearly related to the weight fraction (Fig. 4.1, curve 1). The sensitivity of the analysis may be expressed as the derivative $\partial I_{ij}/\partial c_j$, where I_{ij} is defined in equation (4.3). For the case under discussion:

$$\partial I_{ij}/\partial c_j = K_{ij}/\mu_m^*$$

Case 2. $\mu_j^* << \mu_m^*$

Since μ_j^* is much smaller than μ_m^*, equation (4.3) becomes:

$$I_{ij} \approx \frac{K_{ij}\, c_j}{\mu_m^*\, (1 - c_j)} \tag{4.6}$$

and equation (4.5) becomes:

$$\frac{\partial I_{ij}}{\partial c_j} \approx \frac{K_{ij}}{\mu_m^*(1 - c_j)^2} \tag{4.6a}$$

For low concentrations of j, the dependence of I_{ij} on the weight fraction is almost linear, although the value of the derivative $\partial I_{ij}/\partial c_j$ may be rather low due to the high matrix absorption, and it tends to zero for $c_j = 0$ (Fig. 4.1, curve 2). Thus, the analysis of a small amount of a light phase in a heavy matrix cannot be performed conveniently by this method. For high concentrations of j, I_{ij} is a nonlinear function of c_j, with a fast-changing derivative [from equation (4.6a)].

Case 3. $\mu_j{}^* >> \mu_m{}^*$

For this case, equation (4.3) is transformed into:

$$I_{ij} \approx \frac{K_{ij}\, c_j}{c_j\, \mu_j{}^* + \mu_m{}^*} \tag{4.7}$$

and equation (4.5) into:

$$\frac{\partial I_{ij}}{\partial c_j} = \frac{K_{ij}\, \mu_m}{(c_j \mu_j{}^* + \mu_m{}^*)^2} \tag{4.8}$$

For high c_j, the matrix absorption $\mu_m{}^*$ may be neglected (in comparison with $c_j\, \mu_j{}^*$):

$$I_{ij} \approx K_{ij}/\mu_j{}^* \approx \text{const}$$

and the method becomes very insensitive (Fig. 1, curve 3). For low concentrations of j, the term, $c_j\, \mu_j{}^*$ may be as low as $\mu_m{}^*$, and the latter term cannot be neglected. Thus, equation (4.7) should be used. In this case, I_{ij} is a nonlinear function of c_j.

In the early stages of its development, QXRD was centered on absorption correction. Historically, X-ray methods were first applied for the analysis of two-phase systems by the internal standard method of Alexander and Klug (1948), followed by the diffraction-absorption technique of Leroux et al. (1953), and finally by the complete analysis of an n-phase system by the method of Bezjak (1961). As QXRD has progressed from the analysis of the relatively simple and stable phases like quartz and rutile to much more complicated minerals and man-made crystalline phases, the calibration problem has become more significant. Numerous attempts to cope with it have created new and promising approaches to QXRD, such as whole-pattern analysis and calibration by calculated pattern.

You will certainly be able to follow these important steps in the development of QXRD in the somewhat eclectic division of this chapter in which "absorption" sections are mixed with "calibration sections." Despite some loss of logic in the construction of this chapter, we have chosen this arrangement in light of the fact that the two main factors in QXRD—absorption correction and calibration—are often closely interconnected and should be treated together.

4.2 Analysis of samples with a known mass absorption coefficient (diffraction-absorption technique)

By the term "known," we assume that the mass absorption coefficient has been acquired—by any relevant technique—independently of diffraction measurements. These techniques (which are discussed in full in Chapter 5) may include simple calculation from the chemical composition of the sample [equation (2.13)] or any experimental technique, such as direct absorption measurements, measurement of Compton scattering, etc. The special case of constant μ^* also falls in this scheme. This apparent separation of diffraction and absorption measurements inspired Leroux et al. (1953) to call this method the diffraction-absorption technique.

4.2.1 Multiphase system with a constant absorption coefficient

This case is not as rare an occurrence as we may think . It includes phase transitions in any system in which the chemical composition remains unchanged during the transition. The well-known polymorphic systems, SiO_2, ZrO_2, and austenite-martensite, are typical examples. Another example is the crystallization phenomenon involving the growth of the crystalline phase in glass-ceramics, polymers, or metallic glasses.

Very often the condition of constant μ^* is not strictly met, but it may be assumed. When analysis of minor phases is attempted, one can safely assume that the absorption coefficient is constant. One example of this situation is the analysis of different forms of calcium sulfate (gypsum, hemihydrate, and anhydrite) in Portland cement. The mass absorption coefficient of the matrix (Portland cement clinker) is ~100 cm^2/g (CuK_α radiation), and that of the calcium sulfate phase(s) ranges from ~64 cm^2/g for gypsum to 78 cm^2/g for anhydrite. Since the weight fraction range of calcium sulfate is 0%-6%, the variation of the value of the mass absorption will be less than 2%.

The calibration constant K_{ij} may be obtained via measurements of the pure *jth* phase. Applying equation (4.2) to the pure phase ($c_j = 1$), we define:

$$(I_{ij})_0 = K_{ij}/\mu_j^* \quad \text{or} \quad K_{ij} = (I_{ij})_0 \mu_j^* \tag{4.9}$$

In the light of equation (4.9) and condition $\mu_j^* = \mu^*$, the basic equation (4.2) transforms into:

$$c_j = I_{ij}/(I_{ij})_0 \tag{4.10}$$

In this procedure, the pure phase serves as an analytical standard for calibration purposes. The linear equation (4.10) may be expressed in more general terms as:

$$c_j = \chi_{ij} I_{ij} \tag{4.11}$$

where χ_{ij} is a constant that can be determined by analyzing any mixture with a known weight fraction of phase j.

This procedure is preferable to calibration via measurements of the pure phase $(I_{ij})_0$. Although the condition μ^* = constant mostly excludes microabsorption effects, there are other sources of nonlinearity, such as preferred orientation. The effect of preferred orientation in a pure phase may be quite different from that in the same phase diluted several times by a matrix. Thus, the best calibration procedure is to prepare and measure a set of analytical standards, each containing a known amount of analyte in the same matrix, and thus to obtain the constant χ_{ij} as a linear regression slope. For example, a set of standards containing 2.5%, 5%, 15% and 30% of austenite in ferrite was prepared by the National Bureau of Standards (NBS standard reference materials 435a, 486, 487, 488) and may be used for analysis of retained austenite in hardened steel.

4.2.2 *Two-phase system with a variable absorption coefficient*

Equation (4.3) is easily transformed into an analytical equation for a two-phase system by assuming that phase j is the first phase and the matrix m is the second phase:

$$I_{i1} = K_{i1}\, c_1/[c_1(\mu_1^* - \mu_2^*) + \mu_2^*] \tag{4.12}$$

Alternatively, expressing K_{i1} in terms of equation (4.9), we obtain:

$$c_1 = [I_{i1}/(I_{i1})_0]\, \frac{\mu_2^*}{\mu_1^* - [I_{i1}/(I_{i1})_0]\,(\mu_1^* - \mu_2^*)} \tag{4.13}$$

Usually, μ_1^* and μ_2^* are known or can be calculated. The implementation of equation (4.13) requires experimental measurements of the intensities $(I_{i1})_0$ and I_{i1} for only one phase of a two-phase mixture; this may be important, for example, if the second phase is unavailable.

Alexander and Klug (1948) verified this equation by measuring the intensity of the $(10\bar{1}1)$ peak of quartz mixed in various proportions with a low-absorption phase, BeO ($\mu_2^* = 8.6\ cm^2/g$, $\mu_1^* = 34.9\ cm^2/g$, $\mu_2^*/\mu^*_1 = 0.25$); a high-absorption phase, KCl ($\mu_2^* = 124\ cm^2/g$, $\mu_2^*/\mu_1^* = 3.6$); and a phase with equal absorption, i.e., cristobalite ($\mu_2^*/\mu_1^* = 1$). Observed $I_{ij}/(I_{ij})_0$ ratios corresponded well to the same ratios calculated from equation (4.13). Application of equation (4.13) may be extended to a multiphase system provided that the rest of the sample (matrix) remains unchanged through the set of samples to be analyzed and its absorption coefficient can be measured or calculated. For instance, the concentration of pyrite in sandstone can be analyzed by this method provided that the proportions of the other phases in the sandstone remain constant. It is interesting to note that in this situation the mass absorption coefficient of the whole sample varies with variation of pyrite abundance.

Although the validity of equations (4.4) and (4.13) was confirmed in the pioneering work of Alexander and Klug (1948), its subsequent application revealed significant deviations from the theoretical curve. One example—that for samples of mixtures of cassiterite SnO_2 and quartz SiO_2—taken from the work of Ivoilov and Denisova (1962) is given in Fig. 4.2. In this example, the ratio $\mu_2^*/\mu_1^* = 5.4$ (FeK_α radiation). The experimental data deviate from the curve calculated by equation (4.13) as a result of the effect of microabsorption, which enhances the intensity of the low-absorbing (quartz) phase. In the

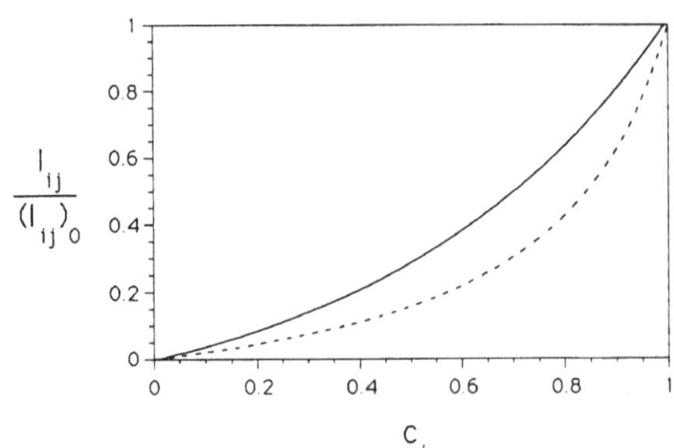

Figure 4.2. Deviation of experimental data (solid line) from the theoretical curve (broken line) due to microabsorption for a two-phase mixture of quartz and cassiterite $\mu_j^* = 0.18\mu_m^*$. Adapted from Ivoilov and Denisova (1962).

presence of significant microabsorption, equation (4.13) does not hold and cannot be applied. A more viable procedure is the creation of an experimental calibration curve of the type shown in Fig. 4.2. In this case, two-phase mixtures used for the preparation of a calibration graph must have a grain-size distribution as close as possible to that of the real samples to be analyzed. Microabsorption correction will be discussed in further detail in the next section.

4.2.3 *General case: multiphase systems with a variable absorption coefficient*

Combining equations (4.2) and (4.9), Leroux et al. (1953) obtained a simple expression for the phase abundance c_j:

$$c_j = \frac{I_{ij}}{(I_{ij})_0} \cdot \frac{\mu^*}{\mu_j^*} \tag{4.14}$$

They suggested direct measurement of μ^* by transmission of monochromatic radiation through the sample and, accordingly, called their method of analysis the *diffraction-absorption technique*. The major advantage of this method is that no special sample preparation (such as mixing with an internal standard) is needed.

For the method of Leroux et al. (1953), pure phases must be available for the measurement of $(I_{ij})_0$. It follows from equation (4.14) that the plot log $[(I_{ij})/(I_{ij})_0]$ versus log (μ^*/μ_j^*) at constant c_j should be a straight line with slope –1. The very first application of this equation revealed a significant deviation from the theoretical slope, although the regression line was linear. Let us take another example from the work of Ivoilov and Denisova (1962). Two-phase test samples were prepared by mixing quartz in various proportions with either a heavy or a light second phase in order to provide a range of μ^*/μ_j values from 0.16 to 5.4; the intensity of the $(10\bar{1}1)$ quartz peak was then measured in all the samples. Equation (4.14) may be rewritten as:

$$\log c_j - \log [I_{ij}/(I_{ij})_0] = \log (\mu^*/\mu_j^*) \tag{4.15}$$

and one can expect a linear regression if the left-hand side of equation (4.15) is plotted against the right-hand side. Indeed, the experimental data corresponded well to the linear regression, although the slope of the line deviated significantly from –1, as can be expected from theoretical equation (4.15) (the broken line in Fig. 4.3).

If the real slope of the regression line is designated α, then equation (4.14) can be rewritten as:

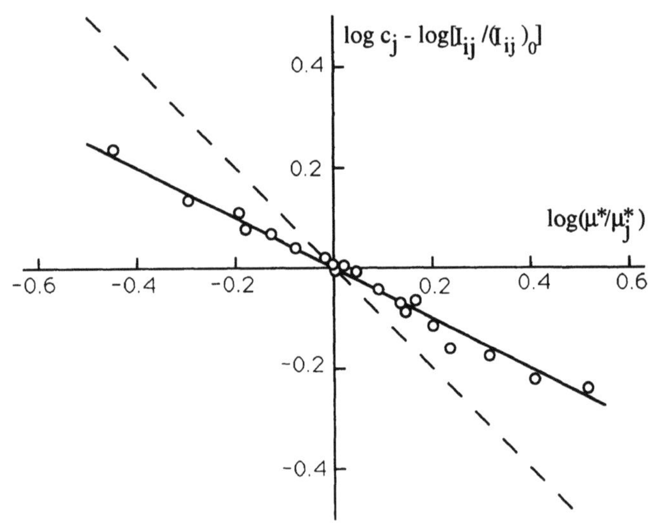

Figure 4.3. Evaluation of the coefficient α: Solid line—experimental regression line; Broken line—theoretical line [equation (4.15)]. Adapted from Ivoilov and Denisova (1962).

$$c_j = \frac{I_{ij}}{(I_{ij})_0} \cdot \left(\frac{\mu^*}{\mu_j^*}\right)^{\alpha} \tag{4.16}$$

The main reason for the observed deviation is the microabsorption effect discussed in Chapter 2. It should be noted that the value of the power α was always found to be less than 1. It follows from an elementary analysis of equation (4.16) that the value of $\alpha<1$ implies an enhancement of the intensity diffracted by the low-absorbing phase in the heavy matrix and a depression of the intensity diffracted by the heavy-absorbing phase in the light matrix. This finding corresponds well to the effect of microabsorption discussed earlier.

The power α in equation (4.16) is not a universal constant but depends on the grain size of the analyte and other phases in the sample, whose role in microabsorption is as important as the difference in absorption coefficients between the analyte and the matrix. Indeed, a range of values have been reported for α in the analysis of the mixtures of the same analyte (quartz) in different matrices:

$\alpha = 1$ (Alexander and Klug 1948)
$\alpha = 0.78$ (Leroux and Mahmud 1960)
$\alpha = 0.54$ (Zav'yalova and Ivoilov 1965).

Gonzalez (1987) expressed α as a function of the mean grain size of the analyte and the other component phases, the volume fraction of solid in a powder specimen, and the ratio of the absorption coefficients of the analyte and the sample. Thus, calculations of α inevitably include poorly known terms, like mean grain size for all phases, and can hardly be carried out in practice.

Nevertheless, despite the semiempirical character of expression (4.16), it may be useful in practical applications that include detection of microabsorption effects and sometimes even in the correction for this effect. If one wants to use the corrected equation (4.16) for quantitative analysis, the power α must be determined experimentally. For this purpose, several test mixtures of analyte with matrix similar to the real matrix in the samples to be analyzed should be prepared and measured. The power α is obtained as the linear regression slope $\log I_{ij}/(I_{ij})_0 - c_j$ versus $\log (\mu^*/\mu_j^*)$, as in Fig. 4.3.

Microabsorption effects are especially severe in the above-discussed method for understandable reasons. The two samples used in the analytical procedure—the pure phase j and the multiphase sample—might have quite different absorption characteristics and grain-size distributions. It seems that for calibration purposes it would be more appropriate to use a sample whose composition resembles that of the sample to be analyzed rather than a pure phase. The "calibration" sample could, for example, comprise a mixture of known proportions of the analyte j with a matrix similar to the matrix of the real sample. Applying equation (4.2) to both the unknown sample and the mixture of known composition, we obtain the following equation for c_j:

$$c_j = [I_{ij}/(I_{ij})_n] (m^*/(m_n^*) (c_j)_n \qquad (4.17)$$

where the subscript n relates to the prepared mixture. If we suppose that $(c_j)_n$ is not far from the expected c_j in the sample and that the grain-size distributions of the mixture n and the examined sample are similar, we may expect that equation (4.17) will not be dependent on microabsorption effects.

The next step, which will give more accurate results but is labor intensive, is, of course, the preparation and measurement of several mixtures covering the expected range of analyte variations.

Theoretically, the weight fraction c_j is a linear function of the product $I_{ij} (\mu^*/\mu_j^*)$:

$$c_j = \gamma_{ij} I_{ij} (\mu^*/\mu_j^*) \qquad (4.18)$$

where γ_{ij} is a constant, and the absorption coefficient of the phase j, μ_j^*, is included simply for convenience and may be ignored if desired. The constant γ_{ij} is determined as the linear regression slope of c_j versus $I_j(\mu^*/\mu_j^*)$. Significant deviation from linearity or a broad scattering of points around the linear regression line is a sign of considerable uncorrected microabsorption effects.

4.2.4 *Examples*

To deepen our understanding of the previous sections, let us consider the following three examples, which are based on the method outlined in this chapter.

4.2.4.1 Analysis of rutile in glass ceramics

TiO_2 (rutile) is added to glass to facilitate the crystallization of the major phases during heat treatment of the glass. If the weight fraction of TiO_2 exceeds some critical value, it will appear as a separate fine-grained phase. Since rutile influences the physical properties of the glass, it is important to be able to measure its abundance. To this end, several mixtures of fine-grained (~1 μm) rutile with noncrystallized glasses were prepared. Although the mass absorption coefficient of the glasses varied from 40 cm²/g to 80 cm²/g vs. 127 cm²/g for TiO_2, a value of $(\mu_j - \mu)d < 0.01$ was to be expected due to the extreme fineness of rutile. Thus, no microabsorption effects were to be expected.

In our example, the chemical composition of the glasses is known, and the mass-absorption coefficients of the test samples and examined samples were calculated by equation (2.13). A calibration graph was prepared with the coordinates $\frac{I_{ij}}{(I_{ij})_0} \cdot \frac{\mu^*}{\mu_j^*}$ vs. the weight fraction of rutile c_j (Fig. 4.4). The linear regression slope was very close to 1, i.e., in accordance with expectations, no microabsorption effects were detected in this case. The method is precise (~5% relative error) and has a low detection limit (no more than a few tenths of a percent).

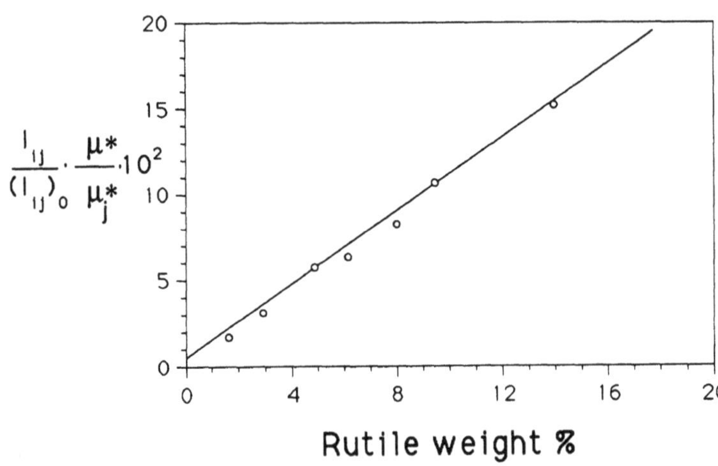

Figure 4.4. Analysis of rutile in glass ceramics. Adapted from Zevin and Zav'yalova (1974).

4.2.4.2 Relative abundance of calcite and quartz in sedimentary rocks

In this example, the mass absorption coefficient of sedimentary rock samples was ~65 cm^2/g, almost double that of quartz (34.9 cm^2/g). The rock was ground and sieved through a 300-mesh sieve, yielding a mean grain size of ~20 μm. Since the microabsorption effect in quartz analysis may be significant, equation (4.16) was chosen for the analytical procedure. The mass absorption coefficient of calcite (75 cm^2/g) is rather close to that of the sample, and no significant microabsorption effects are thus to be expected.

The slope α was determined by measuring the intensity of the $(10\bar{1}1)$ quartz peak in four artificial mixtures containing the quartz and the matrix in weight fractions similar to those in the sample to be analyzed. All measurements were made with CuK_α radiation. The results of the measurements are given in Table 4.1.

The graph according to equation (4.15) yields the slope α = 0.53. The results of analysis of one of the samples are shown in Table 4.2. Phase abundances were calculated by equations (4.14) and (4.16). For quartz, the corrected result deviates significantly from the phase abundance calculated by theoretical equation (4.14).

Table 4.1. Raw data for the calculation of the weight fractions of quartz and calcite in sedimentary rock

c_j	μ^*/μ_j^*	$I_{ij}/(I_{ij})_0$
0.25	1.76	0.185
0.75	1.24	0.67
0.75	0.88	0.81
0.50	0.75	0.59

Table 4.2. Results of analysis of a rock sample

			Weight fraction %	
Sample	Phase	hkl	4.14*	4.16**
Pure phase	Quartz	$10\bar{1}1$	100	100
Pure phase	Calcite	$10\bar{1}4$	100	100
Rock	Quartz	$10\bar{1}1$	9.3	6.9
Rock	Calcite	$10\bar{1}4$	42.4	45.6

*Calculated according to theoretical equation (4.14).
**Calculated according to corrected equation (4.16).

4.2.4.3 Analysis of β-celsian in glass ceramics

β-celsian (barium feldspar $BaO.Al_2O_3.2SiO_2$) is a major constituent of glass ceramics with special dielectric properties (Bogdanova et al. 1966) that are related to the abundance of the crystalline phase. The chemical composition of the glasses is known from the formula for the raw mixture. Thus, the mass absorption coefficient of the sample can be easily calculated.

Pure β-celsian was prepared by repeated sintering of a stoichiometric mixture of components until a single-phase material was obtained. The β-celsian was then micronized and mixed with fused quartz and one of the raw glasses in order to simulate matrices of various compositions. Mass absorption coefficients of known mixtures varied from 112 to 148 cm^2/g ($\mu_j^* = 149\ cm^2/g$). The total integrated intensity of seven celsian peaks (1$\bar{3}$2, 222, 114, 220, 040, 004, 204) was measured with CuK_α radiation. The calibration graph with coordinates $I_j(\mu^*/\mu_j^*)$ versus c_j (Fig. 4.5) did not reveal any significant deviations from linear dependence, and thus the absence of microabsorption effects was confirmed in this particular case. The standard deviation of the experimental points from linear regression σ (c_j) was ≈ 2%. Other errors, such as errors in intensity measurements, and in μ^* calculations, together with the uncertainty of the calibration graph, gave an estimated relative error of analysis of ~5%.

4.3 Internal standard method

4.3.1 *Basic principles*

The internal standard method is the oldest method for correcting for matrix absorption. In this method, the sample to be analyzed is doped with a known amount of reference material, known as the *internal standard*. We will retain this historical name for this doped material, although the word "standard" would be more applicable to a pure phase or mixture of pure phases used for calibration purposes. We will call the latter type of material an *analytical standard*. The weight fraction of the unknown phase j in the doped sample is given by $c_j' = c_j\,(1 - c_s)$, where c_s is the weight fraction of the internal standard and c_j is the weight fraction of phase j in the nondoped sample. Suppose that both phase j and the internal standard have at least one resolved (nonoverlapping) diffraction peak designated i for the unknown phase and h for the internal standard, then two obvious expressions can be written for the intensities of these peaks:

$$I_{ij} = K_{ij}\, c_j\, (1 - c_s)/\mu_d^*; \qquad I_{hs} = K_s\, c_s/\mu_d^* \tag{4.19}$$

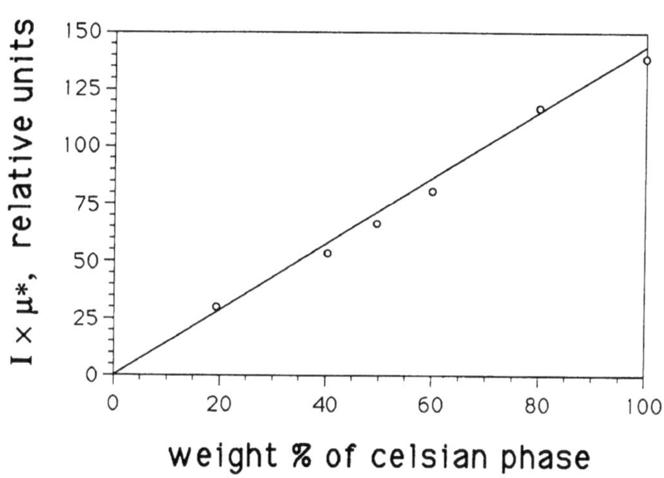

Figure 4.5. Calibration graph for analysis of β-celsian in glass ceramics. Adapted from Bogdanova et al. (1966).

where the subscript s relates to the internal standard, and $\mu_d{}^*$ is the mass absorption coefficient of the doped sample.

The ratio of intensities I_{ij}/I_{hs} is independent of the sample absorption:

$$S_{js} = I_{ij}/I_{hs} = K_{ij}\,(1 - c_s)\, c_j/K_{hs}c_s = \beta_{js}\,(1 - c_s)\, c_j/c_s \qquad (4.20)$$

where $\beta_{js} = K_{ij}/K_{hs}$.

It is clear from equation (4.20) that weight fraction c_j is a linear function of the intensity ratio, namely:

$$c_j = QS_{js} \qquad (4.21)$$

where $Q = c_s/[\beta_{js}\,(1 - c_s)] = \text{const.}$

Equation (4.21) constitutes the basic equation of the internal standard method.

A very important feature of the internal standard method is that it can be implemented to analyze any crystalline phase in multiphase samples without analyzing the remaining phases. For example, the presence of an amorphous phase or any other undetected phase does not interfere with the analysis of any detected phase. Moreover, if all the detected phases (say, n) are analyzed, then the mass balance:

$$\sum_{j=1}^{n} c_j = [c_s/(1 - c_s)] \sum_{j=1}^{n} S_{js}/\beta_{js} \approx 1 \qquad (4.22)$$

will hold if all the phases are detected. Significant deviation of the mass balance from 1 indicates that some phases have not been detected.

4.3.2 *Evaluation of calibration constants*

Calibration constants β_{js} and Q from equations (4.20) and (4.21) can be evaluated by carrying out measurements on a two-phase mixture of a pure phase j and an internal standard. If the composition of this mixture is 1:1, then we can easily obtain an expression for β_{js} from equation (4.21):

$$\beta_{js} = (S_{js})_a \tag{4.23}$$

where the subscript a indicates that the intensity ratio $(S_{js})_a$ relates to a 1:1 mixture. However, it would be methodically better to prepare several artificial mixtures with compositions somewhat close to that of the sample to be analyzed and the weight fraction of the analyte falling in the range of its variation in the analyzed samples. A known amount of the internal standard must be added to each mixture. After the intensity measurements have been made, the constant Q from equation (4.21) is obtained as the slope of the regression line. Due to experimental errors it is better to use the general form of the linear regression:

$$c_j = B + QS_{js} \tag{4.24}$$

rather than equation (4.21).

If we exclude the case of overlapping peaks, which naturally leads to a regression line of the type (4.24), a nonzero coefficient B appears when the experimentally determined background line lies above or below the exact level. Indeed, when the background was chosen below its actual level, the observed intensity of the diffraction peak was greater than the real intensity of this peak by the constant amount i:

$$(I_{ij})_{obs} = (I_{ij})_{real} + i \tag{4.25}$$

Repeating the development of equation (4.21), we will reach the general form of the linear regression (4.24). Coefficient B is usually small, and it may be positive or negative depending on whether the background is overestimated or underestimated.

Inevitable scattering of the experimental points around the regression line brings some degree of uncertainty into the estimation of Q. The standard deviation $\sigma(Q)$ is important in estimating the precision of the analysis. Some examples are given in the next section.

Sometimes β_{jr} values, where the phase r is the potential internal standard, are to be found in the literature. The reference intensity ratio

Table 4.3. Quantitative analysis of quartz‡

Sample	Sample description	μ^* (cm^2g^{-1}) determined by Compton scattering	Weight fraction of quartz in samples $c_{(DA)}$‡‡	$c_{(IS)}$‡‡‡	Difference $c_{(DA)} - c_{(IS)}$
H-77 KSP		35.2	1.6	1.8	−0.2
Kaolinite F	Kaolin	35.5	37.5	39.5	−2.0
H-77 K-T	Kaolin (local)	36.1	3.3	4.1	−0.8
H-77 K-35	Kaolin	36.4	3.9	3.4	+0.5
H-77 BC-75	Ball clay	36.5	16.8	18.0	−1.2
Hycost VC	Ball clay	36.9	14.1	12.3	+1.8
H-17 BC 55	Ball clay	37.5	16.4	17.7	−1.3
H-77	Fine clay (local)	38.4	25.2	25.0	+0.2
H-77 BC TA	Ball clay	39.1	33.4	35.0	−1.6
NB 79-2	Loess soil	51.8	28.1	27.3	+0.8
BS-79-439	Limestone	53.9	1.1	1.3	−0.2
BS-78.731	Iron-rich limestone	122.4	10.5	11.0	−0.5

‡ Taken from Bar-On et al. (1981).
‡‡ Analysis by the diffraction absorption technique.
‡‡‡ Analysis by the internal standard technique.

(RIR) = I_j/I_c is an example in which c stands for corundum. Further details on the implementation of RIRs are to be found in Section 4.7 of this chapter.

4.3.3 Precision of the internal standard method

Starting from equation (4.21), we can estimate the experimental precision in determining the weight fraction c_j. The standard deviation $\sigma(c_j)$ is given by:

$$\sigma(c_j) = \sqrt{[Q\,\sigma(S_{js})]^2 + [S_{js}\,\sigma(Q)]^2} \qquad (4.26)$$

where $\sigma(S_{js})$ and $\sigma(Q)$ are the standard deviations of S_{js} and Q, respectively. The standard deviation $\sigma(S_{js})$ may be defined as an error of reproducibility. It is estimated by repeated measurements of the intensities for one of the samples, preferably one containing a median weight fraction of the analyte. Each measurement is repeated by reloading the sample holder. If N measurements are performed, then the variance is given by:

$$\sigma(S_{js}) = \sum_{f=1}^{N} [(S_{js})_f - \langle S_{js}\rangle]^2 / (N-1) \qquad (4.27)$$

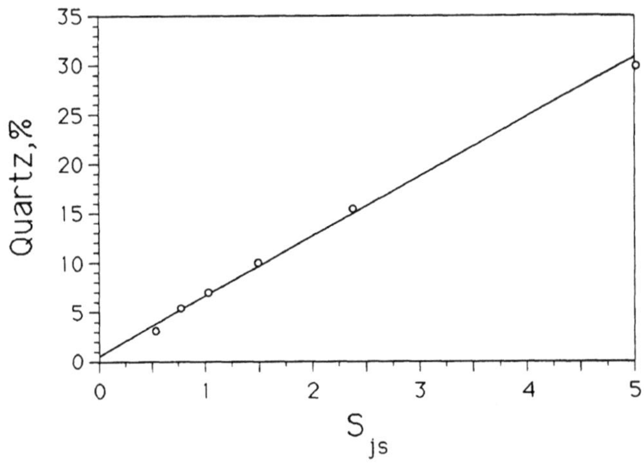

Figure 4.6. Analysis of quartz in geological samples by the internal standard method. Taken from Bar-On et al. (1981).

where $(S_{js})_f$ is the intensity ratio for measurement f ($1 \le f \le N$), and $<S_{js}>$ is the mean value of N measurements equal to $(1/N) \sum_f (S_{js})_f$. Q is the slope of the regression line in the plot of S_{js} vs. c_j. Formulation of $\sigma(Q)$ is given in Chapter 5.

Let us estimate the precision of the analysis of quartz in geological samples (Bar-On et al. 1981). The quartz abundance in these samples is given in Table 4.3. The calibration curve (Fig. 4.6) is linear, and the regression line is:

$$c_j\ (\%) = 0.58 + 6.05\ S_{js}$$

The standard deviation for the slope of the regression line $\sigma(Q) = 0.15$. Repeated measurements of the intensities revealed $\sigma(S_{js}) \approx 0.1$ for a quartz content of 15%-30%. For $c_j = 20\%$, the corresponding $S_{js} = 3.2$, and using these numbers in equation (4.26), we can estimate the precision of analysis to be $\sigma(c_j) \approx 0.8\%$.

4.3.4 Comparison of the internal standard method with the diffraction-absorption technique

The main advantage of the internal standard method is its almost universal character. It is applicable to any multiphase sample, irrespective of whether the matrix is a high or a low absorber. The obvious condition is that the sample may be mixed with the internal standard. Thus, this method is useful only for powder samples and not for coherent mass solid samples such as metals or coatings.

The internal standard method also assumes an automatic correction for some instrumental effects, such as small variations in alignment and long-time instability (drift) of the diffractometer, because these factors affect the intensities of the internal standard and the analyte to the same extent.

The diffraction-absorption technique described in the previous section—like the internal standard method—is also suitable for the analysis of an individual phase or phases in a multiphase sample, and it is thus worthwhile to compare these two techniques. In the former, additional time is needed for the measurement of μ^*, although the time can be shorter than that for mixing the sample with the internal standard. Thus, the diffraction-absorption technique may be faster than the internal standard method. However, it lacks the self-adjusting ability that is associated with the application of intensity ratios (I_{ij}/I_{hs}; β_{js}) rather than of the intensities themselves (I_{ij}; K_{ij}) in analytical equations. The fields of application may also be different for these two techniques: the diffraction-absorption technique is generally not good for heavy absorbers, whereas the internal standard method is not suitable for coherent mass solid samples. It may be said that these techniques are complementary, and circumstances will dictate the method of choice.

The internal standard method can be used not only as an intensity standard but also as a calibration standard for peak position. A possible shift of the internal standard peak due to instrumental factors can be easily detected and used for correction of the angular interval of measuring of integrated or peak intensities for the analyte. It should be remembered, however, that measurements of the peaks of the internal standard and the analyte are usually sequential and are thus separated by a short period of time. The internal standard method can therefore not account for probable short-time instability of the generator output or the electronic system.

The main disadvantage of the internal standard method is the labor required for weighing and mixing the standard with the analyte and measuring the intensity. These additional operations of sample handling could easily increase the duration of analysis by a factor of 2 or 3 in comparison with methods of analysis that do not require a mixing operation. In addition, some uncertainties are associated with this procedure, such as nonuniform mixing, errors in weighing, etc.

The two techniques should have approximately the same precision because additional errors introduced by measurements of an internal standard may well be compensated for by the uncertainties in measuring or calculating the mass absorption coefficient. Direct comparison of the two methods was carried out by Bar-On et al. (1981) for determining the concentration of quartz in a variety of rocks and soil. The mass absorption coefficient needed for the diffraction-absorption technique was determined by Compton scattering. Fluorite (c_S = 20%) was used as the internal standard. Comparison of results of analysis by the two methods shown in Table 4.3 did not reveal any statistically significant differences. The difference between the weight fractions obtained by the two methods for all but two samples fell within the limits of standard deviations $2\sigma(c_j)$ estimated earlier in this section.

4.3.5 *Internal standard—materials and optimal amount*

There are obviously a number of criteria that must be fulfilled before a substance can be used as an internal standard:

1. The potential internal standard should be chemically stable in air, under X-ray radiation, and in admixture with the sample to be analyzed.
2. The absorption characteristics of this material should match the absorption characteristics of the sample in order to minimize the microabsorption effects. Violation of this principle will lead to a deviation from the linear relationship expressed in the basic equation for the internal standard method (4.24). Significant deviations from linearity were found, for example, by Flinter (1975) in the analysis of a quartz-calcite mixture with copper powder as the internal standard and by Baker (1968) in the analysis of quartz in mine dust with vanadium pentoxide as the internal standard.
3. The diffraction pattern of a potential internal standard should exhibit only a few strong diffraction peaks that do not overlap with the analytical peaks of the sample. This requirement means that internal standards are usually high-symmetry compounds.
4. The intensity of the diffraction peak of a potential internal standard should not be affected by particle statistics, i.e., the grain size should not exceed 10 μm. In this respect, material obtained by grinding high-quality single crystals, like quartz or silicon, should be avoided. Each grain of such material represents a single crystal, and extensive grinding to a size of 1–2 μm is needed to produce a material with reproducible intensity. Materials with lesser crystal perfection are preferable as internal standards. Slight broadening of the diffraction peaks is a desired criterion. Materials exhibiting this phenomenon are usually satisfactory internal standards.
5. Materials that exhibit any kind of preferred orientation should be avoided as internal standards.

A research program undertaken at the NBS in a search for possible internal standards was completed several years ago, and the five reference materials listed in Table 4.4 qualified for this purpose. These

Table 4.4. Certified X-ray diffraction intensity standards

Phase	μ, cm^{-1} (CuK$_\alpha$)	I/I_{cor} (RIR)
α-Al_2O_3	124.9	1.00
ZnO	288.0	5.43
TiO_2	549.5	3.44
Cr_2O_3	952.1	2.16
CeO_2	2082.9	14.1

materials meet the demands discussed above and cover a wide range of absorption coefficients.

de Wolff and Visser (1964, 1988) suggested corundum (α-AlO_3) as a universal internal standard. This material meets the demands outlined above and can be obtained commercially in a fine-grained state. By *universal character* for an internal standard, we mean that the calibration constant β_{js} defined in equation (4.23) can be determined and tabulated ready for interlaboratory implementation. The diffraction peak in the ratio $(S_{js})_a$ for a 1:1 mixture of corundum and any phase j is specified by the strongest diffraction line of phase j.

In subsequent years, corundum became accepted as the reference material, and RIRs data [RIR $\equiv (S_{js})_a$, with s meaning corundum] are appearing in increasing numbers in the literature and in the Powder Diffraction File (PDF) issued by the International Centre for Diffraction Data. Taking into account the importance of this development, we will discuss determination and implementation of RIR in Section 4.7.

To determine the optimal amount of internal standard that should be added, we have to look first at how the intensity of the analyte peak is affected by doping the sample with the internal standard. The intensity I_{ij} decreases simply because the fraction of analyte in the sample is reduced due to the addition of the internal standard. If c_s is low, then the intensity of the internal standard peak is also low. Thus, there must be an optimal amount of internal standard that minimizes the coefficient of variation of the phase abundances (c_j). An analysis undertaken by Zevin (1979) leads to the following approximate expression for the optimal weight fraction of the internal standard:

$$c_s \approx 1/(1 + 2\,\mu_s^*/\mu^*) \tag{4.28}$$

If $\mu_s^* \approx \mu^*$, then $c_s \approx 1/3$. In fact, the minimum of $\sigma(c_j)$ as a function of c_s is rather flat, and deviation from the condition (4.28) will not decrease the precision substantially. Nevertheless, c_s = 0.2-0.3 will also give a uniform mixture with the material sample, and that amount of internal standard may be considered as optimal.

4.3.6 *Application of the internal standard method*

As an example of the application of the internal standard method, we will consider the analysis of pyrite in phosphate ore. During calcination of phosphate ore, pyrite FeS_2 may be transformed into the acid-soluble sulfide—pyrrhotite FeS—which will cause problems in subsequent processing of the ore. Sulfur occurs in three different forms in sedimentary phosphate rock—as sulfate, sulfide, or organic sulfur. Chemical methods of pyrite determination, which are based on separation of these three forms of sulfur, are tedious and time consuming. Direct application of QXRD is also impossible, because the weight fraction of pyrite in phosphate rock is no more than 5000 ppm in bituminous phosphate and may be as low as 100 ppm in low organic matter phosphate. This latter figure is well below the detection limit of QXRD. In our example, the weight fraction of pyrite was increased by dissolving out the major mineral of the rock, fluorapatite, in an excess of 0.9 N HCl. In order to prevent generation of CaF_2, which interferes with analysis of pyrite, 3% of silica gel was added to each sample before acid treatment. In addition to pyrite, the residue contains quartz, clays, organic matter, and some other acid-insoluble minor constituents of phosphate rock. Neither the amount nor the chemical composition of the residue was predictable. In these circumstances, the internal standard method seems to be the most appropriate. The internal standard (20% of MgO) was mixed with the residue. Analytical peaks of pyrite (200, 311) and the reference peak of MgO (200) were step-scanned with a 0.05° step and 40 or 20 sec counting time. Approximately 30,000 counts were accumulated for the pyrite peak if its abundance in phosphate was 0.1%. For this sample, counting statistics alone account for ~4% error in the ratio of intensities I_{py}/I_{MgO}.

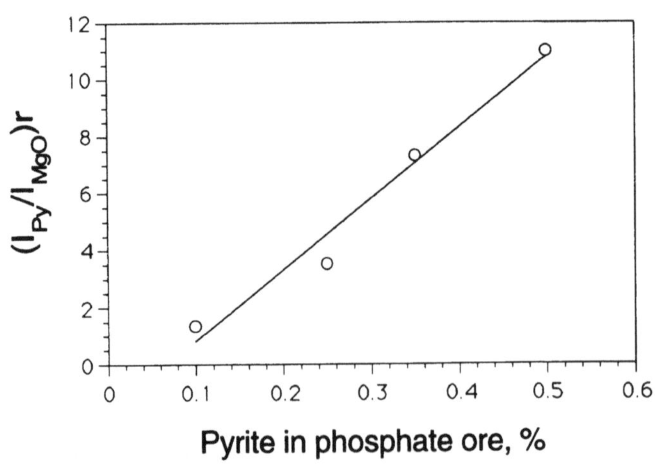

Figure 4.7. Calibration graph for the determination of pyrite in phosphate ore.

For a pyrite abundance of 0.01%, this error increases to ~20%. Calibration standards were obtained by doping pyrite-free phosphate rock with 0.1%, 0.25%, 0.35% and 0.5% of pure pyrite. These mixtures were acid treated (as described above), and the internal standard was added to the residue. For a sample with a constant concentration of pyrite in the untreated phosphate, the ratio of intensities I_{py}/I_{MgO} is not constant but depends on the amount of residue. The invariant characteristic is the product $(I_{py}/I_{MgO})r$, where r is the percentage of residue and varies between 2% and 5%. The calibration graph (Fig. 4.7) shows an unusually large scattering of points around the regression line. The source of the observed deviations lies not in counting statistics but mainly in errors in sample treatment and handling. Nevertheless, the technique was efficient in the quantitative estimation of a pyrite abundance greater than 100 ppm.

4.4 Doping method in quantitative X-ray diffractometry

4.4.1 *Introduction*

Copeland and Bragg (1958) described a technique in which the sample to be analyzed is doped with a known amount of the same material. Ever since it was realized that this method was as good as the popular internal standard method in solving the absorption problem, various modifications of the method of Copeland and Bragg have been suggested. These modifications will be discussed and compared in this section.

First of all, we have to express the weight fraction of the analyte in the doped sample and its mass absorption coefficient. Let us suppose that c_j is the weight fraction of the unknown phase j in the undoped sample. If G is the mass of this sample, the mass of phase j is c_jG. We then dope our sample with a mass G_{jd} of the same phase j, and the weight fraction of phase j in the doped sample becomes equal to:

$$(c_j)_d = (c_j\,G + G_{jd})/(G + G_{jd}) = c_j\,(1 - c_{jd}) + c_{jd} \qquad (4.29)$$

where $c_{jd} = G_{jd}/(G + G_{jd})$ represents the weight fraction of the dopant.

We may consider the doped sample as consisting of two fractions: the doped fraction c_{jd}, having a mass absorption coefficient μ_j^*, and the remaining fraction $(1 - c_{jd})$, having the mass absorption coefficient of the (initial) sample, μ^*. Thus, the mass absorption coefficient of the doped sample is composed of these two terms:

$$\mu_d^* = \mu^*\,(1 - c_{jd}) + \mu_j^* c_{jd} \qquad (4.30)$$

Now, by substituting the values $(c_j)_d$ and μ_d^* in the Klug-Alexander equation (4.2), we arrive at the basic equation for the doping method of QXRD:

$$(I_{ij})_d = \frac{K_{ij}[c_j(1-c_{jd})+c_{jd}]}{\mu^*(1-c_{jd})+\mu_j c_{jd}} = \frac{K_{ij}[c_j(1-c_{jd})+c_{jd}]}{\mu^*[1+c_{jd}(\mu_j^* - \mu^*)/\mu^*]} \tag{4.31}$$

Several variations of the doping method have been proposed to simplify the above-mentioned equation.

4.4.2 *Constant absorption approach*

If the mass absorption of phase j is close to the mass absorption of the matrix (i.e., $\mu_j^* \approx \mu^*$) and/or the dopant is used in a low concentration, i.e., $c_{jd} << 1$, then it is true to say that $c_{jd}(\mu_j^* - \mu^*)/\mu^* << 1$. In brief, this means that the doping does not change the mass absorption coefficient of the sample (i.e., $\mu_d^* \approx \mu^*$).

Thus, basic equation (4.31) can be simplified as follows:

$$(I_{ij})_d = K_{ij}[c_j(1-c_{jd})+c_{jd}]/\mu^* = (K_{ij}/\mu^*)c_j + (K_{ij}/\mu^*)(1-c_j)c_{jd} \tag{4.32}$$

If we consider a set of samples with various proportions of dopant c_{jd} (including, of course, $c_{jd} = 0$, the undoped sample), we will find that the intensity of the diffraction peak in accordance with equation (4.32) is a linear function of the dopant c_{jd}, as is shown schematically in Fig. 4.8. Two points in this diagram, c_0 and I_{ij}, are of particular interest. At c_0, the interception point, $(I_{ij})_d = 0$ and, from equation (4.32), we obtain the following expression for the weight fraction c_j in the undoped sample:

$$c_j = c_0(1-c_0) \tag{4.33}$$

The analysis ends at this point if our task is to analyze only one sample. The experimental work comprises measurements of the intensity I_{ij} in the initial sample and of the intensities $(I_{ij})_d$ in one sample or in a number of doped samples, and construction of a graph similar to that shown in Fig. 4.8. In practice, the case discussed here has a wide application, not necessarily restricted to individual samples. Let us examine, for example, the analysis of the minor phases—calcite, quartz, dolomite, and gypsum—in phosphate ore (Zevin et al. 1988). The major phase, carbonate-fluorapatite, constitutes more than 80% of the ore, with 5%-15% calcite and usually no more than a few

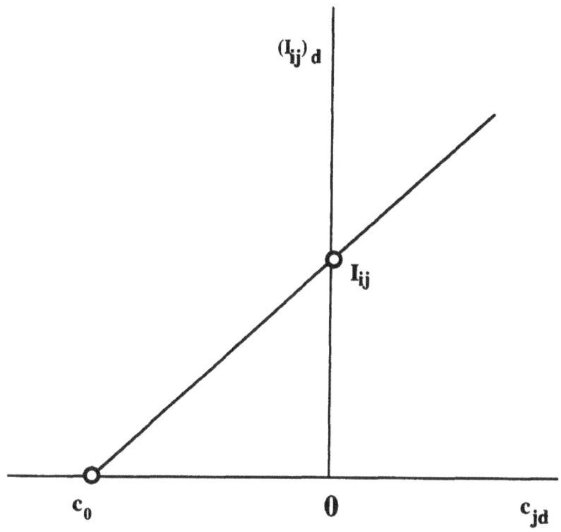

Figure 4.8. Doping method with constant absorption.

percent of the other phases. Under these circumstances, the condition $\mu_d^* \approx \mu^*$ holds for all samples of the batch. Thus, equation (4.32) and Fig. 4.8 take on a more universal character than would be the case for the analysis of a single doped sample. Indeed, if one sample j is analyzed by the doping method and weight fraction c_j is obtained, then we can easily calculate the term (K_j/μ^*), which is constant for all our samples:

$$K_j/\mu^* = I_{ij}/c_{ij} \tag{4.34}$$

This constant term can then be used in the analysis of any sample l of the same batch:

$$(I_{ij})_l = (K_{ij}/\mu^*)(c_j)_l \tag{4.35}$$

Thus, if we dope only one sample of the batch, all we need for analysis of the others are measurements of the intensities $(I_{ij})_l$. If the doping method is applied to a batch of samples, then linear regression (4.32) requires more accurate definition: to this end, a range of proportions of the dopant should be added to the chosen sample, and evaluation of the linear regression by the least-squares procedure should then be performed. When one particular sample is chosen out of a batch for the doping procedure, preference should be given to the sample with the minimal amount of analyte, ensuring proximity of the intercept of the calibration graph to the origin. It should be emphasized that the graph illustrated in Fig. 4.8 is an auxiliary graph used to obtain the value of (K_j/μ^*) and not to determine the weight fraction of the sample to be analyzed. The latter must be calculated from equation (4.35).

From this discussion, we see that measurement of the pure phase is not needed. Calibration is achieved by doping of the sample with a small amount of pure phase. The sample(s) used for calibration are very similar to the real samples, making the intensity measurements and the analysis very reliable, provided, of course, that the dopant is identical to the analyte in the sample.

4.4.3 *General case*

In the general case, the condition that μ^* = constant does not hold, i.e., doping changes the mass absorption coefficient significantly. By measurement of the intensities of the analyte peak in the doped and undoped samples, we create two equations: equation (4.31) for the doped sample and equation (4.2) for the undoped sample. There are three unknowns in these two equations, c_j, K_{ij}, and μ^*. The required third equation is obtained either by measurement of the intensity of peak i in the pure phase or by preparation of a second doped sample with a different amount of doped material.

If the pure phase is used, then the third equation is:

$$(I_{ij})_0 = K_{ij}/\mu_j^* \tag{4.36}$$

Combining equation (4.31) with equations (4.2) and (4.36), we obtain:

$$c_j = c_{jd}/(1 - c_{jd}) \frac{1 - (I_{ij})_d/(I_{ij})_0}{(I_{ij})_d/(I_{ij}) - 1} \tag{4.37}$$

Thus, three different samples are measured: undoped (for I_{ij}), doped [for $(I_{ij})_d$], and the pure phase j [for $(I_{ij})_0$]. From an initial inspection of equation (4.37), we can see that experimental errors in measurements of intensities will undermine the method if:

a. $(I_{ij})_d/(I_{ij})$ is close to 1 (too little doping), or
b. $(I_{ij})_d/(I_{ij})_0$ is close to 1 (too much doping).

Obviously, there is an optimal doping concentration, and this will be discussed in the next section.

Measurement of the pure phase is not obligatory if we prepare and measure an additional doped sample. In this case, equation (4.2) and two equations (4.31) for the two doped samples constitute the system of three equations, which are then solved for the three unknowns c_j, μ^*, and K_{ij}. Ignoring the elementary algebraic transformation, we arrive at the following expression for the phase abundance:

$$c_j = \frac{I_{ij}/(I_{ij})_{d''} - I_{ij}/(I_{ij})_{d'}}{[(1 - c_{d'})/c_{jd'}]\ [1 - I_{ij}/(I_{ij})_{d'}] + (1 - c_{jd''})/c_{jd''}]\ [1 - I_{ij}/(I_{ij})_{d''}]} \quad (4.38)$$

where subscript d' refers to the first doped sample and subscript d" refers to the second doped sample. This technique, although time-consuming, may sometimes be preferable to conventional measurements of the pure phase, because in the former the absorption characteristics of the three samples may be close.

Consider now a batch of samples with a range of absorption coefficients where condition $\mu^* =$ constant does not hold. In order to carry out the analysis, each sample must be doped and the weight fraction c_j calculated according to equation (4.37) or (4.38). It immediately becomes obvious that there are differences between the general case and the simplified case ($\mu^* =$ const), discussed in the previous section, in which only one sample needs to be doped in order to analyze the whole batch. The general doping method ($\mu^* \neq$ const) is not suitable for large-scale analysis because of the time and labor required for doping each sample and measuring the doped and undoped samples for each analysis. In addition, it is inferior to the internal standard method, which requires the measurement of only one—the mixed—sample. However, one can imagine situations in which the general doping method would be useful, for example, when it is difficult to find a suitable internal standard.

4.4.4 *Two-phase approach*

We now consider that doping affects not only the analyte, phase j, but also the other phases in the sample. The intensity of diffraction by any phase except the analyte will decrease in the doped sample, simply due to dilution by doping. However, the extent of the decrease depends on the relationship between the mass absorption coefficients of the doped and undoped samples.

Consider any second phase r with weight fraction c_r in the undoped sample. By measuring the intensities I_{ij} and I_{hr} (where h is the diffraction peak of the phase r), we obtain the ratio:

$$I_{hr}/I_{ij} = (K_{hr}/K_{ij})\,(c_r/c_j) \quad (4.39)$$

The concentration of phase j in the doped sample is defined by equation (4.29), and the concentration of phase r is obviously equal to $c_r(1 - c_{jd})$. By measuring the intensities of the same peaks in the doped sample, the following obvious equation is generated:

$$\frac{(I_{hr})_d}{(I_{ij})_d} = \frac{K_{hr}\ c_r(1 - c_{jd})}{K_{ij}\ [c_j(1 - c_{jd}) + c_{jd}]} \quad (4.40)$$

Equations (4.39) and (4.40) are then solved for c_j:

$$c_j = \frac{c_{jd}}{1 - c_{jd}} \cdot \frac{1}{1 - (S_{rj})_d / S_{rj}} \tag{4.41}$$

where $(S_{rj})_d = (I_{hr})_d / (I_{ij})_d$, and $S_{rj} = I_{hr}/I_{ij}$ are intensity ratios.

In the evaluation of c_j by equation (4.41), we do not have to measure the intensity of the pure phase j, but (at the expense of additional labor) we do have to measure peak h of phase r in two samples, doped and undoped (initial).

In practice, all nonoverlapping peaks of the unknown phase j and the second phase r may be measured and utilized in the calculation of c_j from equation (4.41), thus facilitating multiple determinations of c_j. A mean value would, of course, be more precise.

Bezjak and Jelenic (1971b) suggested a graphic solution for this particular case. It follows from equation (4.41) that:

$$S_{rj} = (S_{rj})_d \, [1 - c_{jd}/c_j(1 - c_{jd})] \tag{4.42}$$

If we take various pairs of peaks i and h of phases j and r, then equation (4.42) represents a linear regression with slope $[1 - c_{jd}/c_j(1 - c_{jd})]$. An example of such a graph, taken from the work of Bezjak and Jelenic (1971b), is shown in Fig. 4.9. The sample is a two-phase mixture of calcite and magnesite, where the analyte is calcite. The initial sample was doped with c_{jd} = 23% of calcite. Each experimental point on the graph represents the ratios of intensities of two particular magnesite

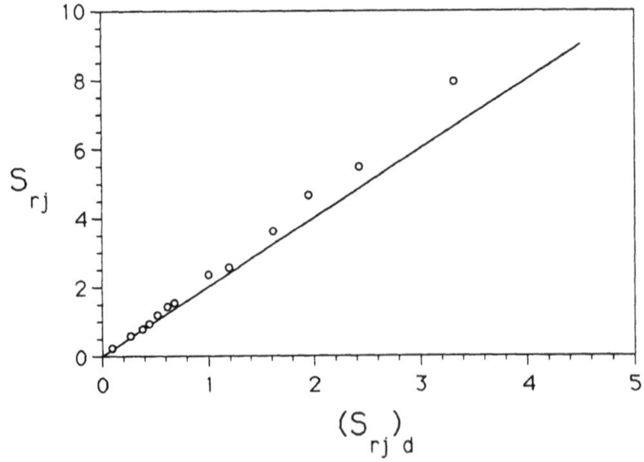

Figure 4.9. Interdependence of intensity ratios before doping S_{rj} and after doping $(S_{rj})_d$ for the system calcite-magnesite. Adapted from Bezjak and Jelenic (1971b).

and calcite peaks. The abundance of calcite, calculated from the slope of the graph c_j, was found to be 29.8%, a value that was very close to the known value of 30%.

4.4.5 *Precision of the analysis and optimal doping*

We define the optimal doping as that which minimizes the relative error of analysis, i.e., the coefficient of variation:

$$\varepsilon(c_j) = \sigma(c_j)/c_j \tag{4.43}$$

where $\sigma(c_j)$ is the standard deviation of c_j.

In the general case of the doping method, we regard the weight fraction c_j as dependent on two intensity ratios, $(I_{ij})_d/(I_{ij})_0 \equiv S_{d0}$ and $(I_{ij})_d/I_{ij} \equiv S_{dj}$. The standard deviation of c_j is given by:

$$\sigma(c_j) = \sqrt{[(\partial c_j/\partial S_{d0})\,\sigma(S_{d0})]^2 + [(\partial c_j/\partial S_{dj})\,\sigma(S_{dj})]^2} \tag{4.44}$$

where $\sigma(S_{d0})$ and $\sigma(S_{dj})$ are the standard deviations of the intensity ratios. In the following calculations, we take into account only the intensity errors due to counting statistics. The rules for calculating $\sigma(S_{d0})$ and $\sigma(S_{dj})$ are given in Chapter 5. Ignoring the tedious intermediate steps, we will arrive at the following expression for $\varepsilon(c_j)$:

$$\varepsilon(c_j) = \sqrt{2}\,\varepsilon(I)\sqrt{\left[\frac{(I_{ij})_d/(I_{ij})_0}{1-(I_{ij})_d/(I_{ij})_0}\right]^2 + \left[\frac{(I_{ij})_d/I_{ij}}{1-(I_{ij})_d/I_{ij}}\right]^2} \tag{4.45}$$

where $\varepsilon(I) = \sigma(I)/I$ is the coefficient of variation in the intensity measurements, which is assumed to be equal for measurements of I_{ij}, $(I_{ij})_0$, and $(I_{ij})_d$. The precision of the intensity measurements $\varepsilon(I)$ defines the lower limit of precision attainable in the analysis $\varepsilon(c_j)$. Thus, the ratio $\varepsilon(c_j)/\varepsilon(I)$ may be considered as an indicator of the quality of the chosen experimental conditions. We will try to find the minimal value of this ratio as a function of $(I_{ij})_d/I_{ij}$, the ratio of intensities in the doped sample $(I_{ij})_d$ and undoped sample I_{ij}. This ratio is closely related to the amount of doped material.

Let us choose three levels of the initial concentration of analyte in the sample:

1. A low concentration corresponding to a very low intensity I_{ij}, such that I_{ij} could be regarded as negligible in comparison to the intensity of the same peak in the pure phase, i.e., $I_{ij}/(I_{ij})_0 \approx 0$;
2. Any intermediate concentration corresponding to the ratio $I_{ij}/(I_{ij})_0 \approx 0.20$; and

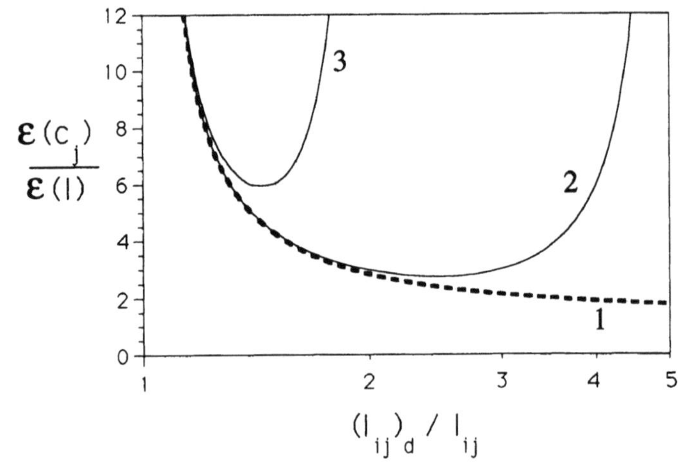

Figure 4.10. Optimal doping: 1—low phase abundance $I_{ij}/(I_{ij})_0 \approx 0$; 2—intermediate phase abundance $I_{ij}/(I_{ij})_0 = 0.2$; 3—high phase abundance $I_{ij}/(I_{ij})_0 = 0.5$.

3. A high concentration corresponding to the ratio $I_{ij}/(I_{ij})_0 \approx 0.50$.

Three curves of $\varepsilon(c_j)/\varepsilon(I)$ for the three levels of the initial concentrations are shown in Fig. 4.10. The minima of these curves correspond to the optimal doping, which we will express conveniently as the optimal ratio $(I_{ij})_d/I_{ij}$.

For the low concentration of analyte (curve 1), the error $\varepsilon(c_j)$ approaches its minimum value $\sqrt{2}\varepsilon(I)$ at very large $(I_{ij})_d/I_{ij}$ ratios, i.e., for a very large amount of dopant. Unfortunately, high concentrations of dopants are unacceptable in this particular case because such doping may violate the very important condition that makes this case exclusively attractive, namely, $\mu^* \approx \mu_d^*$. On the other hand, the error in the weight fraction increases greatly if $(I_{ij})_d/I_{ij}$ approaches 1, i.e., for a very small amount of dopant. Some compromise is thus essential, and the amount of dopant that doubles the initial intensity I_{ij} [i.e., $(I_{ij})_d/I_{ij} \approx 2$] seems to be optimal. In this case $\varepsilon(c_j) \approx 2\sqrt{2}(I_{ij})$, and the relative error in weight fraction would be 6% for a 2% error in the intensity measurements.

Curve 2 for the intermediate level of phase abundance has a minimum at $(I_{ij})_d/I_{ij} \approx 2.5$, although for all values of $(I_{ij})_d/I_{ij}$ between 2 and 3, the error ε (c_j) does not exceed $3\varepsilon(I)$, which is also acceptable. Thus, the amount of dopant that increases the intensity of the analytical peak by 2-3 times in the doped sample may be regarded as optimal.

Curve 3 has a sharp minimum at $(I_{ij})_d/I_{ij} \approx 1.5$, but even the minimal value of $\varepsilon(c_j)$ is rather high—six times greater than the relative error in the intensity measurements. This result confirms the premise that the doping method is applicable only to the analysis of samples with a low or intermediate level of analyte abundance. The upper limit seems to be about 20%.

4.4.6 *Implementation of the doping method*

The method discussed in Section 4.4.2, which assumes constant absorption, is very popular and may be considered the best method for analyzing minor phases, provided, of course, that the conditions for its implementation are fulfilled. The great advantage of this "μ^*= constant" method is that only one sample must be doped in order to determine the constant required for calibration. This method is thus suitable for large-scale analysis.

In the general case, in which μ^* is not constant, the doping technique becomes less practical. Not only does it demand the doping of each sample, but two samples, one doped and one undoped, must be measured, which considerably increases the time required for analysis. Doping methods can thus only be implemented in special cases in which only one sample or a small batch of samples is to be analyzed. In this application, one should remember that calibration is usually very good in doping methods. Indeed, if we consider those doping techniques in which only original and doped samples are used (such as the two-phase approach or general method with double doping), we remember that the pure phase is not measured and that all the samples used have very similar absorption characteristics. Thus, samples analyzed by the doping method may be used afterwards for purposes of calibration in other QXRD techniques, e.g., the diffraction-absorption method. The latter method may be much more precise if a sample with a known abundance of analyte is used for calibration rather than a pure phase. Such a sample may be obtained by doping. Let us suppose that we can, by some technique (experimentally or by calculation), acquire the mass absorption coefficient of the original sample μ^* and that of the doped sample μ_d^*. Then, combining equations (4.2) and (4.29), we obtain the following expression that permits calculation of the phase abundance c_j:

$$(I_{ij})_d / I_{ij} = \frac{\left[c_j(1 - c_{jd}) + c_{jd}\right] \mu^*}{c_j \, \mu_d^*}$$

or

$$c_j = \frac{c_{jd}}{[(I_{ij})_d / I_{ij}]\,(\mu_d^* / \mu^*) - 1 + c_{jd}} \tag{4.46}$$

Doping methods could be very useful in combination with another technique, i.e., complete analysis of the n-phase system discussed in Section 4.6. The important feature of the latter is that imperative normalization is imposed on the sum of the weight fraction $\sum_j c_j = 1$. However, as soon as one phase is determined by any independent (say doping) method, this normalization can be dropped. This phase can then be regarded as an internal standard for the analysis of any other phase in the analyzed sample. Thus, all the phases can be analyzed by implementation of the internal standard technique (Section 4.3). The mass-balance equation can now be used for a routine analytical check of the completeness of the analysis. If the sum of the determined weight fractions is significantly lower than unity, then one has to look for an undetected phase.

Further integration of these two methods has been proposed by Popovic and Grzeta-Plenkovic (1979). Their technique comprises doping of all but one phase in the sample to be analyzed. Although this technique is theoretically valid, it is time-consuming and does not confer any advantage over the conventional method of analysis of the n-phase system, in which one calibrating mixture is prepared by mixing n pure phases in any optimal proportions.

4.5 Dilution Method

4.5.1 *Basic aspects*

In this method, the sample is diluted with any inert material, crystalline or amorphous, preferably one with a known mass absorption coefficient. In contrast to the requirement for the internal standard method, the diffraction peaks of the diluent do not have to be measured, and unlike the doping method the diluent should not contain any of the phases to be analyzed. Measurements of undiluted and diluted samples replace measurements (or calculation) of sample absorption. The mathematics of this method is quite simple.

Let G_d be the mass of the diluent and G the mass of the original sample; then the weight fraction of the diluent is given by:

$$c_d = G_d / (G + G_d) \tag{4.47}$$

and the weight fraction of the unknown phase in the diluted sample is given by:

$$\frac{c_j G}{G + G_d} = \frac{c_j (G + G_d - G_d)}{G + G_d} = c_j - c_j c_d = c_j (1 - c_d) \tag{4.48}$$

The mass absorption coefficient of the diluted sample is defined by the rule of additivity:

$$(\mu)_d^* = \mu_d^* c_d + (1 - c_d)\,\mu^* \tag{4.49}$$

where μ_d^* is the mass absorption coefficient of the diluent and μ^*, as usual, is the mass absorption coefficient of the original sample.

We have now defined all the terms needed to apply the basic equation (4.2) to the two samples, i.e., the original sample and the diluted sample. They are:

$$\begin{cases} I_{ij} = K_{ij}c_j / \mu^* \\ (I_{ij})_d = K_{ij}c_j(1-c_d)/[\mu_d^*\, c_d + \mu^*(1-c_d)] \\ (I_{ij})_0 = K_{ij}/\mu_j^* \qquad \text{(when } c_j = 1) \end{cases} \tag{4.50}$$

where I_{ij} and $(I_{ij})_d$ are the intensities of the *ith* diffraction peak of the phase j in the original and diluted samples. The third equation in the system of equations (4.50) is obtained by measuring the pure phase. There are three unknowns, μ^*, K_{ij}, and c_j, in the system of three equations. Solving these equations for the weight fraction c_j in the original sample we obtain:

$$c_j = \frac{c_d\,\mu_d^*}{(1-c_d)\,\mu_j^*} \cdot \frac{(I_{ij})_d\,/(I_{ij})_0}{1-(I_{ij})_d\,/I_{ij}} \tag{4.51}$$

Diffraction measurements are performed on the diffraction peak (or peaks) of the analyte in the three samples (the original sample, the diluted sample, and the pure phase). The latter measurement is carried out only once for each batch, but the first two measurements should, of course, be carried out for each sample to be analyzed.

In this method, the pure phase is used for purposes of calibration in order to obtain the calibration constant K_{ij}. For the same purpose, we may use an artificial mixture with a known proportion of analyte in the matrix, provided that the composition of the mixture approaches that of the real samples. Suppose that this mixture is diluted in the same proportion by the same diluent as is added to the sample to be analyzed. Then, the term $Q = c_d\mu_d^*\,[1 - c_d)\mu_j^*(I_{ij})_0]$ remains constant, and equation (4.51) transforms into:

$$[I_{ij}\,(I_{ij})_d]\,/\,[I_{ij} - (I_{ij})_d] = Q\,c_j \tag{4.52}$$

where Q is the constant term defined above. Equation (4.52) expresses the linear regression with c_j as the argument and Q as the slope. In order to obtain Q, we have to measure one standard mixture or, preferably, several standard mixtures with a known phase abundance of the analyte. As soon as the slope Q has been determined, equation (4.52) can be used to calculate the phase abundance c_j in the analyzed samples. Of course, two samples—the original and the diluted—should be measured in order to obtain I_{ij} and $(I_{ij})_d$. Good examples are to be found in the work of Clark and Preston (1974).

4.5.2 *Optimal dilution*

Dilution of the sample decreases the intensity of the analytical peak, and thus the amount of diluent should not be too great. On the other hand, the phase abundance c_j calculated from equation (4.51) is subject to large errors if the intensities of the diffraction peaks in the original sample I_{ij} and the diluted samples $(I_{ij})_d$ are close. Thus, there must be an optimal dilution, which minimizes the error in the concentration c_j. The general expression for the standard deviation $\sigma(c_j)$ is:

$$\sigma(c_j) = \sqrt{\{ (\partial c_j/\partial I_{ij})\ \sigma(I_{ij})\ \}^2 + \{ [\partial c_j/(\partial I_{ij})d]\ \sigma(I_{ij})d\}^2} \tag{4.53}$$

In the following exposition, we suppose that $\sigma(I_{ij})$ and $\sigma(I_{ij})_d$ are defined by counting statistics, according to the rules discussed in Chapter 5. We ultimately obtain the final expression for the coefficient of variation $\varepsilon(c_j) = \sigma(c_j)/c_j$.

$$\varepsilon\,(c_{ij}) = \varepsilon(I) \sqrt{\frac{1+[(I_{ij})_d/I_{ij}]^2}{1-\ (I_{ij})_d/I_{ij}}} \tag{4.54}$$

where $\varepsilon(I)$ is the coefficient of variation of intensity measurements, which is assumed to be equal for I_{ij} and $(I_{ij})_d$. Because $(I_{ij})_d$ is smaller than I_{ij}, and in order to obtain the equality in $\varepsilon(I_{ij}) = \varepsilon[(I_{ij})_d]$, the time of counting should be increased during the measurements of the diluted sample by an amount proportional to the ratio $\sqrt{I_{ij}/(I_{ij})_d}$. The ratio $(I_{ij})_d/I_{ij}$, which expresses the level of dilution, is a very convenient argument to study the variation of the error $\varepsilon(c_j)$ as a function of dilution. The corresponding curve for which the function is the ratio $\varepsilon(c_j)/\varepsilon(I)$ is shown in Fig. 4.11. This curve has a distinct minimum at $(I_{ij})_d/I_{ij} \approx 0.33$, i.e., when the dilution decreases the primary intensity I_{ij} by threefold. If the mass absorption coefficient of diluent is close to that of the sample, the condition $(I_{ij})_d/I_{ij} = 0.33$ means that one part of the sample to be analyzed must be diluted with two parts of diluent

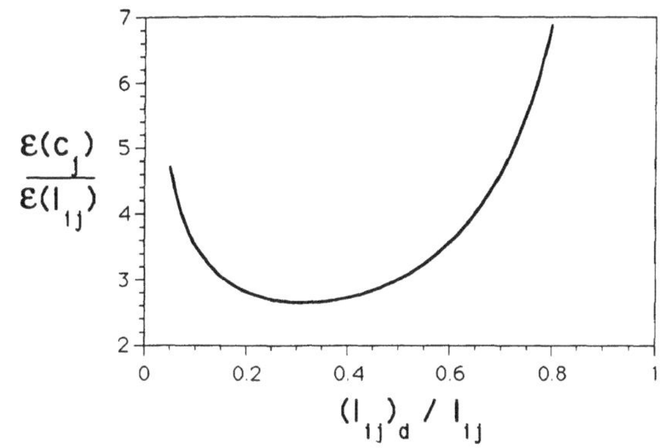

Figure 4.11. Optimal dilution taking expected errors into account.

($c_d = 0.67$). If the condition $\mu_d^* = \mu^*$ does not hold, the optimal dilution corresponding to $(I_{ij})_d/I_{ij} \approx 0.33$ should be found by trial and error.

Figure 4.11 indicates that at the optimal dilution the precision of analysis is $\varepsilon(c_j) = 2.5\ \varepsilon(I)$. Therefore, the precision of the intensity measurements $\varepsilon(I)$ should not fall below 2%-3% in order to keep the relative error of analysis at an acceptable level of $\varepsilon(c_j) < 10\%$.

4.5.3 Potential diluents

Potential diluents must meet the following criteria:

1. Diffraction from the diluent must not interfere with the analytical diffraction peak. Thus, a good diluent would be an amorphous material, provided that scattering by the diluent does not significantly increase the background under the analytical peaks. Silicate glasses, for example, exhibit strong scattering in the 10°-30° 2θ range, if CuK_α is used.
2. The diluent should have a low level of fluorescence.
3. The diluent should be chemically stable, be available as a fine powder, and show good mixing ability with the sample.
4. The mass absorption coefficient of the diluent should be close to that of the sample. This requirement is essential in order to prevent unwanted microabsorption effects.

In defining the amount of diluent, we have to bear in mind the optimal ratio $(I_{ij})_d/I_{ij} \approx 0.33$ (Fig. 4.11). For low-absorbing diluents (organic matter), this means that a very high degree of dilution should be used. The opposite situation will be true when a high-absorbing diluent is used: a small amount of diluent must be mixed with a large mass of sample. In both cases, it is difficult to obtain uniform mixtures.

Fewer restrictions are imposed on the potential diluent in comparison with the internal standard, and usually there are no problems in finding an appropriate diluent.

4.5.4 *Use of a heavy absorber—constant-absorption approach*

Analysis of a high-absorbing phase in a light matrix is difficult due to rapidly changing sample absorption, as discussed in Section 4.1 [equation (4.7) and curve 3 in Fig. 4.1]. At high concentrations of the phase to be analyzed, the gain in intensity obtained by increasing c_j is almost canceled out by increased sample absorption. At lower concentrations, any variations of c_j are associated with sharp variations of sample absorption. A typical example is the analysis of a heavy phase in an organic matrix. In both cases mentioned earlier, dilution of the sample with a suitable (heavy) diluent may result in a situation in which absorption of the diluted sample is almost constant and similar to that of the original sample.

We come, therefore, to the simplified QXRD case of constant sample absorption, and the analysis can then be carried out by one of the methods discussed earlier, for example, the doping technique. Suppose, for example, that the weight fraction of the heavy phase varies within the limits $0<c_j<0.2$. We can arrange that the mass absorption coefficients in two diluted marginal cases ($c_j = 0$ and $c_j = 0.2$) differ by no more than 10%. The mass absorption coefficient is given by expression (4.49), and the condition of approximate constancy of the mass absorption coefficient leads to the equation:

$$\mu_d^* c_d + \mu_m^* (1 - c_d) \approx 0.9 \left[\mu_d^* c_d + 0.8 \mu_m^* (1 - c_d) + 0.2 \mu_j^* (1 - c_d)\right] \quad (4.55)$$

where μ_m^* and μ_j^* are the mass absorption coefficients of the light matrix and heavy analyzed phase, respectively. It follows from equation (4.55) that for $c_d = 2/3$, $\mu_d^* \approx \mu_j^*$, i.e., the heavy absorber must have the same absorption coefficient as the heavy phase to be analyzed.

Dilution has an inevitable side-effect—a significant decrease of intensity due to the combined effects of the dilution itself and the increase in absorption. The ratio of intensities in the original and diluted samples is given by:

$$I_{ij}/(I_{ij})_d = \frac{c_j}{c_j(1 - c_d)} \cdot \frac{\mu_d^*}{\mu^*} = \frac{1}{1 - c_d} \cdot \frac{\mu_d^* c_d}{\mu_j^* c_j + \mu_m^* (1 - c_j)} + 1 \quad (4.56)$$

In the case in which absorption of the heavy phase is overwhelming [$\mu_j^* c_j >> \mu_m^* (1 - c_j)$] and dilution is thus the method of choice, the ratio $I_{ij}/(I_{ij})_d \approx 11$, supposing that $\mu_d^* \approx \mu_j^*$, $c_d = 2/3$, and $c_j = 0.2$. This tremendous decrease of intensity is not surprising, because in the original sample with negligible matrix absorption:

$$I_{ij} = K_{ij}c_j/[\mu_j^* c_j + \mu_m^* (1 - c_j)] \approx K_j/\mu_j^* = (I_{ij})_0$$

i.e., its intensity is almost equal to the intensity diffracted by the pure phase, even if the phase abundance c_j is much less than 1.

The practice of sample dilution with a heavy absorber is very popular in X-ray fluorescent analysis. The same approach was used for QXRD by Flinter (1973, 1975). If the analyzed phase has a very strong reflection, then dilution, even very extensive dilution, might not bring the counting statistics error above the acceptable level of ~1%. In this case, dilution of the samples can generally be used to equalize sample absorption. We found this technique to be very useful in the analysis of quartz in various geological and industrial materials having a quartz weight fraction above 10% and an absorption coefficient ranging from 25 to 80 cm^2/g. The internal standard technique is impractical in this case due to the wide variation in sample composition and the necessity to search for an internal standard for each sample. Samples were diluted five times with corundum.

4.5.5 *Implementation of the dilution method*

Even if we use the dilution method in its original form, which means moderate dilution of each sample, a decrease in the intensity in the diluted sample is unavoidable. Thus, this method is appropriate for analysis of phases with medium or high abundance. This technique may be considered complementary to the doping method; the latter is suitable for the analysis of minor phases. Both dilution and doping methods are more labor-intensive than the internal standard method. They are not recommended for large-scale analysis but rather for the analysis of individual samples.

An example may be taken from the work of Zav'yalova and Ivoilov (1970). Quartz was analyzed in two-phase test mixtures with calcite or barite. Each sample was diluted twofold (c_d = 0.5) with fluorite. The results of the analysis, shown in Table 4.5, are quite satisfactory.

Table 4.5. Results of analysis of quartz in a two-phase mixture by the dilution method

		Quartz, %		
Second phase	μ^*, cm^2/g	Real	Found	Δ
Calcite	43.0	80.0	79.7	+0.3
Barite	70.0	70.0	70.9	−0.9
Sphalerite	90.0	90.0	90.8	−0.8

4.6 Full-phase analysis of the n-phase sample

4.6.1 *Basic equations*

Here we consider the case in which our sample contains n phases and all of them must be determined (*complete analysis*). Of course, the abundance of each phase may be determined by one of the techniques discussed in the previous sections by applying equation (4.2) to each of the n phases. If all the phases were to be analyzed, then an additional mass-balance equation would be available:

$$\sum_j c_j = 1 \tag{4.57}$$

This equation may be implemented in one of two ways:

1. The weight fractions of all of the detected phases are determined by the one of techniques discussed earlier, such as internal standard, doping, etc. The mass balance equation is then used to estimate the accuracy of the analysis, as is acceptable analytical practice. Significant deviation of Σc_j from unity indicates the presence of an undetected phase or an appreciable systematic error in the analysis of the detected phase.
2. The additional equation (4.57) is used to exclude one of the unknowns, i.e., the mass absorption coefficient, from the calculation of weight fractions.

The contribution of this equation was realized in the early years of QXRD in the works of Copeland and Bragg (1958) and Drits (1958). The ratio of the weight fractions of the two phases j and r is expressed as:

$$c_j/c_r = (K_{hr}/K_{ij})(I_{ij}/I_{hr}) = \beta_{rj}S_{jr} \tag{4.58}$$

where $\beta_{rj} = K_{hr}/K_{ij}$ and $S_{jr} = I_{ij}/I_{hr}$. We will retain only the phase subscripts j and r in the ratios for calibration constants or intensities (or elsewhere) and will ignore the peak subscripts i and h. Thus, β_{12} means the ratio of the calibration constants for the first and second phases without reference to the specific peaks used.

For an n-phase sample, (n–1) equations (4.58) could be written, each time taking the ratio of a *jth* phase to the same *rth* phase. These equations, together with equation (4.57), constitute the full system of n linear equations with n unknown phase abundances c_j, provided that the ratios of the calibration constants β_{jr} are known. In practice, we should not even bother to solve equations (4.58) but simply use the basic equation (4.2) and ignore the mass absorption coefficient μ^* or give it any approximate value. Thus, we can calculate a set of provisional c_j' values:

$$c_j' = I_{ij}\,\mu_{ap}^*/K_j \tag{4.59}$$

where μ_{ap}^* is any approximate value of the mass absorption coefficient. Afterwards, the values of c_j' calculated by equation (4.59) are normalized according to equation (4.57) to give the true weight fractions c_j. In this procedure, all K_js should be brought to a common relative scale, but a common multiplier is not essential because it would be normalized together with the absorption coefficient μ_{ap}^* according to mass balance equation (4.57). This approach was first used by Jahanbagloo and Zoltai (1968), and we will use it extensively here and in the following sections.

Another elegant mathematical derivation of the same idea was developed by Bezjak (1961), Péter and Kálmán (1964), and Náray-Szabó and Péter (1964) and was later resurrected by many workers (Chung 1974b; Hooton and Giorgetta 1977; Biederman et al. 1974). By rewriting the basic equation (4.2), we obtain:

$$c_j = \mu^* I_{ij}/K_{ij} \tag{4.60}$$

By summing both sides of this equation over all the phases in the sample and by taking into account the mass-balance equation (4.57), we obtain the following equation for the mass absorption coefficient:

$$\mu^* = \left(\sum_{j=1}^{n} I_{ij}/K_{ij}\right)^{-1} \tag{4.61}$$

A combination of equations (4.60) and (4.61) gives an expression, free of the absorption coefficient, for the concentration of any phase r:

$$c_r = \frac{I_{hr}/K_{hr}}{\sum_{j=1}^{n} (I_{ij}/K_{ij})} = \frac{1}{\sum_{j=1}^{n} (\beta_{rj}/S_{rj})} \tag{4.62}$$

where $S_{jr} = I_{ij}/I_{hr}$ is the ratio of intensities of peaks i and h belonging to phases j and r, respectively, and $\beta_{rj} = K_{hr}/K_{ij}$.

In matrix form, this equation may be written as:

$$\begin{vmatrix} 1/c_1 \\ 1/c_2 \\ \dots \\ 1/c_n \end{vmatrix} = \begin{vmatrix} 1 & S_{12} & \dots & S_{1n} \\ S_{21} & 1 & \dots & S_{2n} \\ \dots & \dots & \dots & \dots \\ S_{n1} & S_{n2} & \dots & 1 \end{vmatrix} \; \begin{vmatrix} 1 & \beta_{12} & \dots & \beta_{1n} \\ \beta_{21} & 1 & \dots & \beta_{2n} \\ \dots & \dots & \dots & \dots \\ \beta_{r1} & \beta_{r2} & \dots & 1 \end{vmatrix} \tag{4.63}$$

From the definition of the ratios β_{rj} and S_{jr}, it follows that:

$$\beta_{jj} = S_{jj} = 1;$$
$$S_{jr} = (S_{rj})^{-1};$$
$$\beta_{jr} = (\beta_{rj})^{-1};$$
$$\beta_{jr} = \beta_{jk} \cdot \beta_{kr} \qquad (4.64)$$

where k is any intermediate phase.

The last equation is particularly important, since the third phase k, which is not found in the sample (or doped into the sample), plays the role of a mediator in acquiring the β_{jr} ratios. As soon as β_{jr} is acquired, we can implement equation (4.62) and thus compensate for the absorption effect. This "external" character of the third phase k, or external standard, inspired Hubbard et al. (1976) to call equation (4.62) an external standard equation.

Equations (4.62) and (4.63) will retain their form if more than one diffraction peak is available for any or all of the n phases. The simplest way to utilize these additional peaks is to calculate the representative intensity I_j as a weighted sum of the available (in this case resolved) peaks:

$$I_j = \sum_{i=1}^{m_1} w_i I_{ij}; \qquad I_r = \sum_{h=1}^{m_2} w_h I_{hr} \qquad (4.65)$$

where m_1 and m_2 are the numbers of the peaks, and w_i and w_h are the corresponding weights. Of course, in evaluating the calibration constant β_{rj} one should use the same weights as in the definition of intensities:

$$\beta_{rj} = \left(\sum_{h=1}^{m_2} w_h K_{hr}\right) \Big/ \left(\sum_{i=1}^{m_1} w_i K_{ij}\right) \qquad (4.66)$$

where weight w_h is assigned to peak h of phase r and weight w_i is assigned to peak i of phase j. The greater weight should be given to the well-resolved and strong peak less susceptible to preferred orientation and to fluctuations in the number of reflecting crystallites. Unfortunately, in this case no quantitative instruction can be given, and weighting is based on largely heuristic judgment.

The remarkable feature of equation (4.62) is that it contains only ratios and no absolute values of intensities or calibration constants. Thus, the method discussed is less sensitive to small fluctuations of instrumental and sample factors, such as variations of tube current and high voltage or small misalignment of the diffractometer. For samples that cannot be considered as infinitely thick, the ratio of intensities S_{jr} would be less sensitive to the sample transparency provided that diffraction peaks i and h of the phases j and r are close to each other.

The evident stability of the β_{jr} calibration constants created hope for interlaboratory exchange of these constants. This opportunity will be discussed later.

4.6.2 *Application of the method to two-phase mixtures*

Equation (4.62) is particularly simple for a two-phase system:

$$c_1 = (1 + \beta_{12}S_{21})^{-1} \tag{4.67}$$

Let us compare equation (4.62) with equation (4.13), which we investigated extensively in Section 4.2.2. The main difference lies in the experimental procedure. For equation (4.13), only one phase must be measured both in the sample and as a pure phase. For the second phase, we need to know only the mass absorption coefficient. In contrast, implementation of equation (4.67) assumes experimental measurements for both phases. In theory, both equations are equally valid, but in practice the technique discussed in this section and expressed in equation (4.67) has some advantages. As will be shown in the next section, the value of β_{12} in equation (4.67) is usually acquired by measuring one two-phase specimen or a number of specimens somewhat similar to the sample to be analyzed, in contrast to the diffraction-absorption technique, in which a pure phase is used for calibration. The method described here may thus be more accurate but, of course, at the expense of additional experimental work that includes intensity measurements for both phases and preparation of the two-phase mixture needed to determine the calibration constant β_{12}.

4.6.3 *Evaluation of the calibration coefficient β_{rj}*

The coefficient β_{rj} can be evaluated by measuring the intensities of the pure phases:

$$\beta_{rj} = K_{hr} / K_{ij} = [(I_{hr})_0 \mu_r^*]/[(I_{ij})_0 \mu_j^*] \tag{4.68}$$

where μ_r^* and μ_j^* are the mass absorption coefficients of the pure phases, and $(I_{ij})_0$ and $(I_{hr})_0$ are the intensities of peaks i and h of the pure phases r and j. However, this method of acquiring β_{rj} suffers from two serious drawbacks. First, we have to know the mass absorption coefficients for the pure phases. Second, and more important, various side effects of the diffraction procedure, like microabsorption and preferred orientation, in the pure phase could differ considerably from those in the real specimen, which is a mixture of several phases. It is preferable to evaluate β_{rj} from two-phase or multiphase mixtures of pure phases which better reflect the situation in the real samples. From equation (4.58), we have:

$$\beta_{rj} = S_{rj} (c_j / c_r) \qquad (4.69)$$

and for a 1:1 two-phase mixture it follows that:

$$\beta_{rj} = S_{rj} \qquad (4.70)$$

In this case, the necessity of evaluating the mass absorption of the pure phase falls away.

In total, we have to determine n–1 calibration constants β_{rj} ($\beta_{jj} = 1$), and we thus need at least (n–1) two-phase mixtures, one n-phase mixture, or a sample with known phase compositions. In the latter case, (n–1) equations similar to equation (4.69) could be written and solved for β_{rj}. These equations are, of course, equally valid for two-phase or multiphase samples.

Bearing in mind this experimental evaluation of β_{jr}, we should take some precautions in the implementation of the relationship $\beta_{jr} = \beta_{jk}\, \beta_{kr}$. In this procedure, any third phase k serves as an intermediate in the evaluation of β_{jr}. This method will be valid only on the assumption that the basic relationship (4.69) is not biased by any side effects, such as microabsorption. Suppose, for example, that both phases j and r are heavy phases and the intermediate phase k is a light phase; then significant microabsorption effects can be expected in the binary j – k and r – k mixtures and much smaller microabsorption effects or none at all in the j – r mixture. Thus, β_{jr} calculated as the product $\beta_{jk}\, \beta_{kr}$ may be different from β_{jr} measured directly from a mixture of the two phases j and r. In such a case the direct measurement of the calibration constant is, of course, preferable.

4.6.4 Presence of undetected phases in the analytical sample

The fulfillment of the conditions of the mass-balance equation (4.57) is essential in the implementation of the above-discussed technique. However, we must ask the question: what would happen if one or more phases were not detected? A common situation is one that occurs when an amorphous phase is present in a significant amount or some minor phases are present in concentrations below the detection limit. We will now show that in favorable circumstances an appropriate correction can be made that may compensate for the incompleteness of the analysis. Suppose that among (n+1) phases that constitute the sample, only n phases were detected and their respective intensities were measured.

Then, following the development of equation (4.62) and taking into account that now $\sum_j c_j \neq 1$, we will arrive at the following equation:

$$c_r = \sum_j c_j / S_{jr}\, \beta_{rj} = (1 - c_{n+1}) / \sum_j S_{jr}\, \beta_{rj} \tag{4.71}$$

The problem now is to estimate the sum $\Sigma\, c_j = 1 - c_{n+1}$, where c_{n+1} is the weight fraction of an undetected phase. Equation (4.60) may be rearranged as:

$$c_j\, \mu_j^* = \mu^*\, I_{ij} / (I_{ij})_0$$

where K_{ij} is replaced by $(I_{ij})_0\mu_j^*$, $(I_{ij})_0$ is the intensity of the analytical diffraction peak, and μ_j^* is the mass absorption coefficient of the phase j. After summation over n phases, the mass absorption coefficient of the specimen may be expressed as:

$$\mu^* = \sum_{j=1}^{n} c_j\, \mu_j^* / \sum_{j=1}^{n} [I_{ij} / (I_{ij})_0] = (\mu^* - c_{n+1}\, \mu^*_{n+1}) / [\sum_j I_{ij} / (I_{ij})_0] \tag{4.72}$$

It follows from equation (4.72) that:

$$\sum_j [I_{ij} / (I_{ij})_0] = 1 - c_{n+1}\, (\mu^*_{n+1} / \mu^*) \tag{4.73}$$

What we have obtained here is a criterion for completeness of the analysis (Zaslavsky 1977). Indeed, if the sample contains only n detected phases and $c_{n+1} = 0$, then:

$$\sum_j [I_{ij} / (I_{ij})_0] = 1 \tag{4.74}$$

If there is at least one undetected phase, then:

$$\sum_j [I_{ij} / (I_{ij})_0] < 1 \tag{4.75}$$

Let us now look at the merits of this criterion. If the mass absorption of the undetected (n+1)*th* phase is close to the mass absorption of the specimen, i.e.,

$$\mu^*_{n+1} \approx \mu^*$$

then

$$\sum_{j=1}^{n} [I_i / (I_{ij})_0 \approx 1 - c_{n+1} = \sum_{j=1}^{n} c_j \tag{4.76}$$

By combining equations (4.76) and (4.71), we get, in this favorable case:

$$c_r = \sum_{j=1}^{n} [I_{ij}/(I_{ij})_0] \,/\, (\sum_{j=1}^{n} S_{jr}\, \beta_{rj}) \tag{4.77}$$

All the quantities in equation (4.77) can be measured. Intensities $(I_{ij})_0$ may be obtained, using equations (4.68) and (4.69), from the same two-phase or multiphase mixtures prepared to evaluate the coefficients β_{rj}. Absorption of the pure phases μ_j^* should be known (μ_j^* is usually calculated from the known chemical composition of the given phase).

There is another obvious method of checking the completeness of the analysis. Suppose that any phase j in our sample is analyzed by two techniques:

1. The external standard method discussed in this section, giving the phase abundance c_j', and
2. Any other method discussed previously, i.e., internal standard, doping, etc., giving the phase abundance c_j.

It can be expected that if an undetected phase is present in the analyzed sample, the ratio c_j'/c_j will be less than 1. Accordingly, the weight fractions of the remaining phases c_r calculated by equation (4.62) must be corrected by multiplying by the correcting term c_j'/c_j. The weight fraction of the undetected phase (or phases) is equal to $1 - (c_j'/c_j) \sum_{r=1}^{n} c_r$. This correction will be meaningful, of course, only if both c_j' and c_j are determined very accurately.

4.6.5 Comparison of equations (4.62) and (4.77)

An analysis based on equation (4.77) is undoubtedly more complicated and time-consuming because $(I_{ij})_0$ values have to be acquired. Thus, a number of considerations should be taken into account before utilizing this equation. First, it was derived on the assumption that the absorption of the undetected phase is equal to that of the mixture of n detected phases. Suppose that this assumption is not valid and $\mu^*_{n+1} > 2\mu^*$. In this case, the abundances calculated by equation (4.77) will bear a greater systematic error (negative) than the abundances calculated by equation (4.62), i.e., by ignoring the undetected phase. Second, we have to take into account that the sum $\sum_j [I_{ij}/(I_{ij})_0]$ from (4.77) is also influenced by various errors, such as counting statistics and particle statistics. If the standard deviation $\sigma[I_{ij}/(I_{ij})_0]$ exceeds the value of c_{n+1}, then implementation of equation (4.77) becomes meaningless. This equation may be used in preference to equation (4.62) only under the following conditions:

1. There is a sound basis for the supposition of the presence of undetected phase(s).
2. There is a sound basis for assuming that the absorption of the undetected phase is close to the absorption of the sample.
3. The sum $\sum_j [I_{ij}/(I_{ij})_0]$ is, at the very least, lower than (about) 0.9.

4.6.6 *Precision of the analysis*

The standard deviation $\sigma(c_j)$ may be calculated from equation (4.62). Suppose that β_{rj} are known accurately and the only sources of deviation are intensity errors, then

$$\sigma(c_r)/c_r^2 = \sqrt{\sum_j [\beta_{rj}\,\sigma(S_{jr})]^2} \tag{4.78}$$

or after some transformations,

$$\varepsilon\,(c_r) = \sqrt{\sum_j \beta_{rj}^2\, c_r^2\, S_{jr}^2\, [\varepsilon^2(I_{ij}) + \varepsilon^2(I_{hr})]} \tag{4.79}$$

where $\varepsilon(c_r) = \sigma(c_r)/c_r$, $\varepsilon(I_{ij}) = \sigma(I_{ij})/I_{ij}$, and $\varepsilon(I_{hr}) = \sigma(I_{hr})/I_{hr}$ are coefficients of variation.

To estimate $\varepsilon(c_r)$, suppose that $\beta_{rj} \approx 1$, $S_{jr} \approx 1$, and $\varepsilon(I_{ij}) \approx \varepsilon(I_{hr}) = \varepsilon(I)$, then

$$\varepsilon(c_r) \approx \sqrt{2}\,(n-1)c_r\,\varepsilon(I) \tag{4.80}$$

For $\varepsilon(I) \approx 3\%$, $n = 5$, $c_r \approx 0.2$, $\varepsilon(c_i\,) \approx 3.5\%$, i.e., values that are quite acceptable.

The simplest way to estimate $\varepsilon(I)$ is to account for intensity errors only on the basis of counting statistics. However, the counting statistics may be the main source of error only for weak peaks and a high background level. Other sources of error, such as particle statistics, orientation effects, microabsorption, and nonuniformity of the sample, are significant for medium and strong peaks. The experimental estimation of the standard deviation $\sigma(I)$ seems to be preferable. These values are calculated from repeated intensity measurements with a reloading sample holder.

Table 4.6. β_{rj} coefficients determined in various laboratories

	β_{rj}			
Phase pair	Náray-Szabó and Péter (1964)	Zav'yalova and Ivoilov (1965)	Hooton and Giorgetta (1977)	Pawloski (1985)
Calcite $(10\bar{1}4)$ –quartz $(10\bar{1}1)$	0.89	0.88	-	0.65
Calcite $(10\bar{1}4)$ –dolomite $(10\bar{1}4)$	1.05	-	1.36	1.85
Calcite $(10\bar{1}4)$ –albite	1.37	1.74	-	-

4.6.7 *Invariance of β_{rj} values*

The apparent independence of β_{rj} on experimental conditions makes it very attractive for use in interlaboratory practice. It was initially suggested that any stable phase, such as MgO (Náray-Szabó and Péter 1967), be used as a constant r phase in the determination of β_{rj} coefficients. This idea later became accepted practice but with Al_2O_3 (corundum) as the reference phase. This aspect will be discussed in detail in Section 4.7.

Unfortunately, β_{rj} depends on the crystal chemistry of the component crystalline phases r and j, on the degree of preferred orientation, and on microabsorption effects. Thus, variations of β_{rj} determined in different laboratories may be significant. Although a full account of this problem will be given in Section 4.7 (in the discussion of the applicability of the corundum ratio), a few examples taken from various works that cover a period of more than 20 years are given in Table 4.6. The large spread of β_{rj} values given in Table 4.6 emphasizes the necessity for critical analysis of published β_{jr} data before they can be applied in analysis.

4.6.8 *Examples*

The external standard method has been quite popular ever since its unquestionable advantages were fully appreciated. The two examples given below will shed some light on the technique.

Example 1. Analysis of a four-phase sample of calcite, apatite, albite, and quartz

The intensities of the analytical peaks in the analyzed sample and in pure phases are given in Table 4.7.

The coefficients β_{rj} were obtained from intensity measurements of two-phase mixtures with 1:1 compositions and are given below in matrix form:

Table 4.7. Intensity data for analyzing a four-phase sample

	Analytical peak		Intensity, cps	
Phase	hkl	2θ	I_{ij}	$(I_{ij})_0$
Calcite	104	29.4	1800	3460
Apatite		31.8	355	1250
Albite		27.8	480	4240
Quartz	101	26.6	335	8200

Table 4.8. Intensities of the analytical peaks

			Intensity, cps	
Number	Phase	μ_j^*, cm^2/g	I_{ij}	$(I_{ij})_0$
1	ZnO	55.0	4661	13311
2	$CaCO_3$	75.3	2298	6299
3	Al_2O_3	31.7	631	4960

$$\beta_{rj} = \begin{vmatrix} 1 & 2.74 & 1.74 & 0.88 \\ 0.36 & 1 & 0.64 & 0.32 \\ 0.57 & 1.57 & 1 & 0.51 \\ 1.13 & 3.11 & 1.97 & 1 \end{vmatrix}$$

Calculation using equation (4.62) gives the following phase abundances: calcite 0.46, apatite 0.25, albite 0.215, and quartz 0.075.

The test for an undetected phase, carried out by calculating $\sum_j I_{ij}/(I_{ij})_0$, gave a value of 0.961. The difference $1-\sum I_{ij}/(I_{ij})_0$ is not statistically significant, and in this case the application of equation (4.77) is not justified.

Example 2. Analysis of a four-phase sample comprising three crystalline phases and one amorphous phase

As a second example, we will take one of the artificial mixtures prepared and measured by Chung (1974a). This mixture is composed of three crystalline phases, ZnO, $CaCO_3$, and Al_2O_3, and one amorphous phase, silica gel. In this case we will consider corundum as one of the unknown phases and not as an internal standard or a "flushing agent," as assumed by Chung (1974a). Intensity data for the analytical peaks of the three detected crystalline phases are given in Table 4.8.

Table 4.9. Intensities for two-phase mixtures

		Intensity, cps		
Number	Phase	I_{ij}	I_{i3}	β_{jr}
1	ZnO	8178	1881	4.35
2	$CaCO_3$	4437	1491	2.98

Table 4.10. Calculated and known weight fractions of crystalline phases in a four-phase mixture of zincite, calcite, corundum, and silica gel

		Weight fractions		
Number	Phase	Calculated by equation (4.62)	Calculated by equation (4.77)	Known
1	ZnO	0.433	0.364	0.344
2	$CaCO_3$	0.312	0.262	0.280
3	Al_2O_3	0.255	0.215	0.214

The calibration coefficients β_{jr} were calculated from intensity measurements on two-phase 1:1 mixtures of zincite(1)-corundum(3) and calcite(2)-corundum(3), as given in Table 4.9.

The matrix for β_{jr} is given below:

$$\beta_{jr} = \begin{vmatrix} 1 & 1.46 & 4.35 \\ 0.685 & 1 & 2.98 \\ 0.230 & 0.336 & 1 \end{vmatrix} \tag{4.81}$$

We recalculated the intensity data for the pure phases, taking into account that for two-phase mixtures of equivalent composition $I_{ij} = K_{ij}/2\mu^*$ and for a pure phase $(I_{ij})_0 = (K_{ij}/\mu_j^*)$. Thus, $(I_{ij})_0 = 2I_{ij}\,(\mu^*/\mu_j^*)$, where μ^* is the mass absorption coefficient of the two-phase mixture. Recalculated $(I_{ij})_0$ intensities are given in the last column of Table 4.8. Phase abundances (Table 4.10) were calculated from equations (4.62) and (4.77) using the intensity data given in Table 4.8 and the matrix (4.81) for the coefficient β_{jr}. In this case, there is a clear advantage to using equation (4.77), which gives phase abundances close to real values. The test for an undetected phase $\sum_j I_{ij}/(I_{ij})_0$ gives a value of 0.84, which is well below 1. Moreover, the intensity measurements were very accurate in the cited work and the difference $1-\sum_j I_{ij}/(I_{ij})_0$ is statistically significant.

4.7 Reference intensity ratios in quantitative analysis

4.7.1 Introduction

Quantitative phase analysis is based on the measurement of X-ray diffraction line intensities and subsequent scaling against the same line intensities in a pure analyte or in a composite specimen with a known amount of analyte. The latter step can be facilitated if the intensities of the analytical peaks in the different phases are brought to a common scale. All of the QXRD techniques discussed in the previous section fall into one of two groups, depending on the scaling mode:

1. "Absolute" techniques, in which the intensity of the analytical peak in the pure analyte $(I_{ij})_0$ is a factor in the analytical equations. Diffraction-absorption and doping techniques are typical of this approach. Due to the obvious dependence of $(I_{ij})_0$ on experimental conditions [I_0 in equation (2.19)], intensities can be brought to a common scale only for a particular type of instrument, tube setting, or detector efficiency. These values are thus not very useful, not even for intralaboratory implementation, to say nothing of interlaboratory use.
2. "Relative" or "ratio" techniques, in which ratios (and not absolute intensities) are used in the analytical equations. Internal and external standard techniques are typical of this type of approach. Due to the apparent independence of intensity ratios on experimental conditions, we can expect that intensity scaling to any common scale would be useful in this case. A short discussion of this matter has already been given in Section 4.6.

The simplest way to bring diffraction peak intensities of various phases to a common scale is to refer them to the diffraction peak intensity of a certain reference phase. Náray-Szabó and Péter (1964) suggested MgO as a reference material in an external standard method. de Wolff and Visser (1964, 1988), looking for an optimal internal standard, chose corundum (α-Al_2O_3) due to its apparent virtues of chemical stability, ready availability in a fine-grained form, and absence of preferred orientation effects. They showed how the internal standard technique can be implemented with intensities of a pure analyte brought to a common scale. Scott (1968) also suggested corundum as a reference material; he pointed out, however, that great care must be exercised when scaled intensities are used for quantification. Corundum was finally accepted as the reference material, and the relevant RIRs appeared in NBS Monograph No. 25 (1969-1978) and in the PDF issued by the International Centre for Diffraction Data.

4.7.2 Definition and acquisition of reference intensity ratios

An RIR (or I/I_c) is defined as the ratio of the strongest peak of any phase j to the strongest corundum peak (113) in a 1:1 mixture (by weight). This ratio is exactly equal to the ratio of the calibration constants K_{sj} and K_{hc}.

$$(RIR)_j = I_{sj}/I_{hc} = K_{sj}/K_{hc} = \beta_{jc} \tag{4.82}$$

where the subscript c refers to corundum, and subscripts s and h refer to the strongest peak of the phase j and corundum. RIR values are defined for CuK_α radiation and for Bragg-Brentano reflection focusing geometry with a constant divergence slit. In the late 1960s, the RIR was chosen to be the ratio of peak height intensities. This practice continues, although the overwhelming use of computer-controlled diffractometers and data-processing software facilitates utilization of more attractive integrated intensities in RIRs (Davis and Smith 1988; Davis et al. 1989). An interesting intermediate case occurs in a modern QXRD approach in which the pattern of an analyzed sample is presented as a scaled sum of the pattern of the component phases (Smith et al. 1988). In this case, the RIR is defined as a parameter of the best fit for mixtures of known composition. Depending on the integration interval, the RIR value might be closer to its "peak height" or "integrated intensity" value. Further details are given in Section 4.7.4.

There are three basic techniques for the determination of the RIR: measurements of powder mixtures, measurements of pure phases, and calculation of the RIR.

4.7.2.1 Measurements on powder mixtures

If a 1:1 mixture with corundum is used and the analytical peaks are the strongest for corundum and the analyte j, then the RIR is defined by equation (4.82). In the general case, in which any other pair of peaks and any weight fraction ratio other than 1:1 are used, then the RIR is given by:

$$(RIR)_j = \frac{c_c}{c_j} \cdot \frac{a_{kc}}{a_{ij}} \cdot \frac{I_{ij}}{I_{kc}} \tag{4.83}$$

Table 4.11. RIR values for ZnO

ZnO			Al_2O_3			
hkl	2θ	a_{ij}	hkl	2θ	a_{kc}	I/I_c
100	31.78	59	012	25.58	61	4.99
002	34.44	46	104	35.16	92	5.24
101	36.26	100	104	35.16	92	5.21
110	56.62	31	116	57.50	83	5.29

where c_c and c_j are the weight fractions of corundum and phase j in the mixture, I_{ij} and I_{kc} are the observed intensities of the chosen pair of peaks, and a_{kc} and a_{ij} are the relative intensities of the same peaks. Note that the mixture need not necessarily be a binary mixture: any other component phase that does not interfere with measurements of peaks i and k may be present. Determination of the RIR by equation (4.83) using several pairs of peaks and several mixtures and averaging the results is certainly advisable, since this procedure will enable the averaging of probable phase-related effects, such as preferred orientation and microabsorption.

Convergence of RIR values obtained from various pairs is generally good, as was demonstrated, for example, by Hubbard and Smith (1977) for ZnO (Table 4.11). The four measurements given in Table 4.11 give no sign of systematic errors. The mean value of the RIR is 5.2±0.1. However, a brief perusal of the RIRs reported in NBS Special Publication No. 25 (1969-1978) reveals greater deviations—up to 5%-7%.

Details of a multimixture, multipeak technique for acquiring RIRs are to be found in the compilation of Snyder (1992). For obvious reasons, this technique is suitable only for phases available in powder form.

4.7.2.2 Measurements of pure phases

By definition:

$$(RIR) = \frac{K_{sj}}{K_{hc}} \cdot \frac{(I_{sj})_0}{(I_{hc})_0} \cdot \frac{\mu_j^*}{\mu_c^*} \tag{4.84}$$

where $(I_{sj})_0$ and $(I_{hc})_0$ are the intensities of the strongest peaks of phase j and corundum in the corresponding pure phases, and μ_j^* and μ_c^* are the mass absorption coefficients of the phases. This technique is suitable if the analyte is available in bulk (as opposed to, say, a thin layer). Mixtures are, however, preferable for determination of the RIR due to the smaller susceptibility to various sample- and phase-related effects. Davis et al. (1990) found a pure phases technique advantageous in a special specimen preparation technique—deposition on porous filters. Parallel determination of the RIR on mixtures and pure phases revealed good correspondence and absence of statistically significant differences.

4.7.2.3 Calculation of the RIR

Calculation of RIR values from the known crystal structure of phase j and corundum is especially attractive. In this case, equation (4.84) may be applied, with $(I_{sj})_0$ and $(I_{hc})_0$ being calculated from a fundamental knowledge of crystal structure, according to equation (3.8). In this way, we not only reduce the experimental work required for measuring binary mixtures, but we can also acquire the needed β_{jr} constants for phases that are not available in the pure state or not available in powdered form. Sources of structural data are listed in Section 4.9.5. Structural data for corundum can be found in Newnham and de Haan (1962). Computer programs are widely used for the simulation of powder patterns from crystal structure data. Some of them, like, for example, MicroPOWD (Smith 1991), generate a table of relative intensities and calculate an RIR value for a given phase. Others add scaling factors directly related to the RIR (Hubbard and Smith 1977). Calculated RIR values are free of sample errors, such as preferred orientation and microabsorption. The main sources of uncertainty lie in the atomic coordinates, in site occupancies, and in thermal parameters (Hubbard et al. 1976). Unfortunately, comparison of experimental and calculated RIR values shows deviations that exceed the reproducibility of experimental data or expected errors of calculations.

RIRs are defined for CuK_α radiation. If a different radiation is desired, it is not difficult to convert the RIR for CuK_α to the required values. Structure factors are not dependent on the radiation used, because atomic scattering factors depend only on $\sin\theta/\lambda = (2d)^{-1}$, except in the case in which the wavelength is close to the absorption edge of one of the phase components, and correction for anomalous scattering is needed. Thus, only the angle-dependent terms in intensity equation (3.8) and the mass-absorption coefficient must be taken into account. From equations (3.8) and (4.84), $(RIR)_\lambda$ for any wavelength λ may be defined as:

$$(RIR)_\lambda = RIR\left(\frac{(LP)_\lambda}{LP}\right)_j \cdot \left(\frac{LP}{(LP)_\lambda}\right)_c \cdot \left(\frac{\mu_\lambda^*}{\mu^*}\right)_j \cdot \left(\frac{\mu^*}{\mu_\lambda}\right)_c \qquad (4.85)$$

where RIR is the reference intensity ratio defined for CuK_α radiation, LP is the Lorentz polarization factor, subscript λ refers to the wavelength λ, and subscripts j and c refer to phase j and corundum, respectively. If we return to experimental methods for the determination of the RIR, we can imagine a situation in which corundum itself might be an unsuitable reference material. If the absorption coefficients of corundum and the analyte differ significantly, the ratio of intensities will be subject to microabsorption effects, causing erroneous RIR values. In order to match μ of the reference material to that of the analyte, the NBS has certified five intensity standards, covering a wide range of absorption coefficients from 125 cm^{-1} up to ~2083 cm^{-1} (Table 4.4) (Hubbard 1980). These five materials are certified for relative intensities and RIRs (calculated values). Four of them may serve as intermediate materials for acquiring the correct RIR for any phase j. Suppose that the absorption coefficient of the reference material k matches the absorption coefficient of the phase j; then by measurement of the strongest peaks of k and j in a 1:1 mixture, we will generate the coefficient:

$$\beta_{jk} = I_{sj}/I_{hk}$$

which is not affected by microabsorption (s and k are the strongest peaks). According to the obvious relationship:

$$(RIR)_j = \beta_{jk} \cdot (RIR)_k \quad (4.86)$$

where $(RIR)_k$ is a certified (calculated and free of microabsorption) RIR for intensity standard k.

4.7.3 *Implementation of reference intensity ratios*

We should remember that RIRs for analyte phases and any other phase involved in an analytical procedure, such as the internal standard procedure, are known before the analysis itself is begun. RIRs may be obtained from a variety of sources, the PDF, literature data, or calculations, or they may be simply "homemade" (obtained experimentally). RIRs do, however, save the time and labor needed for preparation of mixtures of known compositions and the evaluation of calibration constants.

The most straightforward way of applying RIRs is in the internal standard technique, as was initially suggested by de Wolff and Visser (1964). If corundum is chosen as the internal standard and the strongest peaks of analyte and corundum are analytical peaks, then equation (4.21) transforms into:

$$c_j = \frac{c_s}{1 - c_s} (RIR)\, S_{js} \quad (4.87)$$

where S_{js} is the ratio of the intensities of the analytical peaks of the analyte and corundum in the analyzed sample, c_s is the weight fraction of the internal standard, and the RIR replaces the calibration constant β_{js}.

If the internal standard is not corundum, but any other phase s that is not present in the sample, then, by definition:

$$\beta_{js} = (RIR)_j/(RIR)_s \tag{4.88}$$

where subscripts j and s relate to the RIR for the analyte j and the internal standard s. Using this expression for β_{js}, we obtain the following obvious relationship for the phase abundance:

$$c_j = \frac{c_s}{1 - c_s} \frac{(RIR)_j}{(RIR)_s} S_{js} \tag{4.89}$$

which is the basis of the "matrix flushing method" proposed by Chung (1974a) and illustrates the application of the internal standard technique in quantitative analysis when information on matrix absorption is not available. Further details of the implementation of RIRs in the internal standard technique are to be found in Hubbard and Snyder (1988).

Now let us examine the second "ratio" technique—the external standard method—which is of great importance and significance when full specimen analysis is carried out. The basic analytical equation (4.62) contains calibration constant $\beta_{rj} = K_{hr}/K_{ij}$. The analytical peaks h and i are not necessarily the strongest peaks of the phases r and j. The values of β_{rj} are easily expressed through the corresponding RIR and relative intensities of peaks i and h:

$$\beta_{rj} = \frac{(RIR)_r}{(RIR)_j} \cdot \frac{a_{hr}}{a_{ij}} \tag{4.90}$$

where a_{hr} and a_{ij} are the relative intensities of the analytical peaks.

4.7.4 *Invariance of reference intensity ratios*

An RIR, as defined in Section 4.7.2, is an absolute constant for a given crystalline phase. However, published RIR values exhibit significant variation (Davis and Smith 1988; Davis et al. 1989; Dyakonov et al. 1992). Hubbard et al. (1976) compared calculated RIR and two sets of experimental data, one set taken from the PDF (peak intensities) and the other determined by Chung (1974a; integrated intensities) (Table 4.12). Of the six RIR values reported by Chung, three (LiF, KCl, and ZnO) are in good agreement with the PDF data, whereas substan

Table 4.12. Calculated and experimental RIR values

Compound	Calculated value	PDF	Chung (1974a)
LiF	1.66	1.3	1.32
SiO_2 (quartz)	4.10	3.6	-
$CaCO_3$ (calcite)	3.11	2.0	2.98
KCl	5.59	3.9	3.87
ZnO (zincite)	5.38	4.5	4.35
TiO_2 (anatase)	4.32	4.3	3.39
TiO_2 (rutile)	3.05	3.4	2.62
Cr_2O_3	2.39	1.8	-
Ag	16.9	5.2	-
W	38.3	18.0	-

Taken from Hubbard et al. (1976).

tial deviations are observed for three others ($CaCO_3$ and two TiO_2 polymorphs). Calculated RIRs are typically greater than the experimental values by as much as two- or threefold for heavy metals, such as silver and tungsten. A plausible explanation is that microabsorption, which dampens the intensity of a heavy absorption phase and thus decreases experimental RIR values, is responsible for this discrepancy.

Analysis of recently published data on experimental RIR values for minerals (Davis et al. 1989) revealed substantial deviations, even for quartz, one of the most stable minerals. The mean value of RIR was 4.18, with a standard deviation of 1. The minimal (2.7) and maximal (5.1) values differed by a factor of almost 2. Similar deviations have been found for calcite, pyrite, etc. In this latter case, samples of different origin were used, undoubtedly increasing the variability of the RIR. Analyzing the sources of variations of observed RIR, we can separate them into two groups, intrinsic and extrinsic:

1. Intrinsic sources are related to the crystal chemistry of the phase matching corundum. Due to variability in chemical composition and types of structural effects, phases used for preparation of RIR by different diffractionists can be quite different, thus influencing the observed RIR. In the same group of variations, we will include the morphological characteristics of crystallites, which affect the susceptibility for preferred orientation and line broadening, thus influencing the ratio of peak heights to integrated intensities. Structural and morphological variability is typical for minerals.
2. Extrinsic sources relate largely to experimental techniques, such as sample diminution, which affects microabsorption and primary extinction; sample preparation, which affects preferred orientation; and intensity mode (peak heights or integrated).

Recommendations for the determination of RIRs that enable most experimental errors to be avoided are given by Davis and Smith (1988).

In the evaluation of currently available RIR data, we must agree with Hubbard and Snyder (1988), who noted that tabulated RIR should be considered as semiquantitative and utilized for semiquantitative analysis. Accurate analyses can be obtained if RIR are prepared by the diffractionist himself, using phases similar to the analyte. Dyakonov et al. (1992) came to the same conclusion; they showed in a round robin test that variations of RIR obtained in different laboratories were much greater than variations of analytical results. Further details on the acquisition and implementation of RIR values in QXRD are to be found in the book of Davis (1988), who achieved some success in analyzing complicated mineral mixtures. However, the great hope of wide interlaboratory exchange of RIR values has yet to be realized.

4.8 Diffraction patterns with overlapping peaks—full diffraction pattern approach

4.8.1 *Introduction*

The techniques outlined in the previous sections can be applied when there is at least one resolved diffraction peak for each phase to be analyzed. Quite often, this condition is not met, e.g., in the diffraction patterns of Portland cement, oil shales, and many other complex multiphase compositions. It is obvious that as the sample becomes more complex, the probability that a separate peak will not be found must increase. Similarly, when the full diffraction pattern (or a significant portion of it) is to be used for analytical purposes, the case of overlapping peaks will be encountered. The latter approach is demanding increasing attention as the use of computer-controlled diffractometers becomes more widespread, and today it is probably the most promising direction in QXRD.

When the full diffraction pattern is considered, it is almost impossible to avoid the phenomenon of overlapping (although separate peaks could probably be selected for each phase). There are two different approaches to processing diffraction patterns with extensive overlapping peaks. In the first approach, known as *profile fitting* (or pattern decomposition), the clusters of overlapping peaks are separated into components (or individual peaks) (Parrish et al. 1976; Will et al. 1983; Toraya 1986). This technique is based on some knowledge of peak shape and position but does not require any structural or compositional data. As soon as a cluster has been decomposed, the diffraction peaks have been separated, and integrated intensities have been assigned to each component, all of the conventional methods of QXRD discussed so far are applicable. Thus, the problem here is not a specific QXRD problem but rather one of peak separation (which will be discussed in Chapter 5). In the second approach, we consider the composite pattern to be the sum of the scaled patterns of individual phases and try to extract quantitative information without actual separation of the peak clusters into components.

Counterparts to these two approaches may be found in structure refinement by powder diffraction data. One approach is peak separation and implementation of the conventional methods of structure analysis and refinement developed for single crystal studies. The second approach is the continuous pattern approach developed by Rietveld (1969). This method has been used extensively, and hundred of papers have been published since the first publication of Rietveld. It is difficult to evaluate the merits of these two different approaches in QXRD, since no comparative studies have been published. Although the Rietveld approach is the method of choice in structure refinement, in QXRD the two techniques might be equally useful. In practice, in structure refinement, weak peaks, which are very difficult to separate accurately, bear as important structural information as do strong peaks. In QXRD, on the contrary, weak peaks are less suitable for analysis, and greater weight is thus given to the stronger ones.

Historically, the first attempt to solve the problem of overlapping peaks was made by Copeland and Bragg (1958) in their analysis of Portland cement. The general theory was developed by Karlak and Burnett (1966), who showed that even with complete overlapping of the diffraction peaks of all phases, the problem can be solved, at least analytically. In the mathematical procedure developed by Karlak and Burnett (1966) the number of phases in the sample is strictly equal to the number of analytical peaks. In fact, for the best analysis, the number of analytical peaks may exceed the number of phases. This approach was developed in the works of Fiala (1976), Weiss et al. (1983), and Szabó (1980).

It should be noted here that there is a distinct difference in the QXRD methodologies applied for resolved and overlapping peaks. The former case provides the option of analysis of any individual phase, if only one particular phase is of interest, whereas in the latter case it is impossible to analyze separate phases. If all n phases contribute to the overlapping analytical peaks, then all n phases detected in the sample must be analyzed simultaneously. If these n phases constitute our sample, then the mass balance equation (4.57) is valid and may be used in the analytical treatment.

4.8.2 *Basic principles*

Let us suppose that I_i is the integrated intensity of the *ith* diffraction peak in the pattern of the analyzed sample that is accumulated in a particular angular segment:

$$\Delta 2\theta_i = 2\theta_{2i} - 2\theta_{1i}$$

The intensity I_i is simply the sum of the contributions I_{ij} from all the phases j in the analyzed sample, supposing that phase j has a diffraction peak (or peaks) in the range $\Delta 2\theta_i$:

$$I_i = \sum_{j=1}^{n} I_{ij} \tag{4.91}$$

The intensity of the *ith* peak of the *jth* phase may be expressed as the fraction (relative intensity) a_{ij} of the intensity of the major peak of this phase, say I_{jj} (we assign to this peak a number j equal to the number of the phase).

$$I_{ij} = a_{ij} I_{jj} \tag{4.92}$$

By applying the Klug-Alexander equation (4.2) to the major peak I_{jj} of phase j, we obtain:

$$I_{jj} = K_{jj} c_j / \mu^* \tag{4.93}$$

where K_{jj} is the calibration constant related to the major peak of phase j. Accordingly, for any peak of phase j:

$$I_{ij} = a_{ij} K_{jj} c_j / \mu^* \tag{4.94}$$

Combining equations (4.91) and (4.94), we obtain:

$$I_i = (1/\mu^*) \sum_{j=1}^{n} a_{ij} K_{jj} c_j \qquad 1 \leq i \leq m \tag{4.95}$$

where m is the number of peaks (or clusters) that we intend to measure.

Equation (4.95) constitutes the basic equation for the method discussed. I_{ij} is specified as the intensity diffracted by phase j in the angular range $\Delta 2\theta_i$. In fact, I_{ij} may represent a single peak or a cluster of poorly resolved peaks. In the latter case, a_{ij} from equation (4.95) is actually the sum of relative intensities for each individual peak in the cluster:

$$a_{ij} = \sum_{k} (a_{ij})_k$$

where $(a_{ij})_k$ is the relative integrated intensity of the *kth* component of the cluster of peaks diffracted by phase j in the range $2\theta_i$. This equation is especially important when a_{ij} values are acquired by computation.

The problem now at hand is that there are n unknown concentrations c_j $(1 \leq j \leq n)$ and an unknown mass absorption coefficient μ^*. The calibration constants K_{jj} and the relative intensities a_{ij} are characteristic of the phases and are supposedly known. By measuring m peaks ($1 \leq i \leq m$), we will provide m equations (4.95) to be solved for n unknown weight fractions c_j. The obvious conditions are:

1. Each phase must contribute to at least one of the m analytical peaks, and
2. The number of analytical peaks m must be equal to or greater than the number of unknown phases.

Only n equations of the system (4.95) are independent, and formally speaking only n peaks must be measured to calculate the phase abundances. The intensity of each peak in excess of the first n peaks could be calculated from the defined phase abundances c_j. However, these additional diffraction peaks bear valuable information, and their utilization improves the precision of the analysis. In this discussion, we have so far deliberately ignored the mass absorption coefficient, i.e., the (n + 1)*th* unknown in equations (4.95). The following actions to compensate for the sample absorption depend on the mode of use of the usual constraint (mass balance equation):

$$\sum_j c_j = 1 \tag{4.96}$$

We may now proceed in one of two ways:

1. We may apply this constraint to normalize the provisional weight fractions c_j', as was proposed in Section 4.6.1. In this case, we simply ignore the mass absorption coefficient in equation (4.95) by considering it to be a constant factor. The provisional weight fractions c_j' that we hope to calculate from the new system of equations (4.97) are normalized according to equation (4.96) to transform them into real weight fractions.

$$I_i = \sum_j^n a_{ij} K_{jj} c_j' = 1 \qquad 1 \leq i \leq m \tag{4.97}$$

or simply in conventional form:

$$\mathbf{I} = \mathbf{AKC} = \begin{vmatrix} 1 \\ 1 \\ 1 \\ \cdot \\ \cdot \\ \cdot \\ 1 \\ 1 \\ 1 \end{vmatrix}$$

where $\mathbf{I} = [I_{ij}]_{mx1}$, $\mathbf{A} = [a_{ij}]_{mxn}$, $\mathbf{C} = [c_j']_{nx1}$, and $\mathbf{K}$ (diagonal) = $[k_{ij}]_{nxn}$

$$k_{ij} = \begin{cases} 0 & i \neq j \\ K_{jj} & i = j \end{cases}$$

Taking into account the normalization, we can divide all the terms in the right-hand side of equation (4.97) by any constant divider, e.g., the calibration constant K_{rr} of any phase r that may or may not be present in the sample. Thus, equation (4.97) is transformed into:

$$I_i = \Sigma\, a_{ij}\, \beta_{jr}\, c_j' \tag{4.98}$$

where $\beta_{jr} = K_{jj}/K_{rr}$, or in matrix form:

$$\mathbf{I} = \mathbf{A\beta C}$$

where $\boldsymbol{\beta}$(diagonal) = $[\beta_{ij}]_{nxm}$

$$\beta_{ij} = \begin{cases} 0 & i \neq j \\ K_{jj}/K_{rr} & i = j \end{cases}$$

This dummy phase r may, for example, be corundum, and in this case β_{jr} from equation (4.98) is simply the RIR for phase j. Here, we benefit from all the advantages of using ratios of calibration constants β_{jr} instead of absolute values K_{jj} as discussed in detail in Section 4.6.

2. We may apply constraint (4.96) to check the phase balance, i.e., to check whether all the phases present in the sample were detected. In this case, we have to acquire the mass absorption coefficient by any independent method. Application of an internal standard in this particular case will be discussed later in this chapter, and here we suppose that the value of μ^* is known, i.e., that it is measured or calculated. Then, assuming $\bar{I}_i = I_i\mu^*$, we arrive at the following equation:

$$\bar{I}_i = \sum_j a_{ij}\, K_{jj}\, c_j \qquad 1 \leq i \leq m \tag{4.99}$$

which is:

$$\bar{I} = \mu^* I$$

and in matrix form:

$$\bar{I} = \mathbf{A\,K\,C}$$

where $\mathbf{C} = [c_j]_{nx1}$. In this case, all K_{jj} should be put on the same scale, and only values of K_{jj}, and not ratios β_{jr}, are needed.

In both cases, we will work with the system of equations of the same type, (4.97), (4.98), or (4.99). It is useful to express these equations in matrix form:

$$\begin{vmatrix} I_1 \\ I_2 \\ \cdots \\ I_m \end{vmatrix} = \begin{vmatrix} \alpha_{11} & \alpha_{12} & \cdots & \alpha_{1n} \\ \alpha_{21} & \alpha_{22} & \cdots & \alpha_{2n} \\ \cdots & \cdots & \cdots & \cdots \\ \alpha_{m1} & \alpha_{m2} & \cdots & \alpha_{mn} \end{vmatrix} \times \begin{vmatrix} c_1 \\ c_2 \\ \cdots \\ c_n \end{vmatrix} \qquad (4.100)$$

where $\alpha_{ij} = a_{ij} K_{jj}$ and a_{ij} and K_{jj} are defined in equations (4.92) and (4.93). It is easy to see that each column j in the α_{ij} matrix represents the diffraction pattern of the *jth* phase, and each row i represents the contributions of the phases to the peak i, provided that all the phases are present in the sample in equal fractions. If α_{ij} constants are acquired by measurements of pure phases, then obviously, from equations (4.92) and (4.93), we obtain:

$$\alpha_{ij} = a_{ij} K_{jj} = a_{ij} (I_{jj})_0 \mu_j^* = (I_{ij})_0 \mu_j^* \qquad (4.101)$$

where $(I_{jj})_0$ is the intensity of major peak of phase j, and $(I_{ij})_0$ is the intensity of the *ith* peak of phase j. It is assumed that measurements of all intensities $(I_{jj})_0$ and $(I_{ij})_0$ are carried out under the same experimental conditions, and thus the correct scaling of the α_{ij}*th* coefficients will be ensured.

4.8.3 *Evaluation of phase abundances*

The system of linear equations (4.100) may be solved for weight fractions by conventional methods. If the number of peaks m is equal to the number of phases n, then the solution is given by:

$$\mathbf{C} = \boldsymbol{\alpha}^{-1}\mathbf{I} \qquad (4.102)$$

where $\boldsymbol{\alpha} = [\alpha_{ij}]_{mxn}$, $\mathbf{I} = [I_{ij}]_{mx1}$ and $\mathbf{C} = [c_j]_{nx1}$.

We suppose that the inverse matrix $\boldsymbol{\alpha}^{-1}$ does exist, as will be the case if the columns in the matrix $\boldsymbol{\alpha}$ do indeed represent the patterns of the individual phases in the n-phase system. In other words, the elements of any column in the $\boldsymbol{\alpha}_{ij}$ matrix must not be a linear combination of the elements of the remaining columns.

If the number of peaks exceeds the number of phases (m>n), then the weight fractions c_j are found as the values that determine the best fit of the scaled sum of the individual patterns to the experimental data I_i. The most likely weight fractions c_j are those that minimize the sum of the squares:

$$\sum_i w_i (I_i - \sum_j \alpha_{ij} c_j)^2 \quad \min \tag{4.103}$$

where w_i is the weight assigned to peak i. Then, n normal equations with unknowns c_j are derived in equation (4.103), and the least-squares solution is given in matrix form:

$$\mathbf{C} = (\boldsymbol{\alpha}^T \boldsymbol{\alpha})^{-1} \boldsymbol{\alpha}^T \mathbf{I} \tag{4.104}$$

where $\mathbf{C} = [c_j]_{nx1}$, $\boldsymbol{\alpha} = [\alpha_{ij}]_{mxn}$, $\mathbf{I} = [I_i]_{mx1}$ and the symbols T and –1 are used for the transpose and inverse matrices, respectively.

Solving equations (4.102) and (4.104) gives unconstrained c_{ij} values. However, they are, in practice, constrained because all c_j are positive and do not exceed 1. Thus, the constraints:

$$0 \leq c_j \leq 1 \tag{4.105}$$

may be imposed in the minimization procedure (4.103), and sometimes more reliable $c_j th$ values can be obtained. In this case, the solution cannot be written in a simple analytical form, but the relevant mathematical procedure of minimization under constraints is well documented (Karlak and Burnett 1966; Renault 1987; Smith et al. 1988; Snyder 1992).

The optimal weight w_i, i.e., the weight minimizing the standard deviation $(\sigma_c)_j$, is defined as (Szabó 1980):

$$w_i = 1/\sigma^2 (I_i) \tag{4.106}$$

where σ (I_i) is the standard deviation in intensity measurements. There is some uncertainty in the definition of σ (I_i). Very often $\sigma(I_i)$ is considered as the standard deviation of the counting statistics, mostly as a result of the easy way of calculating it. However, some supplementary considerations must be taken into account. For example, diffraction peaks with a high multiplicity factor are less susceptible to particle statistics and deserve higher weightings. Finally, the nature of the peak overlapping and sample-related effects (e.g., preferred orientation) should be taken into account in assigning the weights. Although this somewhat heuristic judgment is difficult to express in figures, some qualitative guidelines may be drawn. For example, it seems reasonable to assign greater weights to the nonoverlapping peaks and lesser weights to the peaks that exhibit preferred orientation.

4.8.4 *Evaluation of the constants* α_{ij}

In equation (4.100), there are mxn constants α_{ij}, which are subdivided into mxn – n relative intensities a_{ij} and n calibration constants K_{jj}. There are two ways to acquire these constants: either experimentally, by measurements of pure phases or mixtures of pure phases, or theoretically, by calculation from fundamental knowledge of the crystal structure of the phases. The latter method will be discussed separately (see Section 4.9). It should be noted that a_{ij} values given in PDF cards are not adequate here because they apply to peak intensities, whereas the method discussed here is based on integrated intensities. The two rows of a_{ij} values (peak and integrated) can be quite similar in a narrow 2θ range, but not in a wide 2θ range, when the peak broadening starts to increase due to chromatic dispersion.

In the determination of a_{ij} and K_{jj} values, pure phases or mixtures of known composition can be used. There are no special problems in acquiring a_{ij} data from pure phase measurements, although one has to bear in mind that these data must be collected under constant experimental conditions for all phases. However, the α_{ij} coefficient, which combines the relative intensities a_{ij} and the calibration constants K_{jj}, may suffer from sample-related effects. The values of a_{ij}, although relatively insensitive to microabsorption effects, may be affected by preferred orientation. The values of K_{jj} are obviously sensitive to microabsorption.

Mixtures of pure phases will imitate the absorption characteristics of real samples better than the pure phases themselves, and therefore mixtures should preferably be used for determining α_{ij} values. Each sample of known composition will give m equations (4.100) for calculating α_{ij} (note that in this case values of c_j are known). Consequently, to acquire all mn constants, we require at least n samples with known compositions. It is better, of course, if the number of samples p > n. Equation (4.100) holds for each of the p samples. Thus, we have p x m linear equations which can be represented in matrix form:

$$\begin{vmatrix} I_{11} & I_{12} & \cdots & I_{1p} \\ I_{21} & I_{22} & \cdots & I_{2p} \\ \cdots & \cdots & \cdots & \cdots \\ I_{m1} & I_{m2} & \cdots & I_{mp} \end{vmatrix} = \begin{vmatrix} \alpha_{11} & \alpha_{12} & \cdots & \alpha_{1n} \\ \alpha_{21} & \alpha_{22} & \cdots & \alpha_{2n} \\ \cdots & \cdots & \cdots & \cdots \\ \alpha_{m1} & \alpha_{m2} & \cdots & \alpha_{mn} \end{vmatrix} \begin{vmatrix} c_{11} & c_{12} & \cdots & c_{1p} \\ c_{21} & c_{22} & \cdots & c_{2p} \\ \cdots & \cdots & \cdots & \cdots \\ c_{n1} & c_{n2} & \cdots & c_{np} \end{vmatrix} \tag{4.107}$$

Each column l in the **I** matrix represents the intensities of m analytical peaks in sample l. Each column l in the **C** matrix represents weight fractions of n phases in sample l.

The system of linear equations (4.107) is solved by the least-squares procedure:

$$\boldsymbol{\alpha} = \mathbf{I}\,(\mathbf{C}^{T}\ \mathbf{C})^{-1}\ \mathbf{C}^{T} \tag{4.108}$$

where $\boldsymbol{\alpha} = [\alpha_{ij}]_{mxn}$, $\mathbf{I} = [I_{ij}]_{mxp}$, $\mathbf{C} = [c_{ij}]_{nxp}$ and T and –1 in matrix notation have their usual meanings of transpose and inverse matrices, respectively.

In order to ensure reliable values for α_{ij}, we have to demand sufficient diversity of the phase abundances c_{jl} in the mixtures used for evaluating the $\alpha_{ij}th$ constant. This problem will be discussed in full in Section 4.10, which is devoted to standardless methods of analysis. Here, we give only the expression for the coefficient of variations for α_{ij}, which will aid in estimating the necessary diversity of the initial set of mixtures:

$$\varepsilon\,(\alpha_{ij}) = \sigma(\alpha_{ij})\,/\alpha_{ij} \approx \varepsilon(I)\,\sqrt{\lambda_{max}\left(I^{T}\,I\right)\,\lambda_{max}\left(\left[I^{T}\,I\right]^{-1}\right)} \tag{4.109}$$

where $\sigma(\alpha_{ij})$ is the standard deviation of α_{ij}, $\varepsilon\,(I) = \sigma(I)/I$ is the coefficient of variation in intensity measurements, and $\lambda_{max}\,(I^{T}\,I)$ and $\lambda_{max}\,|\,I^{T}\,I\,|^{-1}$ are the maximal eigenvalues of the product matrixes $I^{T}\,I$ and $[I^{T}\,I]^{-1}$, respectively. The error in the ε_{ij} values is represented as a product of two terms: $\varepsilon(I)$, the experimental error in the intensities; and the product of eigenvalues that characterize the diversity of the set of mixtures used for estimation of α_{ij}.

Sometimes it is more convenient to split the determination of the two terms comprising α_{ij}, namely, a_{ij} and K_{jj}, into two procedures. The relative intensities a_{ij}, which are less subject to sample-related errors, such as microabsorption, can be acquired by intensity measurements of pure phases. Alternatively, these values can be calculated, provided that the crystal structure of the phase is well known (Section 4.10). The calibration constants K_{jj}, which are more susceptible to sample-related errors, can be determined by measuring known mixtures of pure phases. In practice, only one mixture is needed to obtain all n K_{jj} constants. We may now present the basic equation (4.97) in the form:

$$I_i = \sum_{j} \Omega_{ij}\ K_{jj}$$

where $\Omega_{ij} = a_{ij}\,c_j$ is the product of two known terms, a_{ij} and c_j. If the number of intensities exceeds the number of phases, we will obtain K_{jj} values by the least-squares method:

$$K_{jj} = (\boldsymbol{\Omega}^{T}\,\boldsymbol{\Omega})^{-1}\,\boldsymbol{\Omega}^{T}\,\mathbf{I} \tag{4.110}$$

where $K_{jj} = [k_{jj}]_{nx1}$, $\boldsymbol{\Omega} = [\Omega_{ij}]_{mxn}$, and $\mathbf{I} = [I_{ij}]_{mx1}$.

4.8.5 Precision of the analysis

The standard deviation of the phase abundance $\sigma(c_j)$ is estimated from equation (4.102). Both α_{ij} and I_i are subject to error and thus contribute to $\sigma(c_j)$. If the relative intensities a_{ij} and the calibration constants K_{jj} for all phases are known with high precision, then from equation (4.102) it follows that:

$$\sigma(c_j) = \sqrt{\sum_{i=1}^{n} \left[\alpha_{ji}^{-1}\ \sigma(I_i)\right]^2} \tag{4.111}$$

where $\sigma(I_i)$ is the standard deviation of intensity I_i. A better way to estimate $\sigma(I_i)$ and thus $\sigma(c_j)$ is repeated measurements of I_i with reloading of the sample holder if the specimen is a powder or with replicate samples if the specimen is a solid. In both cases:

$$\sigma(I_i) = \sqrt{\frac{\sum_{r=1}^{p} [(I_i)_r - \langle I_i \rangle]^2}{p-1}} \tag{4.112}$$

where p is the number of successive measurements, and $\langle I_i \rangle$ is the mean value of the intensity.

4.8.6 Internal standard technique in the case of overlapping peaks

Let us begin with the simplest case—that in which the internal standard has at least one separate peak, say n + 1, and the internal standard does not contribute to any of the analytical peaks i. Then, the intensity of the internal standard peak is expressed as:

$$I_S = (1/\mu^*)\, K_S\, c_S \tag{4.113}$$

where c_S is the weight fraction of the internal standard and K_S is the calibration constant. Using equations (4.99) and (4.113), we obtain the ratio of the intensities $S_{is} = I_i/I_S$, i.e.:

$$S_{is} = \sum_j a_{ij}\, \beta_{js}\, c_j'/c_s \qquad 1 \le i \le m \tag{4.114}$$

where $c_j' = c_j\,(1 - c_S)$ is the weight fraction of phase j in the sample doped with the internal standard and $\beta_{js} = K_{jj}/K_S$. The system of linear equations (4.114) is solved for c_j' in the same way as the system of linear equations (4.100).

Let us now consider the case of general overlapping in which the internal standard contributes to each analytical peak and each phase contributes to the chosen peak of the internal standard. Although this case may not seem realistic, we will consider it in order to evaluate the simpler and more realistic situations of partial overlapping. Let us choose one diffraction peak with a significant contribution of the internal standard and call it tentatively an internal standard peak. In accordance with equation (4.95), the intensity of this peak is expressed as:

$$I_s = (1/\mu^*)\,[\sum_j a_{sj} K_{jj} c_j' + K_s c_s]$$

For any other diffraction peak:

$$I_i = (1/\mu^*)\,[\sum_j a_{ij} K_{jj} c_j' + a_{is} K_s c_s]$$

where a_{sj} is the relative intensity of the peak of phase j overlapping with the peak of the internal standard and a_{is} is the relative intensity of the internal standard peak contributing to the analytical peak i. Dividing I_i by I_s, we obtain the ratio S_{is}:

$$S_{is} = \left[\sum_{j=1}^{n} a_{ij}\,\beta_{js}\,c_j' + a_{is}c_s\right] \Big/ \left[\sum_{j=1}^{n} a_{sj}\,\beta_{js}\,c_j' + c_s\right]$$

where $\beta_{js} = K_{jj}/K_s$.

This equation can be rearranged into a more convenient form in the way suggested by Karlak and Burnett (1966) to give:

$$\sum_{j=1}^{n} c_j'\,\beta_{js}\left(S_{is}\,a_{sj} - a_{ij}\right) = c_s\left(a_{is} - S_{is}\right) \tag{4.115}$$

Running index i from 1 to n, we obtain n linear equations of the type (4.115), which can be solved for n unknown phase abundances c_j', provided that all the constants a_{ij} and β_{js} are known. Note that here we use neither the material balance equation nor any other approximation. This means that the internal standard method is applicable even when all the peaks of the internal standard overlap with the analytical peaks. Equation (4.115) may be simplified if there is a separate peak for the internal standard. Then, the coefficients $a_{sj} = 0$ for all $1 \le j \le n$. Equation (4.114) is thus transformed into:

$$\sum_{j=1}^{n} c_j'\,\beta_{js}\,a_{ij} = c_s\left(S_{is}\,a_{hs} - a_{is}\right) \tag{4.116}$$

where the subscript h is assigned to the nonoverlapping peak of the internal standard and a_{hs} is its relative intensity.

Finally, we consider the special case in which: 1) all the analytical peaks are separate; 2) the peak of the internal standard overlaps with the analytical peak of one of the phases, say phase number 1; and 3) there is a separate peak of the internal standard. In this case, all the phases except phase 1 can be analyzed by the conventional internal standard method discussed in Section 4.3. Analysis of phase 1 requires special treatment. We assign the number 1 to the analytical peak of phase 1 that overlaps with the peak of the internal standard. Taking all the imposed conditions into account, we find that the contribution of phase 1 to the analytical peaks of the other phases $a_{i1} = 0$ for all $i \neq 1$. In addition, $a_{1s} \neq 0$, which means that the internal standard contributes to the intensity of peak 1. Then, equation (4.116) written for the peak $i = 1$ will give:

$$c_1' \beta_{1s} a_{11} = c_s (S_{1s} \cdot a_{hs} - a_{1s}) \tag{4.117}$$

Thus, the weight fraction of phase 1 in the sample before doping with the internal standard may be expressed as:

$$c_1 = QS_{1s} + P \tag{4.118}$$

where $Q = c_s a_{hs}/a_{11} \beta_{1s}(1 - c_s)$ and $P = c_s a_{1s}/a_{11} \beta_{1s} (1 - c_s)$ are constants. Equation (4.118) is similar to the basic equation of the internal standard method, but it includes two constants Q and P, instead of one constant in equation (4.21). If known mixtures are used to evaluate these two constants, then at least two mixtures are needed in the case discussed here.

4.8.7 Continuous pattern approach

We have so far assumed that the quantity I_i, which appears in almost all the equations in this section, is the integrated intensity of the diffraction peak i in the diffraction pattern of the sample to be analyzed. Application of the equations derived in this chapter assumes that at least n diffraction peaks (most probably, n clusters of diffraction peaks) must be selected in the diffraction pattern, and each of the n phases must contribute to at least one of these peaks (clusters). The n peaks must have well-defined boundaries, where the intensity of diffraction falls to the background level. These conditions cannot always be met, especially if we wish to increase the number of peaks to $m>n$. This latter condition is particularly important for multiphase mixtures of low-symmetry phases whose diffraction pattern may be composed of poorly resolved clusters of diffraction peaks. However, all the elaborations in this chapter leading to some basic analytical equations, such as equation (4.100), are valid if we consider the intensity I_i simply to be the intensity diffracted by the sample at any running scattering angle $2\theta_i$ and not as an integrated intensity of a peak cluster. In this case, the new meaning given to the components a_{ij} from equation (4.100) is also clear. Let us now apply equation (4.94) to a pure phase j. The intensity diffracted by the pure phase j at any scattering angle $2\theta_i$ is equal to:

$$(I_{ij})_0 = \alpha_{ij}/\mu_j^* = \alpha_{ij} K_{jj}/\mu_j^*$$

and

$$\alpha_{ij} = (I_{ij})_0 \, \mu_j^* \tag{4.119}$$

The diffraction pattern of our sample can now be presented as a scaled sum of the patterns of pure phases:

$$I_i = \sum_j (I_{ij})_0 \; (\mu_j^*/\mu^*) \; c_j \tag{4.120}$$

where I_i is the intensity diffracted at any scattering angle $2\theta_i$ within the chosen range $2\theta_1$-$2\theta_2$. The number of points I_i depends on the range $2\theta_1$-$2\theta_2$ and the step size. The scaling factor is $(\mu_j^*/\mu^*)\, c_j$. We suppose, of course, that in equation (4.120) the background has been removed from the diffraction patterns of the pure phases and of the analyzed sample. (Methods for background definition and subtraction are discussed in Chapter 5.) If the sets of $(I_{ij})_0$ data are acquired for all phases present and are stored in the computer memory, then the scaling factors are recovered by minimizing the sum of the squares:

$$\sum_{i=1}^{m} \left[I_i - [(I_{ij})_0 \, \mu_j^*] \, c_j/\mu^*\right]^2 \tag{4.121}$$

where m is the number of points i in the range of interest. By minimizing the sum of the squares (4.121) we are, in fact, seeking the best fit of the sum of the scaled patterns of pure phases to the pattern of the analyzed sample. There are some important practical remarks to make in connection with these minimization procedures. First, it is imperative that both I_i and $(I_{ij})_0$ data be collected under identical experimental conditions. The number of points $2\theta_i$ used in the collection of data for I_i and $(I_{ij})_0$ patterns need not be the same as the number of data points in equation (4.121). Summation of collected data over short $\Delta 2\theta$ intervals is essential for reducing the volume of computations without interfering with the precision of the results. A typical step for data collection is 0.02°, whereas the interval of integration may be set at 0.10°-0.15°, thus reducing the number m of points in equation (4.121) by 5-7 times. Retaining a small step in data collection is essential for the analysis of the residue remaining after subtracting the scaled individual patterns $(I_{ij})_0$ from the observed pattern.

If no constraints are imposed on c_j*th*, then minimization of the sum given in expression (4.121) generates a system of normal linear equations with unknowns $c_j\ (\mu_j^*/\mu^*)$. The solution is given by:

$$\mathbf{C}\boldsymbol{\mu} = \left(\mathbf{I}_0^T\, \mathbf{I}_0\right)^{-1} \mathbf{I}_0^T\, \mathbf{I} \qquad (4.122)$$

where $\mathbf{I}_0 = [(I_{ij})_0]$, $\mathbf{C} = [c_j]_{n\times 1}$, and $\boldsymbol{\mu}$(diagonal) = $[\mu_{ij}]$ in which:

$$[\mu_{ij}] = \begin{cases} 0 & i \neq j \\ \mu_j^*/\mu^* & i = j \end{cases}$$

The weight fractions are obtained by multiplying the scaling factors by (μ^*/μ_j^*). If the mass absorption coefficient of the sample μ^* is not known, it can simply be ignored (as has already been done on many occasions), and the true c_j values are then obtained by normalization (4.57). Values of μ_j^*, the mass absorption coefficients of the phases, must be known.

The weight abundances c_j are subject to the obvious constraint:

$$0 \leq c_j \leq 1$$

which may be imposed on the minimization procedure (4.121). Solution (4.122) is not valid in this case, and the best scaling factors $c_j\ (\mu_j^*/\mu^*)$ are obtained by the general minimization procedure with its imposed constraints. The goodness of fit expressed in equation (4.121) is characterized quantitatively by an intensity reliability factor:

$$R = \sum_i [I_i - \sum_j (I_{ij})_0\, c_i\ (\mu_j^* / \mu^*)]^2 / \sum_i I_i \qquad (4.123)$$

where $c_j(\mu_j^*/\mu^*)$ are scaling factors evaluated by equation (4.122). Smith et al. (1987) have emphasized that in good analyses $R \leq 10\%$. Qualitative evaluation of the fit may be carried out by examining the difference pattern:

$$I_i - \sum_j [(I_{ij})_0\, c_j\ (\mu_j^*/\mu^*)]$$

The pattern might reveal, for example, the presence of minor undetected phases. The profile of the difference function in the areas of poor fit is an important tool, enabling us to pinpoint the reasons for the observed discrepancies. For example, the preferred orientation of any phase in the sample being analyzed will be detected in the difference trace as a significant positive value under the reflection from a preferentially oriented plane. In this way one can detect other phase-related systematic deviations, such as peak shift or peak broadening.

As soon as the main procedure expressed in equation (4.121) has been established, we can try to make use of other instrument- and sample-related parameters to improve the fit. The first consideration should be the angular correspondence of I_i and $(I_{ij})_0$. There are at least two reasons why $(I_{ij})_0$ may be shifted relative to I_i.

1. Instrument-related effects, for example, a small sample displacement s in the direction normal to the sample surface, may be present. The corresponding shift of the diffraction peak will be:

$$\Delta 2\theta_i = 2s \cos\theta_i / R = u \cos\theta_i \tag{4.124a}$$

   where R is the radius of the goniometer and $u = 2s/R$ is considered an unknown parameter to be refined by the best-fit procedure. Instrument-related effects can be corrected by using an internal (positional) standard. In this procedure, a known amount of internal standard must be mixed with each sample including, of course, the pure phases.

2. Phase-related effects may be present, for example, if the lattice parameters of the pure phases and the phases being analyzed are not identical. In the simplest case of uniform lattice expansion or contraction, the peak shift is given by:

$$\Delta 2\theta_{ij} = 2\,(\Delta d/d) \tan\theta_{ij} = \gamma_j \tan\theta_{ij} \tag{4.124b}$$

   where γ_j is a refinable parameter.

The general way to compensate for probable instrument- or lattice-related shifts is to vary u and γ and then to seek the best fit of the observed I_i data and the sum of scaled $(I_{ij})_0$. One such technique is implementation of the minimization procedure (4.121), with I_i remaining unchanged but $(I_{ij})_0$ shifted by an amount:

$$\Delta 2\theta_i = u \cos\theta_i + \gamma_j \tan\theta_i$$

where u and γ_j are defined in equations (4.124a) and (4.124b). We start, of course, from $u = 0$. The upper value of u in a correctly operated diffractometer can hardly exceed 0.2°, and the probable phase-related shift should be evaluated from the crystal chemistry of the given phase. The goodness of fit is defined by the residual sum of squares (4.123). The weight fractions c_j corresponding to the best fit are considered to be the most likely values.

We have touched here on a very important issue—the structural correspondence of the pure phases used for scaling and the real phases in the sample being analyzed. The difference between these phases can go far beyond the variations in lattice constants, and it may include factors such as positional and occupational atomic parameters, the degree of lattice imperfection, etc. It is a real challenge to include these parameters into the fitting procedure together with the phase abundances. This branch of QXRD, which is based on Rietveld structure refinement formalism, is growing rapidly and will be discussed in Section 4.9.4. In the current discussion, we have supposed that the bank of pure phase data $(I_{ij})_0$ consists of experimental digitalized patterns of pure phases. The success of full-pattern fitting depends on the quality of the patterns and mostly on the structural correspondence between the model (the pure phase) and the real object (the phase being analyzed). This observation is especially important for natural phases, which are subject to various deviations from ideal structure. In such cases, it is highly advisable to collect pure phase patterns for the minerals extracted from geological samples of the deposit being studied.

There are two other sources of pure phase patterns. The first one is crystal structure data, which can be transformed into calculated digital patterns. (This approach will be discussed in greater detail in the next section.) The second source is the PDF or any similar database.

Thus, the primary database for digitalized patterns of analytical standards may consist of three kinds of patterns: experimental digitalized patterns of the pure phases, patterns generated from structural data, and patterns generated from the PDF or any similar database (Smith et al. 1987). Two further steps are needed to create an integrated base for analytical standards.

1. All patterns must be put on the same scale.
2. Certain shape functions must be assigned to diffraction peaks generated from structural data and PDF data.

The first condition is met if the diffraction patterns are supplied with RIRs. According to the definition of RIRs (Section 4.7):

$$(RIR)_j = [\mu_j^*(I_{jj})_0]/[\mu_c^*(I_{hc})_0]$$

where $(I_{jj})_0$ is the intensity of the major (100%) peak in the pure phase j and $(I_{hc})_0$ is the intensity of the corresponding peak in the reference material (corundum). The term $(I_{ij})_0\,\mu_j^*$ of the minimization scheme (4.121) may then be expressed as:

$$(I_{ij})_0\mu_j^* = a_{ij}(I_{jj})_0\,\mu_j^* = a_{ij}\,(RIR)_j\,\mu_c^*\,(I_{hc})_0 = \chi\,a_{ij}\,(RIR)_j \quad (4.125)$$

where a_{ij} is the relative intensity at step i, and $\chi = \mu_c(I_{hc})_0$ is constant. Thus, $(RIR)_j$ reduces the patterns of different phases to a common scale.

By substituting $(I_{ij})_0\mu_j^*$ from equation (4.121) with its scaled value from equation (4.125), we obtain a general minimization scheme, in which patterns of different origins can be implemented:

$$\sum_{i=1}^{m} [I_i - a_{ij}\,(RIR)_j\,c_j\,(\chi/\mu^*)] \qquad (4.126)$$

The additional constant term χ does not pose an analytical problem. For example, it will be accounted for automatically if c_j values are scaled by mass balance equation (4.57).

Equation (4.126) is also useful if only experimental patterns are utilized for analytical standards. First, RIR values help to account for probable small variations of experimental settings in the acquisition of standard patterns. Second, mass absorption coefficients of phases μ_j are excluded from expression (4.126) [in comparison with expression (4.121)], thus preventing the potentially troublesome calculation or measurement of $\mu_j th$. Techniques for acquiring calculated or experimental RIR values are discussed in Section 4.7.

The method of generating pure phase patterns from crystallographic and instrumental data has had some success (Smith et al. 1988). The continuous-pattern approach in QXRD was pioneered by Frohnsdorff and Harris (1964) and Bezjak and Jelenic (1971a), who applied it to the quantitative analysis of Portland cement. A large set of cement samples with phase compositions determined by chemical balance analyses was used to obtain the pure phase data $(I_{ij})_0$. The method is exactly the same as the method that has already been outlined for the recovery of α_{ij} coefficients and expressed in equations (4.107) and (4.108). This technique, which was unfortunately almost forgotten, has recently reappeared in the works of Smith and his co-workers (Smith et al. 1987, 1988, 1989), who obtained very good results in analyzing multiphase mineral mixtures. In general, the continuous-pattern approach is a powerful and very promising technique in QXRD, being particularly suitable for computer-controlled diffractometers. The apparent disadvantage is the time needed to collect the whole pattern. This time may sometimes be reduced by excluding parts of the pattern with zeroed intensity. However, even in the best case, the collecting time per sample will be no less then 0.5 -1 hour.

4.8.8 Examples

Two examples of the application of the method under discussion are taken from Szabó (1980)—the discrete peak clusters approach—and Smith et al. (1987)—the whole pattern approach.

In the first example, Szabó analyzed mixtures of four fertilizers, ammonium dihydrogen phosphate (j=1), ammonium hydrogen ortho-

Table 4.13. Relative intensities of the phases being analyzed*

Peak number i	d, Å	α_{ij} j = 1	j = 2	j = 3	j = 4	j = 5
1	5.32	50.34	0	0	0	0
2	5.05	0	30.32	0	0	0
3	3.1	62.70	41.87	0	34.69	0
4	2.8	0	16.15	48.72	21.58	0
5	2.5	0	10.91	0	0	21.43
6	2.09	0	0	0	0	24.26
7	1.82	0	0	0	6.03	0
8	1.6	0	0	10.64	4.10	20.76

*Data are taken from Szabo (1980).

Table 4.14. Results of analysis of the data presented in Table 4.13

Phase, j	Concentration, c_j, % Observed	Actual	$\sigma(c_j)$, %
1	14.99	15.0	0.37
2	15.09	15.0	0.50
3	15.37	15.0	0.50
4	34.51	35.0	0.50
5	20.04	20.0	0.49

In the estimation of $\sigma(c_j)$, only counting statistics were taken into account.

Table 4.15. Results of the analysis of six-phase mixtures by the whole-pattern approach

Phase	Actual %	Observed %	Relative error, %
CaF_2	17	19	12
$CaCO_3$	17	17	0
α-Al_2O_3	17	18	6
$Ca(OH)_2$	17	17	0
NaCl	17	15	12
α-SiO_2	17	15	12

Data taken from Smith et al. (1987).

phosphate (j=2), ammonium chloride (j=3), and potassium chloride (j=4), together with corundum (j=5). The peak overlapping is characterized by the following table of α_{ij} coefficients (Table 4.13). The results of the analysis under optimized conditions of time sharing between the peaks are good, as shown in Table 4.14.

Smith et al. (1987) obtained excellent results in analyzing complex six-phase mixtures by the continuous-pattern approach (Table 4.15). The test phases in this mixture were taken from the same stock of reference materials used to produce $(I_{ij})_0$ patterns. Although large derivations are to be expected for samples of natural origin, the potential of the full-pattern approach is clearly demonstrated in this example.

4.9 Implementation of calculated powder patterns in QXRD

4.9.1 *Introduction*

In this section, we will discuss the implementation of calculated diffraction patterns for calibration (or scaling) purposes in QXRD. By a *calculated pattern*, we mean that the intensities of certain diffraction peaks are calculated from crystal structure data. Implementation of this method requires that the structure of the analyte be known with sufficient accuracy.

It seems that the first time calculated intensities were used for calibration in QXRD was in the analysis of retained austenite in hardened plain carbon steel by Averbach and Cohen (1948). Structure calculations are very simple for austenite (Cu-type structure) and martensite (tetragonally distorted α-Fe-type structure). The next step in utilizing calculated patterns was made by Jahanbagloo and Zoltai (1968), who analyzed multiphase mixtures of minerals by using a number of calculated diffraction peaks for each phase for calibration purposes. The implementation of the calculated-pattern approach was further refined by the systematic work on calculated X-ray powder patterns of Borg and Smith (1969) and the intensive development of suitable computer programs (Smith et al. 1983; Parrish et al. 1976).

For obvious reasons, calculated patterns are free of sample-related effects, such as preferred orientation, extinction, microabsorption, and surface roughness. Comparison of calculated and observed patterns usually reveals these effects, and care must be taken for proper sample preparation. In all cases, however, one must bear in mind the probable discrepancies between calculated and observed diffraction intensities.

There are at least two cases in which the implementation of calculated patterns may be useful:

1. When pure phases are not available (in this case, calculated patterns have to be used for calibration purposes); and
2. When calibration by means of calculated patterns may save the time required for preparation and measuring known mixtures of pure phases.

The second situation is open to dispute because calibration by means of pure phases or their mixtures (if these phases are genuine, of course) promises better accuracy than calibration by means of calculated constants. The first situation is not uncommon. Even if we put aside the obvious difficulties of separation or preparation of pure phases (analytical standards) discussed in Chapter 5, calculation of a powder pattern gives us an excellent tool for taking into account possible crystal imperfections, including compositional imperfections (solid solutions) and positional imperfections, such as various types of disorder. A typical example is the substitution of Fe for Mg in olivines. Due to differences in atomic scattering factors for Fe and Mg, this substitution leads to significant variations in diffracted intensities. The possibility of using calculated patterns in order to find the relative intensity for each olivine from Mg_2SiO_4 to Fe_2SiO_4 was given by Jahanbagloo and Zoltai (1968). If the composition of the olivine could be established, for example, by measurements of lattice constants or by microprobe analysis, then the calculated pattern for the real mineral could be of great help in QXRD. Disordered layer silicates is another group of minerals for which calculated patterns are essential in the quantification of multiphase mixtures. Suitable techniques and computer programs have been developed by Reynolds (1985).

Finally, quantification of phase compositions by means of calculated patterns may be accompanied by structure refinement of the phases being analyzed. This novel approach to QXRD, based on the Rietveld refinement procedure, has shown promising results.

Any computer program for calculating diffraction patterns generates integrated intensities of diffraction peaks, which can usually be reduced to a common (for various phases) scale. In conventional QXRD techniques, the integrated intensity of the diffraction peak is a measure of the phase abundance. Thus, calculated integrated intensities for pure phases are employed in the evaluation of calibration constants. There is, however, another novel approach to QXRD, which is based on full pattern fitting of the sum of pure phase patterns and observed patterns of the mixture being analyzed. This technique was discussed in Section 4.8. For implementation of the calculated pattern in this case, an additional step is needed, namely, the simulation of a continuous diffraction pattern. The diffraction peak should be approximated by a suitable analytical function, usually with a set of adjustable parameters. The two ways of implementation of the calculated data will now be discussed.

4.9.2 *Calculation of calibration constants*

QXRD, like many other instrumental analytical techniques, is based on the comparison of the observed response (diffraction peak intensity) produced by an analyzed sample and the corresponding response (intensity) produced by a sample of known composition (pure phase or any analytical standard). In this technique, we match (or calibrate) the observed intensity to the phase abundance, which is our prime interest in QXRD. It is assumed, of course, that the intensity diffracted by the analyte in the analyzed sample and in the standard are comparable, or in other words, that they have been reduced to a common scale. This demand is fulfilled automatically if the intensities diffracted by pure phases or analytical standards are obtained experimentally and both the analyzed and standard samples are run at the same experimental setting. This correspondence does not exist if the diffraction lines of a pure phase are calculated, and special steps have to be undertaken to reduce the intensities to the scale of experimental values, depending on the mode of calibration.

In the analytical equations developed so far in this chapter we have used two types of calibration constant:

1. The single or absolute calibration constant K_{ij} defined in equation (4.2) and characteristic for diffraction-absorption, doping, and diluting techniques.
2. The ratio of calibration constants $\beta_{jr} = K_{ij}/K_{hr}$ (where the subscripts j and r designate the analytical peaks of phases j and r, respectively), which is relevant to the internal and external standard techniques.

Both types of calibration constant can be calculated if the structure of the phase is known, although the first is more troublesome in scaling.

Suppose that we use a conventional diffractometer with Bragg-Brentano focusing geometry and a constant beam-divergence mode. The integrated intensity of the diffraction peak of the pure phase $(I_{ij})_0$ needed to calculate K_{ij} is given by equation (3.8), assuming that the volume fraction of the phase $c_j = 1$ and $\mu = \mu_j$. We now combine the structure-independent terms 1 and 2 in equation (3.8), which incorporates the fundamental physical and instrumental constants into a single constant K.

The calibration constant K_{ij} is expressed as:

$$K_{ij} = \left(I_{ij}\right)_0 \mu_j^* = K\left(LP\right)_i \frac{|F_{ij}|^2 D_{ij} M_{ij}}{2\vartheta_j^2 \rho_j} \qquad (4.127)$$

The meanings of all the terms in expression (4.127) are given in equation (2.19); and the linear absorption coefficient from equation (3.8) is expressed as $\mu_j = \mu_j^* \rho_j$.

The second type of calibration constant β_{jr} is calculated directly from crystal structure data. According to its definition and from equation (4.127):

$$\beta_{jr} = K_j/K_r = (I_{ij})_0\, \mu_j^*/(I_{hr})_0\, \mu_r^*$$

$$= \frac{(LP)_i |F_{ij}|^2 D_{ij} M_{ij} \cdot \vartheta_r^2 \rho_r}{(LP)_h |F_{hr}|^2 D_{hr} M_{hr} \cdot \vartheta_j^2 \rho_j} \qquad (4.128)$$

All the quantities in this equation are easily calculated if the structures of the two phases j and r are known. Note that the β_{jr} constants that are calculated directly by equation (4.128) fit the analytical equation perfectly and do not need any additional scaling.

In contrast, the absolute calibration constant K_{ij} [or the calculated intensity $(I_{ij})_0^{calc}$] has to be adjusted to the scale of experimental intensities. In other words, all the parameters of the experimental settings, including the intensity of the primary beam, the goniometer radius, and the receiving slit dimensions, should be evaluated and used in calculation of the constant K from equation (4.127). This is certainly not a practical task in QXRD. There is, however, another way of scaling K_{ij} by means of any reference phase with a well-known structure, provided that this phase is available for experimental measurements.

Suppose that $(I_{hs})_o^{exp}$ is the experimental intensity of the peak h of the reference phase S, determined at the instrument setting used in a particular QXRD case. Then, according to equation (4.127), the structure-independent term K is given by:

$$K = \frac{2(I_{hs})_o^{exp}\, \mu_s^* \rho_s \vartheta_s^2}{(LP)_h |F_{hs}|^2 D_{hs} M_{hs}}$$

and using this expression for K in equation (4.127), we obtain for K_{ij}:

$$K_{ij} = (I_{hs})_o^{exp}\, \beta_{js}\, \mu_s^* \qquad (4.129)$$

where β_{js} is calculated by equation (4.128), in which subscript r is replaced by s. If the reference phase is corundum, then β_{js} is $(RIR)_j$. The value of K_{ij} defined in equation (4.129) is reduced to the intensity scale used in QXRD measurements.

Whereas the expression for $(I_{ij})_0$ is taken from the basic equation (3.8), the term K encompasses the structure-independent coefficients, and a_{ij} is expressed on a scale of 0-100. Assuming K to be a constant for all the phases, we immediately obtain an expression for the scaling coefficient ω_j:

$$\omega_j = \frac{(LP)_j |F_{jj}|^2 D_{jj} M_{jj}}{200\, \vartheta_j^2 \rho_j \mu_j^*} \qquad (4.130)$$

where all terms in this equation, except μ_j and υ_j, are related to the 100% peak of phase j.

The scaled intensity of any peak i of phase j is equal to:

$$(I_{ij})_o^{SC} = a_{ij}\, \omega_j \qquad (4.131)$$

Scaling factors applied in various computer programs—although they include all structure-related terms—may differ from ω_j expressed in equation (4.130) in that they may or may not include absorption coefficient μ^* or may have different numerical coefficients. Short discussions on scaling factors applied in different computer programs may be found in Hubbard et al. (1976) and Altree-Williams (1978).

The calibration constants K_{ij} and β_{jr} are expressed through the scaling factors ω_j and the relative intensities a_{ij}:

$$\beta_{jr} = \frac{K_{ij}}{K_{hr}} = \frac{\omega_j\, a_{ij}\, \mu_j^*}{\omega_r\, a_{hr}\, \mu_r^*} \qquad (4.132)$$

$$K_{ij} = (I_{ij})_o^{sc}\ \mu_j^* = K a_{ij}\, \omega_j \qquad (4.133)$$

The constant term K in equation (4.133) is evaluated by means of a second reference phase s with a known structure. Taking into account expressions (4.129) and (4.132), we arrive at:

$$K_{ij} = (I_{hs})_o^{exp}\ \mu_j^*\ \frac{a_{ij}\, \omega_j}{a_{hs}\, \omega_s} \qquad (4.134)$$

where $(I_{hs})_o$ is the observed intensity of peak h of the reference phase s and a_{hs} is the relative intensity of this peak.

Scaling of relative intensities can be achieved by means of calculated RIRs. Indeed, according to RIR expression (4.84), the intensity of the strongest peak of phase j is given by:

$$(I_{jj})_o = (I_{hc})_o\, \mu_c^*\, (RIR)_j/\mu_j^* = \alpha (RIR)_j/\mu_j^* \qquad (4.135)$$

where $(RIR)_j$ is the reference intensity ratio for phase j, and $\alpha = (I_{hc})_o$ μ_c^* is a constant that can be ignored in scaling. Thus, the quantity

$(RIR)_j/\mu_j^*$ takes the place of scaling factor ω_j in all the scaling equations (4.130) – (4.134).

Calculated calibration constants K_{ij} and β_{jr} were derived from expressions for integrated diffraction peak intensity. Naturally, implementation of these constants assumes that the integrated intensities of the diffraction peaks are measured experimentally. However, in practical work, peak intensities are sometimes used, mainly as a means of escaping problems of overlap. Unfortunately, there is no simple but accurate way to convert integrated intensity to peak intensity by calculation, because the diffraction peak profile is a complex 2θ-dependent function, which is influenced by a number of experimental and phase-related factors.

We should also note that two diffraction peaks at different scattering angles are involved, either directly or indirectly, in the definition of the calibration constants β_{jr} [equation (4.133)] and K_{ij} [equation (4.135)]. If the width of these two peaks is different, then the ratio of the calculated integrated intensities will differ from the ratio of observed peak intensities. In the case in which peak width of the phases being analyzed is affected by phase-related factors, such as microstrains or small coherent length, calculated calibration constants must be used only for integrated intensities.

If instrumental factors are dominant in peak broadening, then diffraction peak profiles are similar for the various phases. If, in addition, the diffraction angles $2\theta_{ij}$ and $2\theta_{hr}$ for two phases j and r are close, then the ratio calibration constant β_{jr} calculated for integrated intensities might also be suitable for peak intensities.

We can see from the conversion factor established by Swanson et al. (1962) that an error of less than ~10% may be expected in β_{jr} calculated for integrated intensities but used for peak intensities if the two peaks i and h are separated by less than 10°.

Another and perhaps more accurate way to fit β_{jr} and K_{ij} coefficients to peak intensity measurements is via an acquired experimental conversion factor. This factor is obtained by measuring the integrated intensities I_{ij} and I_{hr} and the peak intensities I_{ij}^p and I_{hr}^p for the peaks in the diffraction pattern of the sample being analyzed. The conversion coefficients are deduced from these measurements as follows:

$$\varepsilon_{ij} = I_{ij}^p/I_{ij}; \qquad \varepsilon_{hr} = I_{hr}^p/I_{hr}$$

The conversion of the calibration constants is given by:

$$\beta_{jr}^p = \beta_{jr}\,(\varepsilon_{ij}/\varepsilon_{hr}); \qquad K_{ij}^p = K_{ij}\,\varepsilon_{ij}$$

where β^p_{jr} and K^p_{ij} are calibration constants for which peak intensities are measured, and β_{jr} and K_{ij} are calculated values defined by equations (4.131) and (4.132).

4.9.3 *Full-pattern approach*

Calculated data are very useful in the novel full-pattern approach to QXRD discussed in Section 4.8. The basic principle is the fitting of individual (pure phase) patterns to the sample pattern, when phase abundances are obtained from scaling factors of the best fit. We now have to distinguish between two technically different approaches, which were discussed in Section 4.8.

1. In the first approach, the diffraction pattern is divided into clusters of diffraction peaks. The integrated intensity of the cluster is considered as the sum of scaled integrated intensities of individual phases. Note that separation of the cluster into individual peaks is not needed.
2. In the second approach, the diffraction pattern is not separated into peaks or clusters, but continuous or quasi-continuous patterns of individual phases are fitted to the pattern of the sample being analyzed.

The first technique is suitable for calculated data, as can be seen from the development of the basic equation (4.95) in Section 4.8. The calibration constant K_{jj} for the main peak of phase j is simply equal to $K\omega_j$. Accordingly, all coefficients α_{ij} from equation (4.100) are equal to:

$$\alpha_{ij} = K\, a_{ij}\, \omega_j$$

where ω_j is a scaling factor calculated for each phase, as was discussed earlier, and a_{ij} is the relative intensity of peak i of phase j. By substituting α_{ij} from equation (4.100) with a calculated value $\alpha_{ij}\, \omega_j$, we will obtain an equation that differs from the basic equation (4.100) only in terms of the constant K. This constant can be ignored if we use the mass balance equation $\Sigma c_{ij} = 1$ for scaling provisional phase abundances, as was discussed in detail in Sections 4.6 and 4.8.

We have discussed the continuous-pattern fitting technique in a previous section, assuming that the diffraction patterns of analytical standards are acquired experimentally. We now suppose that the patterns of the analytical standards are calculated. The obvious step in this approach is simulation of a continuous pattern from calculated integrated intensities. Any suitable analytical function may be assumed to approximate the diffraction peak. Due to continuous variation of the peak profile with the scattering angle, the profile function should include adjustable 2θ-dependent parameters, which are determined in such a way as to provide the best fit of the calculated profile to the real peak profile. Thus, we come to the profile-fitting technique, which, in fact, has two applications:

1. Simulation of diffraction patterns, based on a structural model.
2. Decomposition of overlapping peaks into individual components, without referring to the structural model.

In light of the importance of the profile-fitting technique in QXRD, we will devote a special section (Section 5.8) to technical problems associated with this topic, including a discussion of probable profile-fitting functions. Here, we suppose that the profile of any diffraction peak h of the phase j is adequately described by an analytical function $f_{hj} = f(2\theta - 2\theta_{hj})$. The argument of this function is $2\theta - 2\theta_{hj}$, 2θ being the running scattering angle and $2\theta_{hj}$ being the position of peak h of phase j calculated by the lattice constants of this phase.

Function f_{hj} must satisfy the normalization:

$$\int_{2\theta_1}^{2\theta_2} f_{hj}\, d(2\theta) = 1 \tag{4.137}$$

where the limits of integration are wide enough to include the peak tails.

Any profile function includes several shape parameters, such as the full width at half maximum (FWHM) Δ_{hj} and the asymmetry index U_{hj}.

Thus, we finally come to:

$$f_{hj} = f(2\theta - 2\theta_h, \Delta_{hj}, U_{hj})$$

where Δ_{hj} and U_{hj} are usually adjusted in the process of pattern fitting.

By multiplying the function f_{hj} by a_{hj}, the relative integrated intensity of peak h, we define the intensity distribution (profile) of peak h on a relative scale:

$$I_{hj} = a_{hj} f_{hj} \tag{4.138}$$

Note that due to normalization condition (4.137) a_{hj} is indeed the relative integrated intensity of peak h. Values of I_{hj} are reduced to a scale common to all phases by multiplying them by the scaling factor ω_j and the constant K defined earlier in this section:

$$I_{hj}^{sc} = K\, \omega_j\, a_{hj}\, f_{hj} \tag{4.139}$$

The intensity diffracted by phase j at any angle $2\theta_i$ is obtained by accumulating the contributions of all h = 1,2,...,m peaks in the investigated 2θ range:

$$(I_{ij})_o^{sc} = K\,\omega_j \sum_{h=1}^{m} a_{hj}\,(f_{hj})_i \tag{4.140}$$

where $(f_{hj})_i$ is the value of f_{hj} at $2\theta = 2\theta_i$.

The set of $(I_{ij})_0$ values, taken at discrete $2\theta_i$ angles with small steps between adjacent angles, actually represents the diffraction pattern of the phase j reduced to a common scale.

The intensity scattered at any angle $2\theta_i$ by a multiphase sample is made up of contributions of individual phases:

$$(I_i)_{calc}^{sc} = (1/\mu^*) \sum_j c_j\,(I_{ij})_o^{sc}\,\mu_j^* \tag{4.141}$$

where superscript sc indicates that intensity is reduced to a common scale and $(I_{ij})_o^{sc}$ is the calculated intensity of the pure phase j at the angle $2\theta_i$ defined in equation (4.140).

In order to fit the calculated intensities $(I_{ij})_{calc}^{sc}$ to the observed intensities, the latter must be stripped of the background. This operation, which is discussed in detail in Chapter 5, might be problematic for complex mixtures of some phases as a result of multiple peak overlapping. The pattern-fitting technique presents an excellent solution to this problem, namely, generating—rather than stripping—the background. The background is presented as a power series of the scattering angle:

$$(I_i)_b = a_0 + a_1(2\theta_i) + a_2(2\theta_i)^2 + \ldots \tag{4.142}$$

and is added to the calculated intensity.

The crucial step is fitting the calculated intensities to the observed intensities, which is achieved by minimization of the sum of the squares:

$$\sum_i \left[(I_i)_{obs} - (I_i)_{cal}^{sc} - (I_i)_b\right]^2 = \sum_i \left[(I_i)_{obs} - (K/\mu^*)\sum_{j=1}^{n} c_j\,\mu_j\,\omega_j \sum_{h=1}^{m_j} a_{hj}\left(f_{hj}\right)_i - (I_i)_b\right]^2 \tag{4.143}$$

where equations (4.141) and (4.140) are used to express the scaled calculated intensity $(I_{ij})_{calc}^{sc}$. The adjustable parameters of the minimization procedure (4.143)—in addition to the weight fractions c_j—are the parameters of the background expansion a_0; a_1; a_2 . . . and the parameters of the peak profile function Δ_{hj} and U_{hj}. There are two unknown constants in the sum of the squares (4.143): the coefficient K and the mass absorption coefficient μ^*. Thus, the fitting procedure will not provide the expected phase abundances c_j but rather the provisional values $c_j' = (K c_j/\mu^*)$. As usual, there are two ways of scaling $c_{j'}$ to the true weight fraction c_j:

1. By means of the mass balance equation

$$c_j = c_j' / \Sigma c_j' \tag{4.144}$$

2. By means of an internal standard.

If c_S is a known weight fraction of the internal standard and c_S' is the provisional weight fraction obtained by the minimization procedure (4.143), then the scaling coefficient is given by:

$$\xi = c_S / c_S'$$

and the true abundance of any analyzed phase by:

$$c_j = \xi c_j' / (1 - c_S) \tag{4.145}$$

Despite the apparent complexity and the large number of parameters, the method has been successfully used for QXRD by several workers. Werner et al. (1979) implemented this technique in the analysis of two-phase mixtures of relatively simple compounds. Excellent results were reported by Bish and Howard (1988) for three-phase mixtures of the NBS standard materials Al_2O_3, ZnO, and Cr_2O_3. In addition, the technique can be used to obtain other refined parameters, including background coefficients and peak shape parameters.

4.9.4 *Simultaneous QXRD and structure refinement by the pattern-fitting method*

Since Rietveld (1969) published his pioneering work more than 20 years ago, structure refinement from powder data has become a well-established technique. Although the application of this technique to X-ray diffraction data may be difficult as a result of the intrinsic complexity of the peak profile, some excellent examples of structure refinement have been published.

The Rietveld technique can be applied to multiphase samples as well as to a single phase. In the former case, the scaling factors for each individual phase will comprise the phase abundance c_j in addition to other instrument- or phase-related terms. We can now continue along the same lines as in the previous subsection, since the minimization procedure (4.143) is also applicable here. However, we do not consider the product $a_{ij}\, \omega_j$ from equation (4.143) as a calculated constant but express it through adjustable structural parameters, site occupancies, and site positions. In so doing, we take a very important step in QXRD, namely, attempting to create a reference phase structurally identical with the analyzed phase in the sample. In order to express the structural parameters as a sum of squares (4.143), we must remember that the product $\omega_j a_{hj}$ is the scaled integrated intensity of the peak h of phase j:

$$\omega_j a_{hj} = \frac{(I_{ij})_0}{K} = \frac{(LP)_h |F_{hj}|^2 M_{hj} T_{hj} R_{hj}}{2\mu_j^* \rho_j \vartheta_j^2} \tag{4.146}$$

where M_{hj} is the multiplicity factor, T_{hj} is the overall temperature factor, $(LP)_h$ is the Lorentz polarization factor, υ_j is the volume of a unit cell of phase j, and R_{hj} is the correction term for preferred orientation. The structure factor combined with the temperature factor is expressed in the general form as:

$$F_{hj} = \sum_m U_m f_{mh} \exp[2\pi i(Hx_m + Ky_m + Lz_m)] \tag{4.147}$$

where the subscript m indicates the site occupied by the atoms in a unit cell of phase j; $\sum_m$ is the summation over the occupied sites; f_{mh} is the atomic scattering factor including modification for thermal vibration for the scattering angle $2\theta_h$; U_m is the site occupancy parameter of the atoms; x_m, y_m, z_m are the fractional atomic coordinates of the site m; and H, K, L are the Miller indices of the diffraction peak h of phase j.

The adjustable structural parameters must include the lattice constants that are used in the calculation of the scattering angles $2\theta_{hj}$. R_{hj} is usually applied as a function of the angle between the preferentially oriented plane of phase j and any other plane of this phase. Finally, there may be some adjustable instrument-related factors, such as zero goniometer setting or probable sample displacement.

By substituting $a_{ij}\,\omega_j$ in the equation for the sum of the squares (4.143) by expression (4.146), we construct an optimization scheme with the following adjustable parameters:

- scaling factor (Kc_j/μ^*)
- peak shape parameters: FWHM, asymmetry parameters
- lattice constants a_j, b_j, c_j, α_j, β_j, γ_j
- positional parameters x_m, y_m, z_m
- occupational parameters U_m
- overall temperature factor B_j defined by $T_{hj} = \exp[-2B_j(\sin^2\theta_h/\lambda^2)]$
- background expansion parameters a_0, a_1, a_2, . . .
- preferred orientation parameters R_{hj}
- peak position shift due to sample or instrumental aberrations.

There is a full spectrum of cases in which only some of the structural parameters are regarded as variables (for example, only the lattice constants).

Application of this technique requires that the relevant computer software be available (Hill and Howard 1987; Madsen and Hill 1990; Hill 1991; Taylor 1991). Although the principle of scaling of individual phases is the same as that described here, the scaling factor itself may take a form other than the product Kc_j/μ^*. For example, in the pioneering work of Hill and Howard (1987) on simultaneous quantification and structure refinement, the scaling factor produced by a fitting algorithm was:

$$s_j = K\, m_j / Z_j\, M_j\, \upsilon_j$$

where K has the same meaning as in equation (4.127); m_j is the mass of phase j in the sample; and Z_j, M_j, υ_j are the number of formula units in a unit cell, the molecular weight of the formula unit, and the volume of the unit cell, respectively.

The consequent evaluation of the actual phase abundances is carried out in ways that have already been discussed—by application of the mass-balance equation or by employing an internal standard.

We now ask ourselves the logical question: will this procedure work? In such a complex system, we suspect that there will be unwanted interference in various parameters, particularly in the weight fractions (or scaling factors) c_j. It seems that the answer is positive at least if the analysis is carried out by a well-trained crystallographer. Although very few publications on simultaneous QXRD and structure refinement have appeared to date, the number is growing fast.

Hill and Howard (1987) applied simultaneous structure refinement and QXRD to the two-phase mixtures rutile-corundum, rutile-quartz, and silicon-quartz. They used neutron diffraction and obtained excellent results. In their study, atomic positions and unit cell dimensions, refined for mixtures of various compositions, converged to the same values, a finding that verified the reliability of the minimization procedure. Moreover, the quantitative analysis was highly accurate, with a coefficient of variation of ~1%. Madsen and Hill (1990) have recently reported good results for a five-phase mixture consisting of galena, pyrite, sphalerite, chalcopyrite, and quartz, with rutile as the internal standard.

Another analysis of a complex five-phase mixture—the product of calcination of zinc concentrates—was reported by Howard et al. (1988). Taylor and Pecover (1988) carried out an analysis of zeolite-bearing rocks, which included structure refinement of natural zeolites.

It must be noted in a concluding remark that full-scale fitting, which includes all of the adjustable parameters mentioned above, is not obligatory. For example, if occupation parameters, thermal vibration or some of the positional parameters are known with good precision, they can be fixed, thus considerably simplifying the refinement procedure. Examples are to be found in the works of O'Connor and Raven (1988) and Bish and Howard (1988). Many hints and examples can be found in the book edited by Young (1993).

4.9.5 Sources of structural data

The following structural and chemical data are needed to calculate the powder pattern of any phase j:

- chemical composition;
- symmetry (space group);
- lattice parameters;
- positional parameters (atomic coordinates);
- site occupation parameters; and
- thermal vibration parameters.

Complete information on crystal databases may be found in one of the publications of the Data Commission of the International Union of Crystallography. We are particularly interested in databases that provide atomic positions. The following list gives the most complete of these databases:

1. *Structure Reports (published by the Reidel Publishing Company, P.O.B. 17, 3300 AA Dordrecht, the Netherlands).* Each annual volume is published in two parts: A. Metals and Inorganic; and B. Organic. Each volume is indexed, and 10-year indexes are also available.

2. *Inorganic Crystal Structure Database (ICSD).* Currently, the ICSD contains more than 26,000 entries. The database is maintained by Fachinformationszentrum Energie Physik Mathematik, Karlsruhe, Germany, and the Institute for Inorganic Chemistry of the University of Bonn. The database may be accessed interactively using the CRYSTIN retrieval system via a telecommunication link to the host INKA in the Fachinformationszentrum.

3. *Cambridge Structure Data Base (CSD).* The CSD contains structural information on organic and organometallic compounds. It comprises more than 50,000 entries. The database is maintained by the Cambridge Crystallography Data Centre. In order to facilitate use of the stored data, the Centre has developed a search, retrieve, analysis and display program, a system that is available worldwide.

4. *Metals Crystallographic Data File (CRYSTMET).* This database for metallic phases (elemental, binary, ternary, etc.) is maintained by the National Research Council of Canada. Phases are taken as metallic if they are composed of elements to the left of the Zintl line (Al, Ge, Sb,

Po, B, S, Sc, Te). Borides, carbides, nitrides, silicides, phosphides, arsenides, sulfides, selenides, and tellurides are also included.

5. *Powder Diffraction File (PDF).* The PDF is maintained by the Joint Committee for Powder Diffraction Standards (JCPDS). Although the entries in this database do not include much of the needed structural information, such as atomic positions, some of them comprise calculated powder patterns. If applied with a scaling factor, these data might be suitable for use in QXRD. It is also possible that original structural data may be found from the references provided in the PDF. In addition, important sources of structural data are the specialty journals, such as *Acta Crystallographica*, *Journal of Solid State Chemistry*, *Zeitschrift für Kristallografie*, and *Journal of Less Common Metals*. Acquired crystal structure data must be analyzed critically with respect to the crystalline phase being analyzed. The most troublesome factors are likely to be site occupation parameters in phases with considerable atomic (ionic) substitution. This means that the particular chemical composition of the analyzed phase (if known) should be used together with other structural data in calculating the powder pattern. A good example is the olivine series discussed earlier.

4.9.6 *Quality of structural data*

The quality (whether good or bad) of structural data may be deduced from a comparison of the intensities calculated on the basis of the structural data with those observed in the original work on structure determination. The criteria for accepting structural data for QXRD are rather stringent. For example, intensity data with 15%-20% errors may be considered acceptable for determination of the main frame of the crystal structure. A great number of crystal structures were determined in the 1930s and 1940s by visual estimation of the intensities on rotation of Weissenberg photographs. On the other hand, 15%-20% errors in the calibration constant or the intensities of the diffraction peak may be considered unacceptable for QXRD.

There are several qualitative criteria that should be taken into account in judging the quality of structural data. These include the method of data acquisition (film or diffractometer), the volume of data (three-dimensional or two-dimensional), absorption correction, etc. The quantitative characteristic that usually appears in the structure description is the reliability index:

$$R = \Sigma \mid |F_o| - |F_c| \mid / |F_o|$$

where $|F_o|$ and $|F_c|$ are the observed and calculated structural factors, respectively.

We set a somewhat arbitrary boundary between good and bad structural data at $R \approx 10\%$. In contemporary structural analysis, in which anisotropic temperature parameters and the location of light atoms are taken into account in the refinement procedure, the reliability index often drops to 3%-5%. Although these structures are undoubtedly good, we should remember that the intensity is proportional to the square of the structure factor, and the relative error in the intensity is given by $\varepsilon(I) \approx 2\varepsilon\,(F)$. Thus, it would not be unusual to encounter a 5%-10% error in the calculated intensity of any diffraction peak, even for good structural data.

The errors in the calculated intensities are bound to the uncertainty in the atomic occupancy and positions, the thermal vibration and the atomic scattering factors.

Some information on the expected errors in the calculated intensity may be acquired by simple calculation of the intensity with variations imposed on the positional parameters. Hubbard and Smith (1977) carried out similar calculations for Cr_2O_3 and found that a deviation of the Z-parameter of Cr^{3+} by 4σ (=0.0003) (σ being the standard deviation reported in structural studies) caused appreciable (up to 10%) variations in the relative intensities of certain peaks and a 6% deviation of the RIR value.

Thus, structural data that are not precise might lead to incorrect calculated calibration constants and serious analytical errors. However, the same data can be considered as a good starting point for structural refinement with simultaneous quantification. In the near future, development of this novel approach will most likely show the applicability of structural data to compounds more complex than quartz, rutile, and calcite, which are usually taken as the components of test mixtures.

4.9.7 Comparison of calculated and experimental powder data

Both calculated and experimental powder data bear some intrinsic errors. For calculated intensities the errors are caused by uncertainties in the structural data, although the errors should be less than 5%-10% if the structural data are good. Experimental intensities might be fraught with specific powder intensity errors, such as preferred orientation, microabsorption, and nonhomogeneity. Unfortunately, very little work has been done on the comparison of these two types of data. Some work was done by Smith et al. (1987), who reported good correspondence of calculated and observed intensities and good accuracy of analysis of two- and three-phase mixtures of quartz, halite, fluorite, rutile, calcite and hematite. However, considerable systematic discrepancies were also detected and reflected in the RIR values.

Table 4.16. Comparison of calculated and experimental β_{jr} values

Phase j	Diffraction peak, hkl	β_{jr}	
		Calculated	Experimental
Quartz SiO_2	101	46.3	45.3±0.3
Calcite $CaCO_3$	104	35.2	47±1
Sylvite KCl	200	63.3	65±1
Zincite ZnO	101	61.3	53±2
Anatase TiO_2	101	48.8	47±1
Eskolaite Cr_2O_3	104	26.5	22±1
Hematite Fe_2O_3	104	40.7	30.2±0.1
Ag	111	194	122±1

Data taken from Altree-Williams (1978).

To give an impression of the order of magnitude of the deviations of calculated and experimental (powder) intensities, Table 4.16 presents data taken from Altree-Williams (1978). All the data were brought to a common scale by a method similar to that discussed in this section. The second (r) phase in all cases was corundum, and its 113 peak was taken as the reference peak. It can be seen that there is good correspondence between calculated and observed data for quartz, sylvite, and anatase, but there are significant discrepancies for the other phases. The deviations generally increase as the linear absorption coefficient of the phase increases, silver powder being the most extreme case. It is highly likely that the observed deviations result from microabsorption effects of the types discussed in Chapter 2. It is also clear from the table that calculated and observed β_{jr} data would be much closer for high-absorbing phases if one of them were to be used as a second model phase for determination of the scaling factor. This implies that great care should be taken in choosing the reference phase for scaling the calculated data. It thus seems best to use a high-absorbing phase for scaling the calculated data for heavy phases and a low-absorbing phase for scaling light phases.

4.9.8 Summary

Implementation of calculated patterns in QXRD seems to be a very attractive and promising approach, even though it has not yet found wide application. Apparently, this method is much more sophisticated than other methods: it demands not only precise intensity measurements, as does any other QXRD method, but also a great deal of computational work and a deep insight into the crystal structure itself. Extensive experience with crystallographic methods is obviously needed in order to utilize all the advantages of this method.

The major restriction to the use of calculated patterns is imposed by sample effects, mainly microabsorption. We can neither calculate this effect with acceptable precision nor estimate it experimentally if only the analyzed sample—and none of the pure phases—is in our possession. Thus, this effect could drastically bias the results of the analysis. It would be advisable to apply this method to a multiphase system only when the ratio of the linear absorption coefficients of the phases does not exceed 2-3. Communition of the powder would be of great help. Examples of the successful application of calculated data, such as the analysis of retained austenite, iron oxides, and zirconia polymorphs, fall in this category.

4.10 Standardless Methods

4.10.1 *Introduction*

Let us look at the diffraction patterns of Nubian sandstone from southern Israel. This material is composed of two phases, quartz (the major phase in the raw material) and kaolinite (the major phase after beneficiation). Naturally, a change in the abundance of one of the phases (kaolinite) causes a change in the abundance (and hence a change in the peak intensity) of the second phase. We will use this fact to analyze two beneficiated samples of Nubian sandstone, which we will designate samples 2 and 3. Neglecting the difference in absorption between kaolinite and quartz, we can write the following obvious equations:

$$c_{2Q}/c_{3Q} = I_{2Q}/I_{3Q};\ c_{2k}/c_{3k} = (1 - c_{2Q})/(1 - c_{3Q}) = I_{2K}/I_{3k} \quad (4.148)$$

where c_{2Q}, c_{3Q}, c_{2K} and c_{3K} are the phase abundances of quartz and kaolinite in samples 2 and 3; and I_{2Q}, I_{3Q}, I_{2K}, and I_{3K} are the integrated intensities of the corresponding analytical peaks ($10\bar{1}0$ for quartz and 001 for kaolinite). The two equations (4.148) are easily solved for quartz weight fractions and produce c_{2Q} = 16.5% and c_{3Q} = 4.1%. Note that we do not use a pure phase as an analytical standard. The conventional method of analysis with pure quartz as the standard produces similar results: c_{2Q} = 15.0% and c_{3Q} = 5.0%. This example illustrates the principles of the standardless method, which are valid for n-phase mixtures as well as for two-phase samples.

Let us now look more intensively into the simple mathematics of conventional phase analysis, in which pure phases are used as analytical standards. We ignore, for a while, micro- and macro-absorption effects, so that the simple equation holds:

$$(I_s)_{ij} = K_{ij}\, c_{sj} \quad (4.149)$$

where the subscript i refers to the analytical peak, the subscript j refers to the phase, and the subscript s refers to the sample. Let us suppose that there is at least one resolved (separated) analytical peak for each phase. If we assign the same number j to the analytical peak as to the phase, then subscript i can be omitted from the equations that follow. Suppose that the analyzed sample is composed of n phases. Thus, we need only n – 1 pure phases in order to acquire n – 1 calibration constants K_j and to determine the abundances of the first n – 1 phases in the analyzed sample. The weight fraction of the remaining *nth* phase is obtained from the mass-balance equation:

$$\sum_{j=1}^{n} c_{sj} = 1 \quad 1 \leq s \leq n \tag{4.150}$$

The analyzed sample and the n – 1 pure phases constitute a set of n samples. It follows from equation (4.149) that $c_{sj} = (I_s)_j / K_j$, and by applying equation (4.149) to each sample (including the pure phases), we obtain the following set of equations:

$$\sum_{j=1}^{n} (I_s)_j \, \alpha_j = 1 \quad 1 \leq s \leq n$$

where $\alpha_j = (K_j)^{-1}$.

These equations may be written in matrix form as:

$$\begin{vmatrix} (I_1)_1 & 0 & \cdots & 0 \\ 0 & (I_2)_2 & \cdots & 0 \\ \cdots & \cdots & \cdots & \cdots \\ (I_n)_1 & (I_n)_2 & \cdots & (I_n)_n \end{vmatrix} \times \begin{vmatrix} \alpha_1 \\ \alpha_2 \\ \cdots \\ \alpha_n \end{vmatrix} = \begin{vmatrix} 1 \\ 1 \\ \cdots \\ 1 \end{vmatrix} \tag{4.151}$$

Each row s in the intensity matrix represents the intensities for the different phases in the sample s, and each column j represents the intensities of the diffraction peak of phase j throughout the set of n samples. The zeros in the first n – 1 rows reflect the fact that the first n – 1 samples are pure phases, and for them $(I_s)_j = 0$, if $s \neq j$.

Suppose now that the first n – 1 samples are not pure phases but are composed of the same n phases in proportions different from those in the *nth* sample. The system of linear equations (4.151) will thus look like:

$$\begin{vmatrix} (1_1)_1 & (1_1)_2 & \cdots & (1_1)_n \\ (I_2)_1 & (I_2)_2 & \cdots & (I_2)_n \\ \cdots & \cdots & \cdots & \cdots \\ (I_n)_1 & (I_n)_2 & \cdots & (I_n)_n \end{vmatrix} \times \begin{vmatrix} \alpha_1 \\ \alpha_2 \\ \cdots \\ \alpha_n \end{vmatrix} = \begin{vmatrix} 1 \\ 1 \\ \cdots \\ 1 \end{vmatrix} \tag{4.152}$$

The system (4.152) is solved for the constants α_j, and then the phase abundances c_j are obtained by applying equation (4.149). In other words, n equations (4.149) can be written for each phase in each of the n samples. If we write n mass-balance equations (4.150) for each of the n samples, then we have n + n equations to be solved for n unknown phase abundances c_{sj} and n unknown calibration constants K_j. The following conditions must be met in order to obtain reasonable solutions to equations (4.152):

1. All the samples must contain no more than n identical phases to ensure the validity of the mass-balance equation (4.150).
2. The compositions of n samples must be sufficiently diverse to ensure the stability of the system (4.152).

These considerations constitute the basis of the so-called *standardless method*, in which samples whose compositions are not known are actually used as substitutes for the analytical standards (pure phases). This idea came into being more than 50 years ago, even before the pioneering work of Klug and Alexander on quantitative diffractometry. Hermans and Weidinger (1948) used this approach to determine the crystallinity of cellulose and other polymers. The diffraction pattern of the polymer was separated into two parts, "crystalline" scattering and "amorphous" scattering. Afterwards, the abundance of the crystalline phase was determined in exactly the same way as we determined quartz in Nubian sandstone. However, these ideas remained unused for almost 20 years, perhaps due to the rapid development of more conventional methods based on analytical standards and inspired by the work of Klug and Alexander. The standardless method for two- and three-phase mixtures reappeared in the work of Moore (1965). By that time, it was realized that the results of analysis could be significantly influenced by crystal-chemical differences between adopted analytical standards and the real phases in the analyzed samples. The standardless approach was generalized in a series of subsequent works (Bezjak 1975; Zevin 1977; Fiala 1980), and today the full theory can be presented.

4.10.2 Derivation of phase abundances and calibration constants

Starting with the Klug-Alexander equation (4.2), we obtain the ratio of the intensities diffracted by the same phase j in two different samples s and t:

$$\frac{(I_s)_j}{(I_t)_j} = \frac{c_{sj}\,\mu_t^*}{c_{tj}\,\mu_s^*} = \frac{c_{sj}}{c_{tj}} \cdot \frac{\sum_j c_{tj}\,\mu_j^*}{\sum_j c_{sj}\,\mu_j^*} \tag{4.153}$$

The following development depends on a knowledge of sample absorption.

4.10.2.1 Analysis of samples with a known absorption coefficient

If the sample absorption is known from any independent source, such as chemical composition or Compton scattering, then equation (4.153) may be written as:

$$c_{sj} = L_{sj}^{t} \left(\frac{\mu_t^*}{\mu_s^*}\right) c_{tj} \tag{4.154}$$

where $L_{sj}^{t} = (I_s)_j / (I_t)_j$ is the ratio of intensities of the same phase j in different samples s and t (which should not be confused with the intensity ratio S_{rj} of two phases r and j in the same sample, used extensively in the previous sections). By applying the mass-balance equation (4.150) to the left-hand side of equation (4.154) and changing s from s = 1 to s = m, the total number of samples, we obtain a system of linear equations:

$$\sum_{j=1}^{n} \left(L_{sj}^{t} \frac{\mu_t^*}{\mu_s^*}\right) c_{tj} = 1 \qquad \text{for } 1 \leq s \leq m \tag{4.155}$$

with s unknown weight fractions c_{tj} of the phases (j = 1,2,. . . ,n) in the sample t. This system may be written in matrix form as:

$$\mathbf{L} \quad \mathbf{c} = \mathbf{b} \tag{4.156}$$

where **L** is an m × n matrix whose elements are $L_{sj}^{t} = (I_s)_j / (I_t)_j$; **c** is an n × 1 vector whose elements are c_{tj} (the phase abundances in sample t); and **b** is an m × 1 vector whose elements are (μ_s^* / μ_t^*).

The least-squares solution to equation (4.156) may be written as:

$$c_{tj} = \sum_{s=1}^{m} \sum_{k=1}^{n} \left[\mathbf{L}^T \mathbf{L}\right]_{jk}^{-1} \left[\mathbf{L}^T\right]_{ks} \tag{4.157}$$

Equation (4.157) gives a solution to the QXRD problem for sample number t. By changing t from t = 1 to t = m we will obtain the phase abundances for all the samples of our sample set. Samples included in the set of m samples might not all be of equal reliability. For example, an undetected amorphous phase might be suspected in some of the samples, and a lower weighting must be given to such samples in the analytical procedure. On the other hand, samples with one prominent phase, i.e., approaching the ideal pure phase situation, are obviously more reliable. Different weightings may be given to the samples if the **L** matrix is multiplied by an m × 1 vector of weights **w**. Thus, equation (4.156) may be rewritten as:

$$\mathbf{w\,L\,c} = \mathbf{b} \tag{4.158}$$

A further improvement of the solution of equation (4.157) or (4.158) may be achieved by imposing obvious constraints on the value of the weight fractions c_{tj}:

$$0 \le c_{tj} \le 1 \tag{4.159}$$

There is, however, another way of solving the QXRD problem for the discussed case. It follows from equation (4.2) that:

$$c_{sj} = \mu_s^*(I_s)_j K_j^{-1} \tag{4.160}$$

By summing both sides over phase index j and taking equation (4.150) into account, we obtain:

$$\sum_j J_{sj} K_j^{-1} = 1 \tag{4.161}$$

where $J_{sj} = (I_s)_j \mu_s^*$, the intensity corrected for sample absorption.

Varying s from 1 to m will give a system of m equations written in matrix form as:

$$\mathbf{J}\,\boldsymbol{\alpha} = \mathbf{1} \tag{4.162}$$

where **J** is an m × n matrix whose elements are J_{sj}, $\boldsymbol{\alpha}$ is an n × 1 vector whose elements are K_j^{-1} and **1** is an m × 1 unit vector. The least-squares solution of this equation is:

$$\alpha_j = \sum_{s=1}^{m} \sum_{k=1}^{n} \left[\mathbf{J}^T\mathbf{J}\right]_{jK}^{-1} \mathbf{J}_{Ks} \tag{4.163}$$

which will give us n unconstrained calibration constants K_j. The calibration constants are positive, and the relevant constraint $\beta_j \ge 0$ must be imposed on equations (4.162). In this case, the solution to the linear problem is obtained by linear programming methods. As soon as K_j are known, all phase abundances c_{sj} can be easily calculated from basic equation (4.149).

4.10.2.2 Analysis of samples with an unknown mass absorption coefficient

The problem here is to express the ratio of absorption coefficients μ_s^*/μ_t^* in equation (4.156) through measurable quantities, such as intensities of diffraction peaks. Application of the internal standard method will solve the problem. Indeed, the ratio of mass absorption coefficients of the two samples (see Section 4.3) is equal to:

$$\frac{\mu_t^*}{\mu_s^*} = \frac{I_{s,St}}{I_{t,St}} \equiv R_{ts} \tag{4.164}$$

where the subscript St refers to the internal standard. Thus, introduction of the value R_{ts} instead of μ_t^*/μ_s^* in equations (4.156) will enable us to solve them for phase abundances c_{tj}, as was done in equation (4.157).

Another way to compensate for absorption is to mix one chosen sample t with the rest of the samples. If the samples are mixed in a 1:1 proportion and the intensity of the analytical peak of phase j is measured in each of the samples and each of the mixtures, then the mass absorption coefficient for a 1:1 mixture of the two samples t and s is:

$$\mu_{ts}^* = 0.5\,(\mu_s^* + \mu_t^*)$$

and the weight fraction of any phase j in the mixture is given by:

$$(c_{ts})_j = 0.5\,(c_{tj} + c_{sj})$$

The intensity of the diffraction peak of phase j in the 1:1 mixture $(I_{ts})_j$ is expressed as:

$$(I_{ts})_j = K_j\,(c_{ts})_j/\mu_{ts}^* = [(I_t)_j\,\mu_t^* + (I_s)_j\,\mu_s^*]\;/\;(\mu_t^* + \mu_s^*) \tag{4.165}$$

The ratio of the mass absorption coefficients of the two samples s and t, (μ_s^*/μ_t^*), is easily derived from equation (4.166):

$$\frac{\mu_t^*}{\mu_s^*} = \frac{(I_s)_j - (I_{ts})_j}{(I_{ts})_j - (I_t)_j} \equiv R_{ts} \tag{4.166}$$

Substitution of R_{ts} in equation (4.157) will solve the problem. However, mixing of samples is obviously inferior to the internal standard method in this particular application, because the proximity of $(I_{ts})_j$ to $(I_t)_j$ or $(I_s)_j$ would greatly affect the precision of determination of the ratio.

4.10.2.3 Analysis of samples for which the absorption coefficients of the phases are known

One can easily imagine a situation in which the absorption coefficients of the samples are not known, but the absorption coefficients of the phases are known or can be calculated from known chemical compositions of the phases. By multiplying both sides of equation (4.153) by μ_j, the mass absorption coefficient of phase j, and rearranging the terms, we obtain:

$$\frac{(I_s)_j}{(I_t)_j} c_{tj}\mu_j^* = \mu_j^* c_{sj} \frac{\sum_j c_{tj}\mu_j^*}{\sum_j c_{sj}\mu_j^*}$$

By summing the left- and right-hand sides over all the phases, we obtain:

$$\sum_j \frac{(I_s)_j}{(I_t)_j} c_{tj}\mu_j^* = \sum_j c_{tj}\mu_j^*$$

By varying the sample number $1 \le s \le m$ and including the mass-balance equation for sample t, we will arrive at a system of linear equations with n unknown weight fractions c_{tj}:

$$\begin{cases} \sum_j \mu_j^* [1 - (L_s^t)_j] c_{tj} = 0 & 1 \le s \le n \\ \sum_j c_{tj} = 1 \end{cases} \tag{4.167}$$

where $(L_s^t)_j = \frac{(I_s)_j}{V}$.

Table 4.17. Intensities for 13 four-phase test samples

	Quartz		Kaolin		Pyrite		Calcite
Sample	101	100	001	002	311	200	104
1	0	0	56	44	0	0	100
2	12	4.2	51	41	8	7	95
3	20	7	49	38	12	11	99
4	27	9	49	39	19	14.5	83
5	30	12.5	44	35	21	19	69
6	40	14	38	29	26	22	56
7	46	16	34	26	34	28	43
8	55	19	30	23	40	32	32
9	64	23	18	14	48	42	23.5
10	74	26	10	9	54	46	7.5
11	34.5	12	55	45	12.5	11	17.5
12	26	9	52	40	30	24	25
13	17	6	47	37	33	29	46.5

Table 4.18. Phase abundances (%) of 13 four-phase samples

	Quartz		Pyrite		Calcite		Kaolinite	
Sample	Actual	Determined	Actual	Determined	Actual	Determined	Actual	Determined
1	0	0.0	0	0.0	30	26.3	70	66.4
2	7	7.4	6.	46	27	24.3	60	59.8
3	11	12.3	10	8.0	24	26.1	55	56.2
4	15	16.6	14	11.7	21	21.9	50	57.2
5	19	19.5	18	13.9	18	18.3	45	51.11
6	23	24.8	22	16.7	15	14.6	40	43.2
7	27	28.9	26	21.4	12	11.3	35	39.1
8	31	34.6	30	25.2	9	8.4	30	34.2
9	37	39.7	37	31.3	6	6.2	20	21.2
10	44	46.0	43	34.8	3	2.0	10	12.8
11	20	21.3	10	8.4	5	4.6	65	64.9
12	15	16.2	22.5	18.8	7.5	6.9	55	59.8
13	10	10.5	27.5	21.5	12.5	12.2	50	54.9

We applied the technique outlined in this section to 13 four-phase test mixtures of quartz, kaolinite, pyrite, and calcite. The absorption coefficients were obtained by measuring Compton scattering (Bar-On et al. 1981). The intensities corrected by absorption are presented in Table 4.17.

For phases for which two diffraction peaks were measured, the sum of two intensities was used in analytical equations. Application of equation (4.163) gave the following results for the phase abundances (Table 4.18). Most of the observed weight fractions fall within 10% (relative) boundaries around the actual values, although for pyrite the discrepancies are greater. The results reflect the main drawback of the technique—at least in its present state—the fact that precision is only moderate.

4.10.3 Full pattern approach

The theories underlying the standardless technique discussed so far were developed on condition that each phase had at least one separated diffraction peak. Unfortunately, the results of this method may be significantly biased by any systematic intensity variations caused by preferred orientation or any other sample-related factors. In several of the cases discussed in previous sections, we showed that the reliability of the analysis could be greatly improved if a significant portion of the diffraction pattern was involved in the analytical procedure. This also holds true for the standardless technique.

There are two different techniques for processing digitalized patterns. In the first, known as *profile fitting*, overlapping peaks are separated into their components, i.e., into individual peaks (Parrish et al. 1976; Toraya 1986). This technique requires some knowledge of peak shape, but it does not require any structural data. As soon as cluster decomposition has been carried out and more than one peak is available for any phase j, this phase may be represented by combined intensity $(I_s)_j$:

$$(I_s)_j = \sum_k w_k (I_s)_{kj} \tag{4.168}$$

where $(I_s)_{kj}$ is the intensity of the peak k of phase j in sample s and w_k is the weight given to the *kth* peak of phase j. The combined intensities are handled in the conventional way outlined earlier in this section.

In the second technique, known as the *continuous-pattern approach*, the sample pattern is presented as a weighted sum of the patterns of individual phases. For the standardless technique discussed here, this approach is expressed in the following equation:

$$I_{si} = \sum_{j=1}^{n} c_{sj} K_{ji} \tag{4.169}$$

where s (= 1, 2, . . . , m) is the sample index, i (= 1, 2, . . . , t) is the angle index or step number, and j (= 1, 2, . . . , n) is the phase index. Each row s in the **I** matrix represents the diffraction pattern of sample s, and each row of the **K** matrix represents the diffraction pattern of phase j. The matrix **I** can be decomposed into a product of two matrices, **c** and **K**, in an unlimited number of ways, and in the general case, with c_{sj} and K_{ji} not known, there is no unique solution to equation (4.169). However, the equation can be solved for one particular case. Suppose that:

1. The number of samples m = n is equal to the number of phases; and
2. There are at least p zero elements in each row of the **K** matrix.

In other words, each phase does not contribute to at least p diffraction peaks among t diffraction peaks measured (t elements of a row in **K**). The **K** matrix can then be expressed as:

$$\mathbf{K} = \mathbf{W}\,\mathbf{I}$$

where $\mathbf{W} = \mathbf{c}^{-1}$.

Equations (4.170) hold true for each zero element in the **K** matrix:

$$\sum_{s=1}^{n} W_{js} I_{si} = 0 \qquad i = 1, 2, \ldots, p \tag{4.170}$$

The total number of equations (4.170) is p × n (n = number of rows in K). Let us now add n mass balance equations, bringing the total number of equations to $p \times n + n$. If $p \geq n - 1$, then the total number of equations is greater than or equal to n^2, and the total number of unknown elements in **W** is $n \times n$. Inversion of this matrix will give the required weight fractions c_{js}. The system of equations (4.170) is not linear and must be solved by iterations starting from any approximate solution. Data on zero-contribution angles can be acquired from the PDF or by calculation of the powder pattern from known structural data.

Let us now return to the general case expressed in equation (4.169) and consider the difference between I_{si}, the observed intensity at any scattering angle $2\theta_i$, and the sum of the contributions of the individual phases:

$$\varepsilon_{si} = I_{si} - \sum_{j=1}^{n} c_{sj} K_{ji} \qquad (4.171)$$

Our problem may be formulated as the minimization of the sum of the squares (4.172) over all m samples and over all t scattering angles:

$$\min \sum_{s=1}^{m} \sum_{i=1}^{t} \left(I_{si} - \sum_{j=1}^{n} c_{sj} K_{ji} \right)^2 \qquad (4.172)$$

The weight fractions c_{sj} and the coefficients K_{ji} are considered to be parameters of the minimization, subject to the constraints:

$$0 \leq c_{sj} \leq 1 ; \quad K_{ji} \geq 0 ; \quad \sum_{j} c_{sj} = 1$$
$$1 \leq s \leq m ; \quad 1 \leq j \leq n ; \quad 1 \leq i \leq t \qquad (4.173)$$

It is possible that the problems can be solved by iteration, starting from any approximate abundances $c_{sj}^{(1)}$ (Fiala 1980; Zevin et al. 1981). The procedure is repeated up to the point where the difference between the successive iterations $c_{sj}^{(K)} - c_{sj}^{(K+1)}$ falls below the expected error. The greatest uncertainty lies in the set of the initial abundances $c_{sj}^{(1)}$. In order to ensure convergence to correct weight fractions, $c_{sj}^{(1)}$ must be sufficiently close to the global minimum of the sum (4.172). There is always some chance that the solution will yield one of the local minima and not the correct phase abundance. Approximate phase abundances can sometimes be obtained from elemental chemical data, supposing some ideal chemical composition of the phases. Another way is application of conventional X-ray techniques with analytical standards, which do not exactly correspond to the analyzed phases but that can nevertheless be used for scaling and evaluation of the starting values $c_{sj}^{(1)}$.

We applied this technique to the analysis of Israeli phosphorites consisting of three major phases, francolite [carbofluor apatite $Ca_{10}(PO_4)_{6-x}(CO_3)_xF_2$], calcite, and dolomite, and several minor phases, quartz, gypsum, and organic matter. Since the concentration of organic matter was less than 1%, it was ignored. The gypsum was removed by water treatment, and the quartz was analyzed by conventional scaling against a pure quartz sample. The remaining three phases were analyzed by applying the minimization procedure (4.172), as follows. Twelve samples with different proportions of phases were prepared from several samples of phosphate rock by size separation. The absorption correction was acquired by measurements of Compton scattering (Bar-On et al. 1981). The integrated intensity was collected in the 12 scattering angle intervals shown in Table 4.19, together with an overlapping pattern. Initial concentrations were obtained by scaling against pure phases, using separate diffraction peaks.

Bulk chemical analysis was carried out for five samples, and the phase composition was derived from chemical data, taking into account the $PO_4 \rightarrow CO_3$ substitution in francolite (Zevin et al. 1988). Only a few iterations were required to reach the final results given in Table 4.20 for some of the analyzed samples. We found that the minimization converged to the composition very close to that derived from bulk chemistry, even in the cases when the initial weight fraction was rather far from the true values.

The continuous-pattern approach, expressed in equations (4.169), is certainly overburdened with a large number ($n \times t$) of K_{ji} constants, thus increasing the chance of falling into a false minimum during mini

Table 4.19. Angular intervals and contributing phases in the analysis of phosphates

Angular range 2θ, degrees	Phases*
21.0 –23.6	1 + 2
25.0–26.2	1 + 2
26.2–27.0	4
28.5–0.0	1 + 2
30.2–31.2	3
31.2 –34.6	1
35.0–36.4	1 + 2
39.6–41.5	1 + 2 + 3 + 4
42.6–44.0	1
44.0–49.10	1 + 2
49.10–50.2	1
50.2–5.50	1 + 3

*1—francolite, 2—calcite, 3—dolomite, 4—quartz

Table 4.20. Analysis of the major phases in phosphate rock

	Francolite, %			Calcite, %			Dolomite, %		
Sample	Starting value $C^{(1)}$	X-ray anal.	Chem. anal.	Starting value $C^{(1)}$	X-ray anal.	Chem. anal.	Starting value $C^{(1)}$	X-ray anal.	Chem. anal.
1	40	42.0	43.5	30	35.2	36.9	30	21.8	18.5
2	20	18.0	15.2	60	72.7	76.8	20	9.2	7.6
3	70	68.4	69.7	30	29.2	27.5	0	1.8	1.9
4	50	51.4	50.9	30	36.4	37.8	20	10.3	9.0
5	80	81.8	82.5	0	11.7	12.4	20	6.3	7.0
6	50	37.1	-	50	47.5	-	0	12.1	-

mization of the sum of the squares (4.172). In pattern decomposition, the overlapping problem is treated independently of quantification, conferring an advantage on this technique. Unfortunately, very little has been done on the experimental comparison of these two approaches, although full-pattern utilization has to be the main path to further development and to hopes for standardless quantification.

4.10.4 Number of phases in analyzed samples

Although the number and identity of the phases present in an analyzed sample are obtained by well-established manual or automatic identification procedures, it is sometimes useful to apply an independent check. Bezjak (1975) noted that in a set of m samples, each containing no more than $n<m$ identical phases, the intensity diffracted by any sample s at any scattering angle $2\theta_i$ is a linear combination of intensities diffracted at the same angle by n other samples. The number of phases may be established applying these criteria. Relatively simple and easily programmable mathematical tests (factor analysis) are available for this purpose.

The number of factors (phases in our case) responsible for the variability of the intensities in the $\mathbf{I}_{si}$ matrix is equal to the minimum number of linearly independent rows or columns or to the matrix rank. Technically, it is simpler to determine the rank of symmetrical quadratic ($m \times m$ or $t \times t$) matrices $\mathbf{R}' = \mathbf{I}\,\mathbf{I}^T$ or $\mathbf{R}'' = \mathbf{I}^T\,\mathbf{I}$, which is equal to the rank of the original matrix **I**. The rank of the **R** matrices is equal to the number of nonzero rows in the reduced triangular form (Wallace 1960; Wallace and Katz 1964) and to the number of nonzero eigenvalues (Ritter et al. 1976).

Due to inevitable errors in intensities, which contribute to the variability of the $\mathbf{I}_{si}$ matrix, its rank will be equal to m or t, whichever is smaller. This means that major factors (phases) that affect the variability should be separated from insignificant factors caused by experimental errors. The element of the **R** matrix is expressed as:

$$R_{jk} = \sum_{i=1}^{t} I_{ji} I_{ik} \tag{4.174}$$

and the standard deviation $\sigma(R_{jk})$ is obtained by the general rules of error propagation. As a rough estimate, let us suppose that the mean element of the **R** matrix $<R> \approx m <I>^2$, where $<I>$ is the mean intensity in the data matrix and m the dimensionality of the **R** matrix. Then $\sigma(<R>) \approx 2\, m<I>\, \sigma_i(<I>) = 2 <R>\, \varepsilon(<I>)$, where $\varepsilon(<I>)$ is the relative error in the intensity measurements, which can often be assumed to be constant. The **R** matrix is reconstructed from eigenvectors $\mathbf{X}_i$ and eigenvalues λ_i, where i - 1, 2, . . ., r nonzero eigenvectors.

$$\mathbf{R}^r = \sum_{i=1}^{r} \mathbf{X}_i \lambda_i \mathbf{X}_{iT} \tag{4.175}$$

where the superscript r in $\mathbf{R}^r$ indicates that reconstruction was carried out with the full number of nonzero eigenvectors. Now, let us suppose that only h first eigenvectors of the full row of r eigenvectors, arranged in order of decreasing eigenvalues, correspond to statistically significant factors, then the $\mathbf{R}^h$ matrix reconstructed by summation over only h first eigenvectors will be statistically equivalent to the original **R** matrix. If n is the number of detected phases and h>n, then h – n phases are undetected.

We applied this test to the $\mathbf{I}_{si\ 13x7}$ matrix given in Table 4.17. Eigenvalues calculated for $\mathbf{R}_{7x7} = \mathbf{I}^T\mathbf{I}$ are shown graphically in Fig. 4.12. The steep drop in eigenvalues from λ_4 to λ_5 emphasizes the presence of four major factors, in our case the four phases. The $\mathbf{R}^4$ matrix reconstructed from the first four eigenvectors is statistically indistinguishable from the original matrix.

This technique will, however, not reveal an amorphous phase that might otherwise go undetected. The amorphous phase will dilute the sample and reduce the intensities of all the crystalline phases in the sample, but it will not change the rank of the intensity matrix. Various amounts of amorphous phase were added artificially to 8 of 13 four-phase samples with the intensity matrix shown in Table 4.17. In one set the abundance of the amorphous phase varied from 2.5 to 15%, and in the other set, from 5 to 30%. The character of the distribution of the eigenvalues did not change (Fig. 4.12), a finding that emphasizes the presence of the same four crystalline phases.

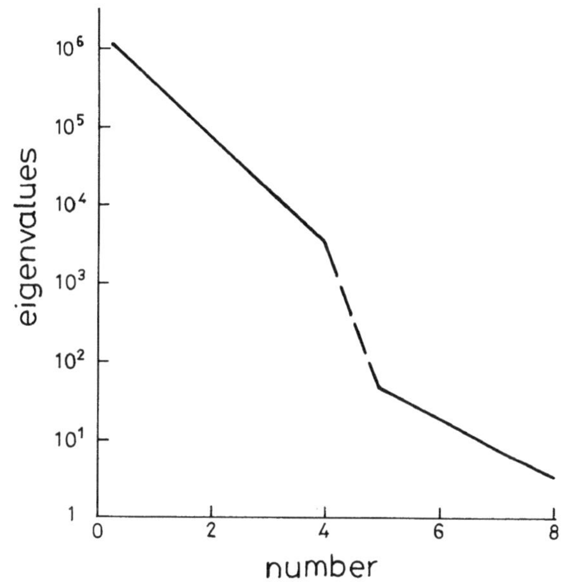

Figure 4.12. Eigenvalues of the product matrix $\mathbf{R} = \mathbf{I}^T\mathbf{I}$. Adapted from Zevin (1992).

4.10.5 Precision of the standardless techniques—optimal sample set

Three factors affect the precision of the standardless technique:

1. Intensity errors,
2. Uncertainty in the mass-balance equation (4.150), and
3. Diversity of the phase composition in the samples.

It appears that all three factors can be brought together in one compact formulation. In the development of the basic equation of the technique discussed in Section 4.10.2, we assumed that the mass-balance equation (4.150) was strictly followed. In fact, due to the probable presence of undetected phases, the sum of the weight fractions $\sum_j c_{sj}$ may be less than 1 and may vary from sample to sample. Following the development of equation (4.162) and taking into account that $\sum_j c_{sj} = r_s < 1$, we obtain a similar expression:

$$\mathbf{J}\boldsymbol{\alpha} = \mathbf{b} \tag{4.176}$$

where $\mathbf{J}$ is a matrix whose elements are $I_{sj}\,\mu_s$, $\boldsymbol{\alpha}$ is a vector whose elements are K_j^{-1}, and $\mathbf{b}$ is a vector whose elements are $b_s = \sum_j c_{sj}$.

Estimation of the error in the elements of the $\boldsymbol{\alpha}$ vector is given by Zevin and Zevin (1989):

$$\frac{\delta\alpha}{\alpha} \leq C(I)\left[\frac{\varepsilon + \Delta b_s/\sqrt{m}}{1-\varepsilon}\right] \tag{4.177}$$

where $\varepsilon = \delta I/I$ is the relative error in intensity measurements, which is supposed to be constant through all $\mathbf{I}_{si}$; $\Delta b_s = \sqrt{\Sigma_s(\delta b_s)^2}$ is the deviation of the actual sum of the weight fractions from 1; m is the number of samples; and C(I) is the condition number of matrix **I**, defined as:

$$C(I) = \sqrt{\lambda_m(\mathbf{I}^T\mathbf{I}) \times \lambda_m([\mathbf{I}^T\mathbf{I}]^{-1})} \tag{4.178}$$

where $\lambda_m(\mathbf{I}^T\mathbf{I})$ is the maximal eigenvalue of the product matrix $\mathbf{I}^T\mathbf{I}$.

The three sources of error noted above are clearly resolved in equation (4.178), with condition C(I) being responsible for the sample's diversity. Condition C(I) takes its minimal value of $\sqrt{(n-1)/n}$ (≈ 1 for large n) in the case of the most diverse samples, when we have one analyzed sample and n – 1 pure phases (analytical standards). Substitution for the pure phases by unknown samples inevitably reduces the diversity of the sample set, increases the condition number C(I), and increases the analytical error.

Equation (4.178) provides us with an excellent tool for checking the "goodness" of the sample set. The following example confirms this premise. The experimental data presented in Table 4.17 were used to calculate the "true" calibration constants (K_{j0}) as slopes of the regression I_{sj} against the true phase abundances c_j. Subsets of four or more samples were chosen at random from the general group of 13 samples, and for each subset calibration constants K_j were calculated from equation (4.163). The relative error in the calibration constants, $\partial K/K = \sum_j |K_j - K_{j0}|/K_{j0}$ was plotted against the condition number of the intensity matrix for the corresponding subset (Fig. 4.13). The correlation of the analytical error with the condition number is obvious. The value $\partial K/K$ remains within the reasonable 10% range if the condition number does not exceed 10. The presence of an undetected (amorphous) phase was simulated by reducing the intensities in randomly chosen rows of the intensity matrix in Table 4.17. This resulted in a dramatic increase of the analysis error. The value of $\partial K/K$ increased from ~0.09 to an unacceptably high value of 0.18 when the presence of an amorphous phase with a maximum abundance of >30% was simulated in 8 of 13 samples.

Intensity errors affect the results of standardless analysis to a lesser degree than variability of the samples. We can expect, however, that

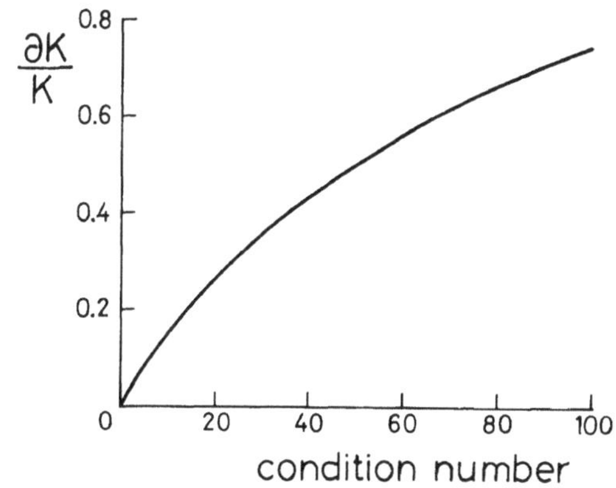

Figure 4.13. Precision of analysis as a function of the condition number of the intensity matrix. Adapted from Zevin (1992).

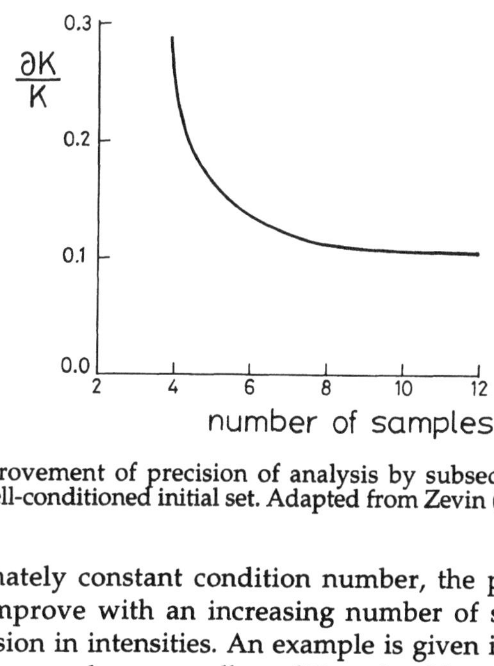

Figure 4.14. Improvement of precision of analysis by subsequent addition of samples to the well-conditioned initial set. Adapted from Zevin (1992).

at an approximately constant condition number, the precision of the analysis will improve with an increasing number of samples due to increased precision in intensities. An example is given in Fig. 4.14. The initial set of four samples was well conditioned, with condition number ~10. Samples added after the initial set did not improve the condition number but rather contributed to the precision of the intensity and improved the results of the analysis.

If the diversity of the sample set is unsatisfactory, it can be improved by using various material-handling procedures known from mineral processing. They include size separation by sifting, gravity separation by sedimentation in various liquids, magnetic separation, flotation, and chemical treatment. These procedures are usually selective for certain phases and produce samples with different proportions of the same phases.

4.11 Combination of X-ray diffraction and chemical data

4.11.1 Phase analysis based on pure chemical data

Knowledge of the chemical (elemental) composition of a sample can be a great help in QXRD. Not only can the absorption be easily calculated, but the main problem—that of phase composition—may sometimes be solved solely in terms of chemical data. The following linear equation is obvious for any chemical component (expressed as elemental or oxide fractions):

$$x_k = \sum_{j=1}^{n} \gamma_{kj} c_j \tag{4.179}$$

where x_k is the weight fraction of component k in the analyzed sample, c_j is the fraction of phase j, γ_{kj} is the fraction of component k in phase j, and n is the number of phases in the analyzed sample. In an m-component sample, equation (4.179) can be written for all m components, and the corresponding set of m equations is given by:

$$\mathbf{X} = \boldsymbol{\Upsilon} \mathbf{c} \tag{4.180}$$

where $\mathbf{X}$ is an $m \times 1$ vector whose elements are x_k, $\boldsymbol{\Upsilon}$ is an $m \times n$ matrix whose elements are γ_{kj}, and $\mathbf{c}$ is an $n \times 1$ vector of the phase abundances c_j.

Our objective—determination of the vector $\mathbf{c}$—can be obtained on the basis of the following premises:

1. The chemical composition of the sample (vector $\mathbf{X}$) is known (measured).
2. The matrix of the component composition of the phases $\boldsymbol{\Upsilon}$ is fixed.

If our system is in thermodynamic equilibrium, then the phase composition of any sample may be defined by an equilibrium phase diagram. In another words, in this case, the set of simultaneous equations (4.180) has a unique solution. If temperature and pressure are constant (the usual case), then at a fixed component composition of the sample, the only variables are the abundances of the components of the phases γ_{kj}. According to the phase rule, the number of degrees of freedom at a particular temperature and pressure in such a system is:

$$f = m - n \tag{4.181}$$

where m is the number of components and n is the number of phases.

The apparent consequence of the phase rule is that $m \geq n$, and the number of equations in system (4.180) is equal to or greater than the number of unknowns c_j. If $m = n$, then $f = 0$ and the system is constant, i.e., the composition s of the phases (the elements of the $\boldsymbol{\gamma}$ matrix) are fixed and are certainly different for various phases. In the case in which $m > n$, there are $m - n$ degrees of freedom, and the composition of one or more phases may vary without changing the number of phases n. This implies that at least one of the phases is a solid solution. The difference between the two systems, $m = n$ and $m > n$, is evident if we try to change the chemical composition of the sample (**X** vector). In the former system, any change in **X** will be accommodated in equation (4.180) by a corresponding change in the phase abundances **c**. In the latter case, the same change in **X** will be accommodated by changes in both **c** and the phase composition matrix $\boldsymbol{\gamma}$. We can even imagine a situation in which the phase abundances will be constant when **X** varies to some extent.

If in the case of $m > n$, the phase composition matrix $\boldsymbol{\gamma}$ is known from a phase diagram or is established by any independent method (microprobe analysis, cell parameters measurements, etc.), then the only unknowns are n phase abundances c_j. The system of m simultaneous linear equations (4.180) is now redundant ($m>n$). In principle, the $m - n$ redundant equations may be removed, and the remaining n equations may be solved for n c_j*th*. However, it is better to implement all m equations (4.180) in a least-squares procedure, in which the solution is found in the following way:

$$\mathbf{c} = (\boldsymbol{\gamma}^T\boldsymbol{\gamma})^{-1}\,\boldsymbol{\gamma}^T\,\mathbf{X} \tag{4.182}$$

Taking into account that $m>n$, we can actually skip the definition of $m - n$ elements of the $\boldsymbol{\gamma}$ matrix and nevertheless solve the system of equations (4.180) for n unknown phase abundances c_j and $m - n$ undefined γ_{kj}. In this case, equations (4.180) are nonlinear.

However, thermodynamic equilibrium cannot be taken for granted for any particular system being analyzed. In many, if not in most cases subjected to quantitative phase analysis, the analyzed sample is not in thermodynamic equilibrium (mainly as a result of kinetic factors), and the phase rule (4.181) does not hold. Equations (4.180) will hold, of course, but in this case the chemical composition of the sample does not always define its phase abundance and the chemical composition of the phases, as in the case of thermodynamic equilibrium. A typical example is the coexistence of polymorphs, such as cristobalite and quartz in a ceramic body or austenite and martensite in quenched steel.

Under non-equilibrium conditions, the number of phases can simply exceed the number of components, thus precluding the use of equation (4.180) for phase analysis. If, however, condition $m > n$ does hold true, the phase composition matrix γ_{kj} must be known for the successful solution of equations (4.180).

We can now formulate the basic requirements for carrying out phase analysis on the basis of chemical data in the general case:

1. the number of phases must not exceed the number of components;
2. the chemical compositions of the phases must be known; and
3. all phases must have different compositions.

The last requirement emphasizes the fact that polymorphs cannot be analyzed by the method under discussion. In mathematical language, this requirement implies that the rank of the γ matrix must be equal to the number of phases n. Even if these requirements are met, the real success of the analysis depends on the variability of the phase compositions with a well-defined γ. In the most favorable situation, some rows in the γ matrix have all the elements equal to zero, except one, $a_{k'j'}$. This means that component k' is found only in phase j', and the phase abundance c_j follows from the obvious equation:

$$x_{k'} = \gamma_{k'j'}\, c_{j'} \tag{4.183}$$

independently of the abundances of the remaining phases in the sample. In the most favorable situation, each component corresponds to a single phase, as, for example, in a mixture of quartz, corundum, and rutile. In this case, γ is a diagonal matrix. On the other hand, if all or some of the phases in the sample have similar chemical compositions, we can expect significant errors in estimating c_{kj} by equations (4.180). If δx_k is the error in the analysis of component x_k, then phase abundances calculated by equations (4.182) are also subject to errors $\sigma(c_j)$. Following the exposition given in Section 4.10, the relative error in the estimation of c_j is expressed as:

$$\varepsilon(c) = C\,(\gamma)\,\varepsilon\,(x) \tag{4.184}$$

where $\varepsilon(x) = \sigma(x)/x$ is the relative error in the determination of the elemental composition, which is assumed to be equal for all elements and $C(\gamma)$ is the condition number of the matrix γ defined as:

$$C(\gamma) = \sqrt{\lambda_m\,(\gamma^T \gamma)\ \times\ \lambda_m\,(\gamma^T \gamma)^{-1}} \tag{4.185}$$

where $\lambda_m(\gamma^T\gamma)$ is the maximum eigenvalue of the product matrix $\gamma^T\gamma$. The two terms in the right-hand side of equation (4.184) clearly indicate two major factors affecting the precision of the phase analysis;

namely, the precision of the chemical analysis $\varepsilon(x)$ and the variability of matrix Υ, expressed by its condition number $C(\Upsilon)$.

The condition number decreases with increasing diversity of the matrix and has a minimum value close to 1 for the case of the diagonal matrix of Υ, thus emphasizing the most favorable case of the analysis.

Although we have so far considered x_k as representing fractions of chemical components, this vector may be extended to other phase-related properties. Of course, in the Υ matrix the corresponding *kth* row must be related to the same properties of the phases under analysis. Other x_k properties might include weight loss in a particular temperature range, solubility in certain acids, magnetic susceptibility, etc.

There are many examples of applications of the discussed technique to the quantification of various rather complex mixtures of phases (see, for example, Hodgson and Dudeney 1984; Pearson 1978). Probably the most important application of this method is to be found in the cement industry. It is used to calculate the potential phase composition of cement clinker, when the chemical composition of the raw mixture is known (Bogue 1955; Majumdar et al. 1972). The main drawback of this method is that it is based on only an approximate knowledge of the Υ matrix. The calculations are based on the assumed compositions of four clinker minerals, alite ($3CaO.SiO_2$), belite ($2CaO.SiO_2$), calcium aluminate ($3CaO.Al_2O_3$), and brownmillerite ($4CaO.Al_2O_3.Fe_2O_3$). In fact, all four of the phases have alternative compositions, a fact that sometimes causes significant deviations of the calculated potential composition of the Portland cement clinker from the true composition.

Although the pure chemical approach to phase analysis is used rather widely, one should be aware of the greater generality and universality of X-ray diffraction methods. The requirements for successful application of pure chemical data to phase analysis by the method outlined earlier are not fulfilled unconditionally. For example, a pure chemical technique cannot be applied to analysis of polymorphs and encounters major difficulties in the analysis of phases with similar and/or variable chemical compositions. In some cases, X-ray diffraction and pure chemical techniques can be complementary to one another. The X-ray technique may be auxiliary, leaving the major role for chemistry, or it may be basic, leaving the auxiliary role for chemistry.

4.11.2 Auxiliary X-ray diffraction data

The first step in the procedure to be outlined is to establish the number and identity of the phases present in the sample to be analyzed. For qualitative identification, it is self-evident that X-ray diffraction is the tool of choice. X-ray diffraction is also required as an auxiliary method of analysis when the sample to be analyzed contains polymorphs or phases with similar chemical compositions. Suppose that calculation by equation (4.182) gives the total abundance c of two polymorphs or two phases j' and j" containing the same element k'. If we can separate diffraction peaks of these two phases, then the ratio of intensities will be given by:

$$I_{j'}/I_{j''} = \beta_{j'j''} c_{j'}/c_{j''} \tag{4.186}$$

where $\beta_{j'j''}$ is the calibration constant defined in Section 4.6.

Combination of this equation with the obvious equation:

$$c = c_{j'} + c_{j''}$$

gives the abundance of an individual phase:

$$c_{j'} = c/[I + (I_{j'}/I_{j''})\beta_{j'j''}] \tag{4.187}$$

Engler and Iyengar (1987) used X-ray diffraction data to establish confidence limits of probable variations of phase abundances. For minerals with variable chemical composition (ankerite, illite, feldspars), they included an additional variable parameter in the relevant elements γ_{kj} of the $\boldsymbol{\gamma}$ matrix. The corresponding set of equations (4.180) is nonlinear in this case. Phase weight fractions c_j and variable parameters in γ_{kj} elements are obtained by methods of nonlinear optimization, subject to constraints for the phase abundances, obtained by X-ray diffraction data. These constraints make the optimization procedure more realistic.

Another auxiliary application of diffraction measurements is specification of the $\boldsymbol{\gamma}$ matrix. The usual problem is establishing the exact composition of a solid solution. In many cases, this can be achieved by measurement of unit cell parameters of the solid solution. Typical examples include mullite (Garbauskas and Goehner 1981) and chlorites (Hodgson and Dudeney 1984).

4.11.3 Basic application of X-ray diffraction

The two basic equations for phase quantification—equation (4.100), which is based solely on diffraction data, and equation (4.180), which is based solely on chemical data—are in fact very similar. The weight fraction x_k of the component k in the sample takes the place of the intensity I_i of the diffraction peak i, and the matrix $\boldsymbol{\gamma}$ of the chemical compositions of the phases takes the place of the matrix of the scaled intensities of the pure phases. These two equations may be easily combined into a single equation to be solved for the phase abundances c_j:

$$\begin{vmatrix} I_1 \\ I_2 \\ \cdots \\ I_m \\ x_1 \\ x_2 \\ x_l \end{vmatrix} = \begin{vmatrix} a_{11} & a_{12} & \cdots & a_{1n} \\ a_{21} & a_{22} & \cdots & a_{2n} \\ \cdots & \cdots & \cdots & \cdots \\ a_{m1} & a_{m2} & \cdots & a_{mn} \\ \gamma_{11} & \gamma_{12} & \cdots & \gamma_{1n} \\ \gamma_{21} & \gamma_{22} & \cdots & \gamma_{2n} \\ \gamma_{l1} & \gamma_{l2} & \cdots & \gamma_{ln} \end{vmatrix} \begin{vmatrix} c_1 \\ c_2 \\ \cdot \\ \cdot \\ \cdot \\ \cdot \\ c_n \end{vmatrix} \quad (4.188)$$

A similar relationship was suggested by Snyder (1992) for the internal standard technique. The solution for the phase abundances c_j is found in the conventional way, expressed, for example, in equations (4.105) and (4.182).

Smith et al. (1989) applied a similar idea of the implementation of chemical data in the full-pattern approach to QXRD. In this case, weight functions of chemical elements act as constraints in the least-squares fitting of scaled pure mineral patterns to the pattern of the analyzed sample, as discussed in Section 4.8.7. This technique was applied to seven mineral mixtures characteristic of sedimentary rocks. Although QXRD without chemical constraints showed good correspondence between observed and real compositions, implementation of constraints further improved the accuracy of the analysis. Improvement of phase analysis can be expected in all cases in which precise x_k and γ_{kj} data are used in the way expressed in equation (4.188).

Majumdar et al. (1972) suggested that X-ray diffraction data could be combined with elemental chemical data in a single procedure that would reveal the elemental composition of the individual phases. If, on the contrary, the phase composition was known, similar calculations would establish the calibration constants needed for the quantification of the phases.

Suppose that we have a set of m samples with variable proportions of n phases in each sample. Let us express the basic equation (4.2) in matrix form:

$$\mathbf{J} = \mathbf{c}\,\mathbf{K} \quad (4.189)$$

where **J** is an $m \times n$ matrix whose elements are $J_{sj} = \mu_s^* I_{sj}$, $n \le s \le m$ being the sample index and $1 \le j \le n$ being the phase index; **c** is an $m \times n$ matrix of phase abundances c_{sj}; and **K** is an $n \times n$ diagonal matrix whose elements are the calibration constants K_j. In fact, our objective is the determination of matrix **K**.

It was shown in Section 4.10 that the values of K_j can be obtained from equations (4.189) alone, if the number of samples is not less than the number of phases and there is sufficient variability in the set of samples. However, use of potentially accurate and precise chemical data should improve the accuracy of the K_j values. A chemical balance equation in the form (4.180) is also applicable for our set of m samples, but in this case $\mathbf{X}$ is an m x p matrix, whose elements are x_{sk}, the weight fractions of k elements in samples s, and $\mathbf{c}$ is an m × n matrix of phase abundances c_{sj}. If the matrix $\boldsymbol{\Upsilon}$ is known, then equation (4.180) may be solved for $\mathbf{c}$ by the least-squares method:

$$\mathbf{c} = (\boldsymbol{\Upsilon}^T\boldsymbol{\Upsilon})^{-1}\,\boldsymbol{\Upsilon}^T\mathbf{X} \tag{4.190}$$

and calibration constants K_j are obtained from equation (4.189) using solution (4.190) for $\mathbf{c}$.

$$K_j = (\mathbf{c}^T\,\mathbf{c})^{-1}\,\mathbf{c}^T\,\mathbf{J} \tag{4.191}$$

This type of approach was used by Renault (1987) in analyzing three-phase mixtures of quartz, microcline, and calcite, generally with good correspondence between known and observed phase compositions. There were two unacceptable results for microcline and quartz among 12 samples. It is interesting to note that these "bad" results might be expected by the deviation of $\sum_j c_j$ from 1 as well as by the deviation of the chemical composition calculated by equation (4.189) from the composition obtained experimentally.

A further development of this X-ray diffraction-bulk chemistry approach was based on a special statistical technique—target transformation factor analysis (Starks et al. 1984). Equation (4.180), written for the set of m samples, is the model equation for factor analysis, in which the elements c_{sj} of the matrix $\mathbf{c}$ can be regarded as the loads and the elements γ_{kj} of the $\boldsymbol{\Upsilon}$ matrix as the scores (Malinowski and Howery 1980). The basic factors are the phase abundances c_{sj}. However, we have more information about the basic factors than in a common factor analysis problem. The number of phases (factors) is usually well established by qualitative X-ray analysis. The directions of the column vectors in the $\mathbf{c}$ matrix are also known, at least approximately, because these vectors are parallel or approximately parallel to the directions of the column vectors in the $\mathbf{J}$ matrix [equation (4.189)]. Finally, the chemical compositions of some of the analyzed phases might be known, and thus some of the elements of the $\boldsymbol{\Upsilon}$ matrix can be fixed.

The technique developed by Starks et al. (1984), which may be used to estimate both the $\mathbf{c}$ matrix and the $\boldsymbol{\Upsilon}$ matrix, was successfully applied to the analysis of phosphate rock. One of the results is a very realistic composition for the major phosphorus-bearing mineral, carbofluorapatite (francolite), as shown in Table 4.21.

Table 4.21. Composition of francolite

	P_2O_5	CaO	F	MgO	SiO_2
γ_{kj}, %	35.8	52.45	4.00	0.00	-0.2
Standard deviation, %	0.87	1.5	0.26	0.10	0.10

Taken from Starks et al. (1984).

4.12 Crystallinity of polymers

Polymers are usually made up of two components with different degrees of order: a crystalline component, with some degree of long-range order in the spatial distribution of the chain molecules, and an amorphous component, possessing only short-range order. Although a sharp interface between these components does not exist and a rather diffuse transition is more likely, this two-phase concept is generally accepted. Since the physical properties of polymer materials, such as mechanical strength and transport properties, are well correlated with the degree of crystallinity, there is a need for a reliable method for determining polymer crystallinity. As soon as we separate diffraction by crystalline and amorphous components, the problem seems to be an ordinary problem of analysis of two-phase systems, which can be solved by many of the techniques discussed at length in this book. However, a closer look reveals that the problem is not that simple, being complicated by the nature of the X-ray scattering by the nonorganized amorphous component, the intrinsic disorder of the crystalline component, and the very important fact that both phases have the same chemical composition.

Scattering by an amorphous material is defined by the distribution of interatomic vectors in the first few coordination spheres and consists of one or more very broad diffraction bands (or halos). The extension of the halo may be tenths of degrees 2θ. Incoherent scattering, which is substantial for light atoms, contributes significantly to the total specimen scattering and must be removed, together with contributions from any other sources of parasitic radiation, such as air scattering. The practical aspects of this first step in data processing are discussed in Chapter 5.

The next step, the separation of the smooth amorphous halo from the diffraction peaks of the crystalline matter, is the real problem. A simple and rather naive technique is delineation of the amorphous halo by drawing by hand a smooth curve under the diffraction peaks. This procedure becomes more reliable if the diffraction pattern of the pure amorphous sample is available. The shape of the amorphous halo obtained from this material can be followed in any manual or computer-fitting procedure in the two-phase sample (Vonk 1973; Polizzi et al. 1990). Suppose now that the amorphous and crystalline contributions to the total diffraction pattern have been successfully separated. Then, the integrated intensity of the amorphous scattering I_a is a measure of the amorphous content. c_c, the degree of crystallinity (fraction of the crystalline phase), is given by:

$$c_c = 1 - I_a/(I_a)_0 \tag{4.192}$$

where $(I_a)_0$ is the integrated intensity of the amorphous halo in a totally amorphous polymer sample. If the latter is not available, but series of specimens with various degrees of crystallinity are, then the standardless method proposed by Hermans and Weidinger (1948) is applicable. Their approach to the problem is exactly the same as that described in the analysis of Nubian sandstone at the beginning of Section 4.10. If we have two polymer samples with different fractions of amorphous and crystalline phases, then the degree of crystallinity is defined (Zevin and Messalem 1982) by:

$$(c_c)1 = [L_c(1 - L_a)]/(L_c - L_a) \tag{4.193}$$

where $L_c = I_{c1}/I_{c2}$; $L_a = I_{a1}/I_{a2}$; I_{c1}, I_{c2} are the integrated intensities of the crystalline peaks in specimens 1 and 2, and I_{a1}, I_{a2} are the integrated intensities of the amorphous phases in the two specimens. It is supposed that the density and composition of both the amorphous phase and the crystalline phase are constant in all the investigated samples. If the number of specimens exceeds two, then the usual least-squares minimization outlined in Section 4.10 may be employed. Attention must be paid to the values L_c and L_a in equation (4.193), i.e., the ratio of intensities in the different specimens. Thus, we have to take the probable absorption correction into account if the transmission technique is used, or we have to carefully maintain "infinitely thick" conditions for reflection techniques. Once crystallinity has been established in one series of specimens, the series can serve as a secondary standard in the analysis of subsequent batches.

In the application of the conventional QXRD method outlined in this section, we have tacitly assumed that the spatial distribution of the interatomic vectors in the amorphous material does not vary between specimens. In other words, we have assumed that the shape of the amorphous halo does not change. This is not necessarily true, and the correct representation of the amorphous scattering must be based on the law of conservation of intensity:

$$\int_0^\infty I(s)\, dV_s = 4\pi \int_0^\infty s^2 < f^2 > ds \tag{4.194}$$

where $s = 2 \sin \theta/\lambda$; dV_s is the volume element in reciprocal space, $<f^2> = \sum_k N_k f_k^2 / \sum N_k$, f_k is the atomic scattering factor of atom of type k, and N_k is the number of atoms of type k in the stoichiometric formula. For a specimen with "isotropic" scattering, equation (4.194) may be rewritten as:

$$\int_0^\infty s^2 I(s)\, ds = \int_0^\infty s^2 <f^2>\, ds \tag{4.195}$$

For a specimen with *anisotropic* scattering (preferred orientation), the same equation holds, but $I(\bar{s})$ must be averaged over all orientations of s:

$$I(\bar{s}) = (1/4\,\pi) \int_0^{4\pi} I(s)\, d\omega$$

where ω is a solid angle in reciprocal space.

It follows from the intensity conservation law that for a purely amorphous specimen the quantity:

$$\int_0^\infty I(s)\, s^2\, ds = \text{const} \tag{4.196}$$

irrespective of any redistribution of molecules and variations in the pattern of interatomic vectors. Therefore, it is more appropriate that I_a and $(I_a)_0$ in equation (4.192) should not be integrated intensities but integrals of the type (4.196), where the lower and upper limits, s_0 and s_p, are chosen to cover one or more amorphous halos. The two representations are, of course, equivalent, if the shape of the $I_a(s)$ curve remains constant throughout the batch of investigated specimens.

The techniques outlined earlier usually overestimate the amount of amorphous phase and underestimate the crystallinity. The reason for this lies in the diffuse scattering of the crystalline phase distributed between reciprocal lattice points and generated by thermal motion and static displacements of molecules. Diffuse scattering is added to the continuous amorphous scattering, apparently increasing the latter. Diffuse scattering by the crystalline phase may be separated by a fitting procedure. Diffraction by the crystalline phase (diffraction peaks) decays rather rapidly with the distance s in reciprocal space, usually due to the Debye-Waller factor, which is significant for polymers, and to the inevitable presence of lattice imperfections. The isotropic thermal vibrations factor has the form $\exp(ks^2)$. Static molecular displacements that do not violate the long-range order (structural imperfections of the first kind) will give the same form of intensity decay. Structural imperfections of the second kind, those that do violate long-range order, produce a more complicated form of intensity decrease; these can, however, also be approximated by a Gaussian function. We assume here, of course, that both kinds of lattice imperfections are isotropic. Assuming a Gaussian form for the decay of crystalline scattering and a monotonous increase of diffuse scattering, Polizzi et al. (1990) suggested the following expression for the latter:

$$I_{dif} - c_c < f(s)^2 > [1 - \exp(-ks^2)] \tag{4.197}$$

where k is a disorder parameter. There is no complete theoretical justification for this equation, although it seems to be correct for high s values, for which diffuse scattering is indeed significant and for which many overlapping peaks create quasi-continuous scattering by the crystalline phase. If equation (4.197) is accepted, then the total scattering of a two-phase specimen may be represented by three terms (Polizzi et al. 1990): continuous scattering by the amorphous phase, diffuse scattering by the crystalline phase due to thermal vibrations of atoms and lattice defects in the crystalline phase. The profile fitting is given by equation (4.198), using an appropriate function for a single line shape (Lorentzian or Gaussian):

$$I_{fit}(s) = (1 - c_c)(I_a)_0(s) + c_c < f^2 > [1 - \exp(-ks^2)] + \sum_m F_m(s, H_m, \alpha_m) \tag{4.198}$$

where $(I_a)_0$ is the scattering by a purely amorphous specimen, F_m is an analytical function fitted to the *mth* crystalline peak, H_m is the height of the *mth* peak, and α_{mn} are parameters defining the width and shape of $F_m(s)$. The unknown parameters, i.e., the crystallinity c_c, the disorder factor k, and the peak-fitting parameters are obtained, as usual, by minimizing the sum of the squares:

$$U = \sum_{i=1}^{N} \{[I_{fit}(s_i) - I_{obs}(s_i)]^2\} \tag{4.199}$$

where I_{fit} is expressed in equation (4.198), and I_{obs} is the total observed scattered intensity after removing incoherent and other parasitic scattering. The observed intensities $(I_a)_0$ and I_{obs} must be transformed to an absolute scale. The scaling coefficient ξ is obtained by application of the intensity conservation law:

$$\xi = \left[\int_{s_0}^{s_p} I_{obs}(s)\, s^2 ds\right] \Big/ \left[\int_{s_0}^{s_p} <f^2> (s)\, s^2 ds\right] \tag{4.200}$$

where the lower limit s_o must be close to s = 0 and the upper limit s_p must be extended over the range where the diffraction peaks disappear.

All the methods discussed earlier are based, in some way, on comparisons with an analytical standard, usually a pure amorphous material. The standardless method of Hermans and Weidinger (1948) is, in fact, one type of commonly used QXRD approach. However, the intensity conservation law (4.196) gives us a clue to an absolute method for the determination of crystallinity. This law, expressed for a crystalline phase, may be written as:

$$\int_0^{\infty} I_{cr}(s)\, s^2 ds = c_c \int_0^{\infty} <f\ (s)^2>\, s^2 ds \tag{4.201}$$

The degree of crystallinity c_c is, however, severely underestimated by this equation, because considerable intensity is removed for I_{cr} (s) due to crystal imperfections. Vonk (1961) suggested that the equation should include a crystal disorder function D(s), which takes into account dynamic (temperature) and static crystal disorder:

$$\int_0^{\infty} I_{cr}(s)\, s^2 ds = c_c \int_0^{\infty} <f\ (s)^2>\, s^2 D(s)\, ds \tag{4.202}$$

In order to convert equation (4.202) into a form that can be used for analytical applications, we have to define the function D(s) and scale the observed intensities $I_{cr}(s)$. The latter aim is achieved by measurement of the total scattering intensity and application of intensity conservation law (4.194). Thus, from equations (4.194) and (4.202) we obtain:

$$c_c = \frac{\int_0^{\infty} I_c\,(s)\, s^2 ds}{\int_0^{\infty} I(s)\, s^2 ds} \cdot \frac{\int_0^{\infty} s^2 <f^2> ds}{\int_0^{\infty} s^2 <f^2> D(s) ds} \tag{4.203}$$

The Gaussian form $\exp(-ks^2)$ is usually assumed for D(s), as has been discussed earlier in this section. The integration limits in equation (4.203) are substituted by any reasonable lower limit s_o and a large enough but finite upper limit s_p. Thus,

$$c_c = \frac{\int_{s_o}^{s_p} I_c(s)\, s^2 ds}{\int_{s_o}^{s_p} I(s)\, s^2 ds} \cdot L \tag{4.204}$$

where $L = \int_{s_o}^{s_p} s^2 \langle f^2 \rangle ds \Big/ \int_0^{s_p} s^2 \langle f^2 \rangle \exp(-k^2)\, ds$. This equation still has two unknowns, c_c and the distortion parameter k. Ruland (1961) proposed that a number of upper limits s_p be chosen to find a k value such that c_c calculated by equation (4.204) would not vary with s_p and could be considered a true value.

Vonk (1973) suggested the following strategy for solving equation (4.204). Combining equations (4.203) and (4.204), we obtain:

$$R_s = 1/c_c + (L/c_c)\, s_p^2 \tag{4.205}$$

where $R_s = \int_0^{s_p} I(s)\, s^2 ds \Big/ \int_0^{s_p} I_c(s)\, s^2 ds$.

A plot of R_s vs. s_p should oscillate about a straight line defined by:

$$y = 1/c_c + (L/2c_c)\, s_p^2 \tag{4.206}$$

c_c and L are obtained from the intercept and slope, respectively, of the straight-line approximation of the R_s curve.

Polizzi et al. (1990) carried out a comparative analysis of four polyethyleneterephthalate samples by three basic techniques: the standardless technique of Hermans and Weidinger (1961a), the absolute method of Ruland (1961), and Vonk (1973), and their own profile-fitting technique. The results are given in Table 4.22. There is surprisingly good correspondence between the crystallinity data obtained by the three different methods. However, there is complete disagreement in the values for the distortion parameter k. One cannot, of course, expect complete correspondence of parameters defined in different ways [in an integral equation (4.202) by Ruland and in a functional nonintegral equation (4.197) by Polizzi et al. (1990)].

Table 4.22. Crystallinity c_c and disorder parameter k for four polyethyleneterephthalate specimens

	Standardless technique*	Absolute method**		Profile-fitting technique***	
Specimen	c_c	c_c	k	c_c	k
A	0.21	0.23	7	0.23	2.0
B	0.32	0.33	6	0.4	2.0
C	0.40	0.41	6	0.42	2.1
D	0.46	0.48	5	0.50	1.9

* Hermans and Weidinger (1961a)
** Ruland (1961) and Vonk (1973)
*** Polizzi et al. (1990)

Nevertheless, as a result of the large discrepancies observed in the distortion parameter k, which is a very important by-product of crystallinity determination, this parameter must be considered a comparative value useful for specimens analyzed by one specific technique.

4.13 Analysis of low-mass samples

The discussions in the previous sections of this chapter were based on the Klug-Alexander equation (4.2) on the tacit assumption that the amount of sample material was sufficiently large to meet the requirements for an infinitely thick sample demanded by this equation. In practice, in a large number of cases requiring quantification, the sample mass is limited. This limitation can be one of two kinds:

a. The available sample mass is limited, such as in mineral inclusions, corrosion products, and rare chemicals.
b. The available sample mass may be unlimited, but only a small mass of sample is irradiated to produce a diffraction pattern, because the sample material is distributed as a thin layer or film.

4.13.1 *Critical sample masses*

For low-mass samples, special attention must be paid to the absorption correction, which will depend on the volume or mass of the sample. Of equal importance to the sample mass G is another characteristic of material abundance [which is especially suitable for the *b* (see above) type of sample], i.e., A, the surface density (loading) or the sample mass per 1 cm^2 of sample surface S:

$$A = G/S$$

For Bragg-Brentano reflection geometry and a sample that cannot be considered infinitely thick, equation (3.17) serves as the basic intensity equation. Note that the absorption correction in this case depends not only on the Bragg angle θ but also on the loading A. Let us calculate the first critical sample mass (loading) that defines the boundary between an infinitely thick sample and a low-mass sample. Suppose that the sample can be considered infinitely thick if the exponent in equation (3.17) is less than 0.02, then the critical mass of the sample per 1 cm^2 of irradiated area (critical loading) is given by:

$$A_{cr} = \frac{\ln 0.02}{2\,\mu^*} \sin\theta = 1.85 \sin\theta/\mu^* \qquad (4.207)$$

The irradiated sample surface $S = B/\sin\theta$, where B is the cross section of the primary beam at the location of the sample. The critical sample mass in Bragg-Brentano reflection mode geometry is equal to:

$$G'_{cr} = A_{cr} \times S = 0.85\,B/\mu^* \qquad (4.208)$$

The quantity G'_{cr}, which defines the boundary up to which the infinitely thick sample approach is valid, is known as the *first critical mass*.

If the sample mass is reduced further, then starting from a certain sample mass, we will find that the diffracted intensity is a linear function of the irradiated sample mass, i.e., it behaves as a non-absorbing sample. Suppose that loading A in equation (3.17) is so small that the exponent can be expanded in a linear series:

$$\exp(-2\mu^* A/\sin\theta) \approx 1 - 2\,\mu^* A/\sin\theta$$

then the diffracted intensity is given by:

$$I_{ij} = K_{ij}\, I_0\, A \qquad (4.209)$$

The mass absorption coefficient then vanishes from the intensity equation, which now expresses a linear dependence on the loading A. We will now define the *second critical sample mass* as the mass for which the approximate intensity equation (4.209) deviates from the exact equation (3.17) by a certain small amount, say 2%. In this case, the following equation holds:

$$1 - \exp(2\mu^* A''_{cr}/\sin\theta) = 0.98\,(2\mu^* A''_{cr}/\sin\theta) \qquad (4.210)$$

The approximate solution leads to:

$$(A''_{cr})_2 \approx 0.025 \sin\theta/\mu^* \tag{4.211}$$

and the second critical mass is given by:

$$(G'_{cr})_2 = (A''_{cr})_2 \cdot S = 0.025\, B/\mu^* \tag{4.212}$$

with the same definition of terms as for equation (4.208). The subscript 2 refers to the 2% boundary for the approximate linear equation (4.210). If a 10% deviation is accepted for equation (4.210), then both values $(A''_{cr})_{10}$ and $(G''_{cr})_{10}$ are greater:

$$(A''_{cr})_{10} \approx 0.11 \sin\theta/\mu^* \quad ; \quad (G''_{cr})_{10} = 0.11\, B/\mu^* \tag{4.213}$$

It must be emphasized that in the definition G''_{cr} we assume that the sample material is distributed evenly over the surface, $S = B/\sin\theta$.

Values of G'_{cr} and $(G''_{cr})_2$ and $(G''_{cr})_{10}$ are given in Table 4.23 for light, medium, and heavy materials, assuming CuK_α radiation, a constant divergence mode with a 1° beam divergence, a goniometer radius of 173 mm, and the height of irradiated sample area H = 12 mm. The cross section of the primary beam $B = HR \tan 1° \approx 0.36$ cm^2. If the irradiated sample surface S is kept constant by a continuous-divergent (θ-compensating) slit, the critical mass is calculated as $A_{cr}S$ where the critical surface density A_{cr} is defined by equations (4.207) and (4.211). Although the figures will be different from those given in Table 4.23, the order of magnitude will be the same.

Table 4.23. First and second critical sample masses for reflection mode Bragg-Brentano geometry (CuK_α radiation)

Material	μ^* cm^{2}g^{-1}	First critical mass G'_{cr} (mg)	Second critical mass (µg)	
			$(G''_{cr})_2$	$(G''_{cr})_{10}$
$(CH_2)_n$	4.80	139	1875	8330
$Mg(OH)_2$	23.90	28	376	1670
SiO_2	35.30	19	254	1126
$CaCO_3$	75.50	8.9	120	530
TiO_2	127.40	5.2	70	315
$BaSO_4$	227	2.9	40	175
CeO_2	334	2.0	27	120

In transmission mode diffractometry, the first critical mass is not relevant because the sample cannot be infinitely thick. However, the second critical mass, corresponding to the situation in which no absorption correction is needed for the intensity equation, can be defined. However, in this case, this critical mass is related to the truly nonabsorbing sample. If the exponent in equations (3.34) deviates from unity by no more than 2%, then $\mu^* A/\cos\theta \leq 0.02$, and the critical loading is given by:

$$(A''_{cr})_{tr} = 0.02\cos\theta/\mu^* \tag{4.214}$$

Taking the irradiated surface as $S = B/\cos\theta$, where B is the beam cross section, we arrive at the following expression for the critical mass:

$$(G''_{cr})_{tr} = (A''_{cr})_{tr} \cdot S = 0.02\,B/\mu^*\cos\theta/\mu^* \tag{4.215}$$

In transmission geometry, nonabsorbing conditions are achieved at greater loadings than in the reflection case.

We can thus define a low-mass sample as one with a mass smaller than the first critical mass G'_{cr} for reflection geometry [equation (4.208)] or the critical mass $(G''_{cr})_{tr}$ for transmission geometry [equation (4.215)]. Although G'_{cr} does not exceed 10-20 mg for inorganic materials, except perhaps for compounds of light elements like Be or Li, the amount of material generating the diffraction pattern is much less than G'_{cr} in many important analytical cases and especially in the analysis of thin coatings or environmental samples.

The dependence of the intensity on the loading A is illustrated in Fig. 4.15 (CuK_α radiation, Bragg-Brentano reflection case).

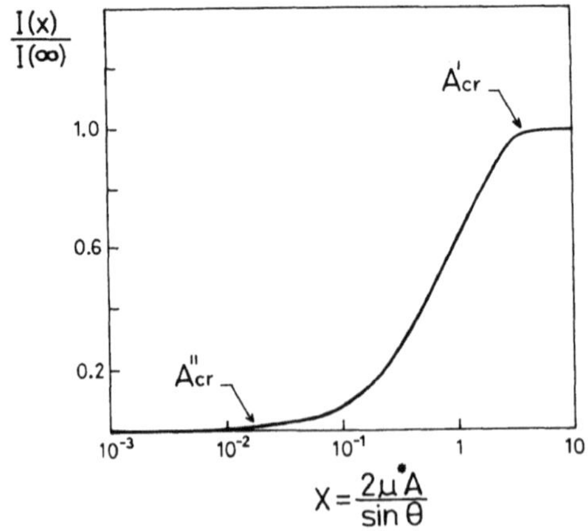

Figure 4.15. Dependence of the intensity on the loading $A = \rho t$.

Three different ranges are apparent:

1. $0 \le A \le A''_{cr}$; the intensity is a linear function of the loading A, and no absorption correction is needed.
2. $A''_{cr} \le A \le A'_{cr}$; the intensity varies with A according to equation (3.34), and a special absorption correction is needed.
3. $A > A'_{cr}$; the sample is infinitely thick, and a further increase of sample mass does not affect the diffraction intensity.

4.13.2 Analytical equations

Suppose now that the sample to be analyzed is a multiphase sample, A_j is the loading (g/cm^2) of phase j, and A (g/cm^2) is the loading of the sample material, then the intensity dependence on A_j retains the character shown in Fig. 4.15. However, we have to remember that the critical loadings A'_{cr} and A''_{cr} are loadings of sample materials and not of phase j. This means that the linear regression $I_j(A_j)$ is valid for a narrower range of A_j than for a single-phase material comprising only phase j.

The analytical equation in the range $A<A''_{cr}$ is:

$$I_{ij} = \frac{K_{ij} C_{ij}}{2\mu^*}\left[1 - \exp\left(\frac{-2\mu^* A}{\sin\theta_j}\right)\right] \rightarrow \frac{K_{ij} C_{ij} A}{\sin\theta_j} = \frac{K_{ij}}{\sin\theta_j} A_j \qquad (4.216)$$

where phase loading $A_j = C_j A$; the calibration constant K_{ij} has its usual meaning, defined earlier in equation (4.2); the sin θ term emphasizes that for low loadings ($A < A'_{cr}$) and constant-divergent geometry the intensity is θ dependent and is enhanced for low-angle reflections; and the θ-dependent term vanishes if a variable divergence slit is used (Kimmel 1987).

Equation (4.216) introduces a new analytical quantity, A_j, i.e., the phase mass per 1 cm^2, instead of weight fraction c_j in the conventional Klug-Alexander equation (4.2). Indeed, in the low-mass range, the intensity does not depend solely on the phase abundance c_j but on the product $A_j = Ac_j$. We can easily imagine that a sample with a very low loading A and a high concentration c_j would generate a diffraction peak of lower intensity than a sample with a lower weight fraction but higher loading A. In both cases, the weight fraction may be calculated if A_j and A are known.

In the sample loading range $A''_{cr} < A < A'_{cr}$, the intensity-weight fraction interdependence is also modified by sample loading A, which appears in the exponent in equation (3.17). In order to retain the common character of the analytical equations in both low-mass ranges of $A < A'_{cr}$, we will slightly modify equation (3.17) to introduce the term A_j, i.e., the loading of phase j, instead of the weight fraction c_j:

$$I_{ij} = \frac{K_{ij}\, A_j}{A\, \mu^*} \left[1 - \exp(-2\mu^* A/\sin\theta_j)\right] \tag{4.217}$$

which is reduced to equation (4.216) if $A < A''_{cr}$. This equation was developed for constant-divergence geometry when the sample intercepts the primary beam and the irradiated sample length shrinks toward higher scattering angles by a factor of $1/\sin\theta$. The irradiated surface might, however, remain constant if the sample is bathed in the primary beam or if the diffractometer is equipped with a variable divergence that maintains a constant irradiated area. In such a case, equation (4.217) should include an additional $\sin\theta_j$ factor:

$$I_{ij} = \frac{K_{ij}\, A_j}{A\, \mu^*} \sin\theta_j \left[1 - \exp(-2\mu^* A/\sin\theta_j)\right] \tag{4.218}$$

4.13.3 Analytical techniques

Let us now consider *b*-type samples (as defined in Section 4.13), i.e., thin coatings or powder depositions on membrane filters. Analytical techniques assuming mixing with a doped phase (such as an internal standard), spiking, or dilution are not relevant in this case. If the layer loading is below the second critical value, no absorption correction is needed. Conventional analytical practice involves preparation of a calibration function (probably a straight line) of diffracted intensity versus phase loading A_j. Samples with known loading A_j are prepared by deposition of a pure phase j from a liquid or gaseous (air) suspension (Altree-Williams et al. 1977). Complete analysis of the samples along the lines discussed in Section 4.6 can also be carried out. The intensity ratio for the thin layer of a 1:1 mixture of two phases is given by:

$$\beta'_{jr} = K_j/K_r\,(\sin\theta_j/\sin\theta_r) = \beta_{jr}\,(\sin\theta_j/\sin\theta_r) \tag{4.219}$$

where β_{jr} is the corresponding ratio for an infinitely thick sample, and the multiplier $(\sin\theta_j/\sin\theta_r)$ appears for constant-divergence geometry. We must remember that the abundance of any phase $c_r = A_r/A$. Then,

$$\sum_r c_r = \sum_r A_r/A = \frac{1}{A} \sum_r I_r \sin\theta_r/K_r = 1 \tag{4.220}$$

where the summation is carried out over all the phases detected in the sample. Going back to the basic equation (4.216), we obtain an expression for the loading of any phase j:

$$A_j = \frac{A}{1 + \sum_r \beta_{jr} (I_r/I_j) (\sin \theta_r / \sin \theta_j)} \tag{4.221}$$

where sample loading A must be calculated from the sample mass G and the area covered by the sample S: A = G/S. If the irradiated sample surface is kept constant, then the $(\sin \theta_j/\sin \theta_r)$ factor vanishes from equation (4.221).

The general case of layer loading, $A < A'_{cr}$, requires a more complicated intensity equation (3.17). In this case, the product $A\mu^*$, which characterizes the layer absorption, must be acquired for each sample.

If the material to be analyzed is deposited on a low-absorbing support, such as a membrane filter, then its absorption may be calculated from the attenuation of a directly transmitted monochromatic beam. The intensity of the transmitted beam is given by:

$$I = I_0 \exp(-\mu^* A) \times \exp[-(\mu_{sp}^* A_{sp})] \tag{4.222}$$

where I_0 is the intensity of the incident beam without an absorber in its path, and μ_{sp}^* and A_{sp} are the mass absorption coefficient and layer loading, respectively, for the supporting material. If $\mu_{sp}^* A_{sp}$ is small (organic filter materials) and constant, then by measuring I and I_0 and using $\exp(-\mu_{sp}^* A_{sp})$ for blank filter measurements, we can easily calculate the product $\mu^* A$:

$$\mu^* A = \ln (I_0/I) - \mu^*_{sp} A_{sp} \tag{4.223}$$

For example, Millipore cellulose nitrate filters had a stable value of $\mu_{sp}^* A_{sp} = 0.040$ for $CuK\alpha$, and filters made of other cellulose derivatives showed $\mu_{sp}^* A_{sp}$ deviating by no more than 0.01 from this value. Fiberglass filters showed much greater absorption ($\mu_{sp}^* A_{sp} \approx 23$ for the filters we used), which was comparable to or even greater than the absorption of the deposited layer. The usefulness of fiberglass filters in conventional analytical procedures is thus questionable.

If the powder sample is supported on a flat polycrystalline surface, then the product $\mu^* A$ can be acquired from attenuation of the reflections from the support. Suppose that (I_{sp}) and $(I_{sp})_0$ are the integrated intensities of any reflection from the support with and without the sample placed on it, respectively, then:

$$(I_{sp})/(I_{sp})_0 = \exp(-\mu^* A / \sin \theta_{sp}) \tag{4.224}$$

where θ_{sp} is the Bragg angle for support reflection. The quantity $\mu^* A$ is easily calculated from this equation. This method was used for powder samples by Zav'lyalova et al. (1964) and more recently by Frevel and

Roth (1982). Altree-Williams et al. (1977) used this technique in the analysis of environmental samples collected on a Nuclepore membrane filter having low loading and low absorption. For diffraction measurements, the Nuclepore filter was placed on a silver filter, and the silver reflection was used for absorption correction. In this case $(I_{sp})_0$ from equation (4.224) is the intensity of silver reflection attenuated by the Nuclepore filter. An in-depth description of the technical details for measuring absorption characteristics of deposited layers is given in Chapter 5.

As soon as the product μ^*A is known, equation (4.217) is transformed into:

$$J_{ij} = K_{ij} A_j \tag{4.225}$$

where $J_{ij} = I_{ij}\,\mu A/\,[1 - \exp(-2\mu^*A/\sin\theta_j)]$, and I_{ij} is the intensity of the analytical diffraction peak.

Subsequent treatment depends on the analytical problem. If only one phase has to be analyzed, the best solution is preparation of a linear calibration graph of J_{ij} vs. A_j. Standard samples needed to construct a calibration graph are prepared by deposition of various amounts of the phase to be analyzed, A_j, in the form of a pure phase or a mixture with other phases. If all the phases have to be analyzed, then by repeating the development of equation (4.221), we will obtain:

$$A_j = \frac{A}{1 + \sum_r \beta_{jr}\,(J_r/J_j)} \tag{4.226}$$

Let us now consider a low-mass sample of the type *a* (as defined in Section 4.13), i.e., a loose powder. If the powder can be evenly distributed over a flat support, then the theory developed earlier will be valid. In the application of the analytical equations developed in Section 4.13.2, we must be aware of the actual geometric scheme employed, i.e., constant beam divergence or constant irradiated area.

Loose powder, even in minute amounts, can be mixed with an internal standard, facilitating application of a powerful QXRD technique. It is appropriate to suppose that the powder is distributed over the small area that is bathed in the primary beam. In this case, the intensity of the analytical peak is related to the peak of the internal standard by the expression:

$$I_{ij}/I_{rs} = \frac{K_j A_j \sin\theta_j\,[1 - \exp(-2\mu^*A/\sin\theta_j)]}{K_s A_s \sin\theta_s\,[1 - \exp(-2\mu^*A/\sin\theta_s)]} \tag{4.227}$$

where A_s is the loading of the standard, and θ_j and θ_s are the Bragg angles of the respective reflections of the analyzed phase and the standard. The exponential terms in equation (4.227) indicate that the ratio I_{ij}/I_{rs} is dependent on sample absorption. However, if the chosen peak of the standard is close to the analytical peak and $\theta_j \approx \theta_s$, the exponential terms vanish, and a simple linear equation holds:

$$I_{ij}/I_{rs} = \beta_{js} A_j/A \tag{4.228}$$

For a reasonable value of μ^*A, I_{ij}/I_{rs} calculated from equations (4.227) and (4.228) will fall within a 10% difference if $[2\theta_j - 2\theta_s]$ is as large as 10°. Absorption compensation expressed in equation (4.228) is valid in a significant range of μ^*A values, making it especially important if the powder cannot be distributed evenly over the support. This situation arises when a sample of less than 1 mg has to be handled and analyzed.

For a more precise analysis, the quantity μ^*A from expression (4.227) must be determined by any independent method. Frevel and Roth (1982) in their excellent work on the analysis of semimicro quantities of materials (5-10 mg) mounted the powder to be analyzed in the recess of an aluminum plate, 1 cm in diameter and 0.01 cm deep. Attenuation of the Al reflections by the sample-internal standard mixture was used to determine the sample absorption coefficient. In a different method that is applicable to high-absorbing materials, the absorption coefficient μ^* may be calculated from the following equation:

$$\frac{I_{rs}}{(I_{rs})_0} = \frac{\mu_s^*}{\mu^*} \frac{A_s}{A} \frac{[1 - \exp(-2\mu^* A/\sin\theta_s]}{[1 - \exp(-2\mu_s^*(A_s)_0/\sin\theta_s]} \tag{4.229}$$

where $(I_{rs})_0$ is the intensity of the internal standard peak in a pure internal standard material filling the sample holder, with loading $(A_s)_0$, μ_s^* is the mass absorption coefficient of the internal standard, and A_s and A are the loading of the internal standard and sample-standard mixture, respectively. Frevel and Roth (1982) achieved remarkable precision for both the absorption coefficient and the phase abundances, sometimes running as high as a few percent.

5 Practical aspects of quantitative phase analysis

5.1 Introduction

In the previous chapters, we discussed the basic techniques and instrumentation for quantitative diffractometry. We now come to the next step in QXRD—the practical realization of the theoretical concepts outlined earlier. In other words, we have to delineate the techniques and conditions required to carry out this task with acceptable precision and accuracy. The quantity that we actually measure is the diffraction intensity, and the objective in this chapter is thus to determine the means of reducing intensity errors (both systematic and random). The pitfalls awaiting us in the practical accomplishment of QXRD can be roughly divided into three groups: instrument-, sample-, and phase-related errors.

Some intensity errors belong strictly to one of these three groups and can be reduced by controlling the relevant instrument-, sample-, or phase-related factors. Typical examples are instrument-related counting statistics errors and sample-related preferred orientation errors. The former can be reduced by increasing the intensity of the primary beam and introducing the correct counting strategy. The latter can be reduced by proper sample preparation. Other errors, for example, particle-statistics errors, are biased, and a number of different factors have to be considered in attempts to reduce them. Such errors can be reduced by sample diminution and by increasing beam divergence or by special movement of the sample during diffraction measurements. Especially important—but not always properly acknowledged—are phase-related errors. All QXRD techniques require scaling of the observed diffracted intensity against suitable reference or analytical standards. Proper scaling assumes full crystallochemical identity of the analyte and the reference phase. This requirement is, of course, not always met, and this fact must be taken into account in the development of any particular QXRD technique.

This chapter is devoted to practical ways of achieving the main goal of QXRD—phase quantification. The chapter includes a discussion of various instrument-, sample-, and phase-related factors and ends with a general discussion of the sensitivity, precision, and accuracy of the analysis.

5.2 Instrumentation

5.2.1 Introduction to instrumentation

The basic principles of powder diffractometer schemes are outlined in Chapter 3. Here, we will discuss some practical aspects of instrumentation in QXRD. Each scheme detailed in Chapter 3 has its virtues and drawbacks; thus, it is advantageous to have access to a variety of diffractometers. The Seeman-Bohlin scheme is the optimal scheme for thin films, for which it is important to exclude reflection from the substrate. Transmission geometry facilitates a decrease in instrumental and chromatic aberrations and may be advantageous in studying materials prone to preferred orientation. Debye-Sherrer geometry in combination with a position-sensitive detector is advantageous for analyzing microgram quantities. However, none of these schemes can compete with the Bragg-Brentano reflection focusing arrangement in terms of versatility. The latter scheme has a number of advantages in practical applications:

1. A wide range of scattering angles (from 2°-3° up to ~160°) is available.
2. The scheme is equally good for thin and bulk samples.
3. The scheme provides a high level of diffracted intensity with acceptable angular resolution.
4. For thick samples, intensity ratios do not depend on sample absorption, thus facilitating application of stored reference intensity ratio (RIR) data for phase quantification.

The number of diffractometers with Bragg-Brentano geometry is overwhelming. Smith (1993) estimated 15,000 such units are in operation, largely due to the efforts invested by Parrish in promoting this type of instrument. The material given in this section is devoted mostly to this versatile scheme, although some basic principles are, of course, applicable to any other diffractometric geometry.

5.2.2 Geometric alignment of the diffractometer

The arrangement given in Fig. 3.2 illustrates the proper alignment of the diffractometer. In the focusing plane, the backbone of the geometric arrangement constitutes a line that makes a particular (usually 6°)

takeoff angle with the anode plane and runs through the center of the focal spot and the center of the goniometer circle. In the zero 2θ position, the divergence slit, the receiving slit, and the antiscatter slit must be placed symmetrically around this central line, which must also pass through the sample surface. All the geometric elements, such as the focal spot and the divergence, receiving, and antiscatter slits, are extended in the axial direction, normal to the focusing plane. In a properly aligned diffractometer, there is a single plane normal to the axial direction that contains the central points of all the elements. In carrying out the alignment procedure, we have to bear in mind that the tolerance limits are rather stringent. If, for example, we want to keep the deviation of the zero position within a reasonable ±0.005°, then for a goniometer with a radius R of 175 mm, the tolerance of locating the receiving slit in the zero position (symmetrical towards the central line in Fig. 3.2) is approximately 2 μm. The tolerances for the divergence slit, the antiscatter slit, and the sample surface are somewhat less critical, although they are still impressive. For example, an angular error of ~0.005° will be produced by a sample displacement of 10 μm in the direction normal to its surface. After the mechanical zero has been established, the goniometer must be calibrated, i.e., the zero of the angular scale must coincide with the mechanical zero.

There are a number of alignment techniques proposed by the manufacturers of diffractometers that require the use of a variety of mechanical and optical tools. Descriptions of the techniques are to be found in the relevant manuals and in the literature (Parrish and Lowitzsch 1962; Schreiner 1986). We would, however, like to describe briefly the brilliant and extremely sensitive method proposed by Tournaire (1954) for finding the zero position of the receiving slit. The intensity of the primary beam is screened out by a nonpenetrating knife edge, which is used as the sample, and is scanned along 2θ angles around the zero point for two knife-edge positions that differ from one another by 180°. The 2θ value that corresponds to equal intensity readings at the two knife-edge positions is the required zero position. A comprehensive review of calibration methods for Bragg-Brentano geometry is given elsewhere (Schreiner 1986).

5.2.3 *Geometric aberrations*

Even in a properly aligned diffractometer, exact focusing of the diffracted rays (convergence to a point) occurs only for the rays parallel to the focusing plane and reflected by crystallites located on the focusing circle. These conditions are violated in real diffractometric schemes as a result of the flat (as opposed to concave) shape of the sample surface, of the penetration of X-rays inside the sample, and of any sample motion that causes the sample surface to deviate from being

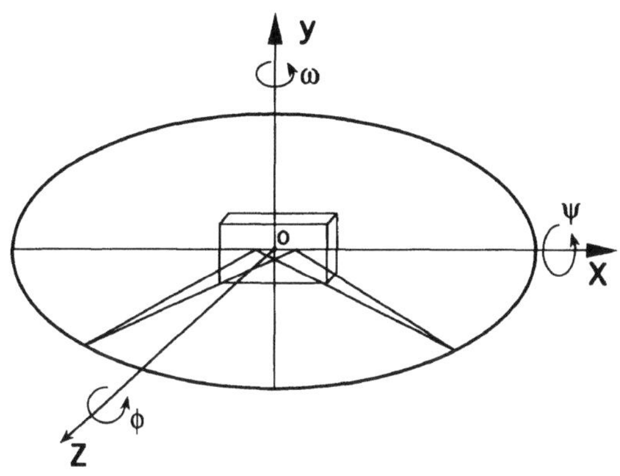

Figure 5.1. Possibilities of sample motion.

tangent to the focusing cylinder (Fig. 3.2b). Movement of the sample surface (Fig. 5.1) includes sample tilt around the OX axis normal to the goniometer axis OY and parallel to the focusing plane; sample oscillation around OY; or simply any misalignment in the 1:2 sample-detector setting. In practice, oblique beams (axial divergence of X-rays) are depressed by the Soller slits (a stack of plates parallel to the focusing plane) (Fig. 3.2a). However, axial divergence remains, causing noticeable defocusing at the low-angle side of the diffraction pattern. The manifestation of a deterioration in the exact focusing condition will be the shift of a peak centroid from its true position and asymmetric or symmetric peak broadening (Wilson 1963).

The full mathematical theory applicable to instrumental aberrations in powder diffractometry is given by Wilson (1963). Each aberration is characterized by a corresponding instrumental function $g(\delta)$, for which an angle $= 2\theta - 2\theta_c$ is the deviation from the position of the centroid $2\theta_c$ of the undistorted diffraction peak. The observed diffraction profile (Fig. 5.2) is the consequent convolution of the ideal profile $f(\delta)$, defined by the real crystal structure, with instrumental functions:

$$F(\delta) = \int_{-\infty}^{\infty} f(\delta - x)\, g(x)\, dx \tag{5.1}$$

It follows from this equation that the moments of $F(\delta)$ are the sums of the moments of the convoluted functions. Thus, the center of gravity (centroid) of the observed profile is given by:

$$(\delta_F)_c = (\delta_f)_c + \sum_i (\delta_{gi})_c \tag{5.2}$$

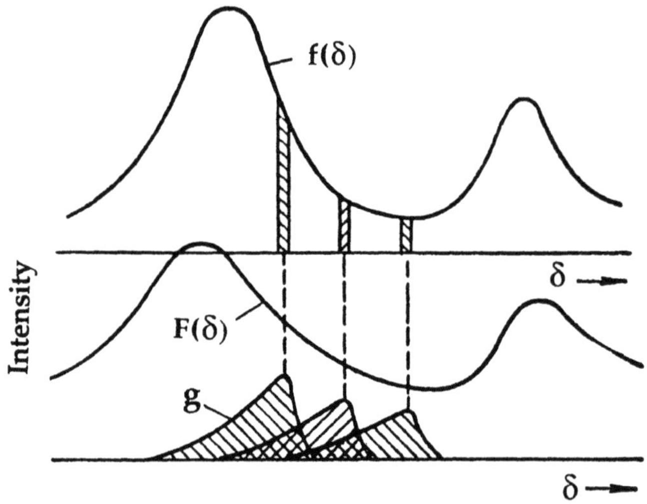

Figure 5.2. Distortion of the diffraction profile. Taken from Kheiker and Zevin (1963).

where $(\delta_{gi})_c$ is the center of gravity of the *ith* instrumental function. The same is true for the variance of F(δ):

$$V_F = V_f + \sum_i V_{gi} \tag{5.3}$$

where V_f is the variance of the ideal profile f(δ) and V_{gi} is the variance of the *ith* instrumental function.

Unfortunately, there is no simple way of calculating the shift of the peak maximum or the increase of its full width at half-maximum (FWHM). We have to bear in mind, however, that strongly asymmetric instrumental functions, such as axial divergence, that produce a significant shift of the peak's centroid might affect the peak maximum only to a small extent. It was found, for example, that under certain conditions the peak shift was five times smaller than the centroid shift (Zevin and Kheiker 1963a).

It might seem that the theory of instrumental peak distortion is relevant to phase identification (based largely on accurate d-values) and not to phase quantification (based largely on accurate intensity measurements), but this impression is not quite correct in light of contemporary trends in QXRD. There are a number of QXRD techniques in which the control of peak displacement and broadening are essential:

1. In the full-pattern approach, the pattern of the multiphase mixture is presented as a scaled sum of the stored individual patterns. It is important to minimize the probable systematic instrumental shift between the component patterns.

2. In the combined technique of structure refinement and quantification, it is important to minimize instrumental peak distortion.
3. In the profile-fitting technique, the number of adjustable parameters can be decreased if instrumental functions affecting the real diffraction profile are known.
4. Cluster separation, if needed, is facilitated by keeping instrumental broadening low.
5. In the quantification of solid solutions, it might be necessary to correlate cell parameters with phase composition. Accurate d-values are needed in this case.

A complete list of the sources of line shift and broadening may be found in the literature (International Tables for X-ray Crystallography, Vol. II 1959; Parrish and Wilson 1962; Wilson 1970, 1980).

Some instrumental functions, such as the width of the receiving slit, are symmetrical with centroid position $(\delta_g)_c = 0$. These functions do not produce peak shifts but do cause peak broadening. Instrumental functions, such as sample displacement or calibration error functions, may be considered Kronecker functions, which produce a peak shift as a whole but do not cause peak broadening. Finally, asymmetric instrumental functions, such as axial divergence, flat sample shape, or penetration functions, produce both peak shift and peak broadening. Jenkins (1983) separated instrumental aberrations into three main categories:

1. Inherent aberrations, such as axial divergence, flat specimen, finite receiving slit and focal spot widths, which are characteristic of the given diffractometer geometry;
2. Alignment errors, such as calibration errors, $\theta/2\theta$ missetting, differences in source-specimen and specimen-receiving slit distances; and
3. Experimental errors, such as specimen displacement and specimen transparency.

The three most prominent individual aberrations are:

- sample displacement in the direction normal to the sample surface
- axial divergence
- sample transparency for low-absorbing (organic) materials.

It seems that sample displacement is a major cause of the poor interlaboratory reproducibility often observed for lattice constants. Axial divergence affects low-angle diffraction peaks, giving them a characteristic asymmetric shape, with tails extended towards lower scattering angles. The extent of existing instrumental aberrations can be determined by running standard reference materials prepared by the NBS.

5.2.4 Variable instrument parameters

Some of the instrument parameters that define the shape and extent of instrumental functions (Wilson 1970, 1980) are not fixed but can be chosen by the diffractionist. The width of the receiving slit b_c and the angular aperture of the divergence slit α are usually among these parameters. In addition, in some diffractometers, the operator can choose the divergence of the Soller slits, and, as a rule, each manufacturer supplies a series of X-ray tubes with different focus dimensions (mainly focal spot width b_f) and, correspondingly, different powers. The question is how to delineate the optimal set of variable instrumental parameters. Before answering this question, we have to remember that the parameters discussed here influence not only instrumental aberrations but also the intensity of the diffraction peak. Thus, the choice of the minimal setting for various apertures can lead to an unjustified decrease of peak intensity and to poor counting statistics. There must, however, be some optimal relationship between the various instrument parameters that gives satisfactory peak intensity together with a reasonable reduction of peak distortion. For example, it is senseless to use a receiving slit width many times smaller than the FWHM of the diffraction peak.

An analytical approach to the problem based on calculating both peak distortion and peak intensity was proposed some time ago (Zevin et al. 1963; Kheiker and Zevin 1963; Zevin 1981). The principle on which this approach is based may be described as follows. Let us suppose that the intensity of the diffraction peak I is a function of n instrument factors a_i:

$$I = \varphi_1 (a_1, \ldots, a_n) \tag{5.4}$$

The distortion of the peak—for example, the variance σ^2—is a function of the same set of instrument factors:

$$\sigma^2 = \varphi_2 (a_1, \ldots, a_n) \tag{5.5}$$

Now, we can define the terms for the minimum peak distortion σ^2 on the condition that I is constant. We thus obtain $n - 1$ equations of the type:

$$\partial\sigma^2/\partial a_i = 0 \qquad 1 < i < n-1 \tag{5.6}$$

From these equations, the optimal ratios between the instrument factors can be derived.

Let us consider two variable instrument parameters, the receiving slit width b_c and the beam divergence in the focusing plane α. The corresponding instrument functions and variances are given as (Klug and Alexander 1954):

i) finite width of the receiving slit

$$g_1(\delta) = \begin{cases} \text{const for } -b_c/2R \leq \delta \leq b_c/2R \\ 0 \text{ for } -b_c/2R \geq \delta \geq b_c/2R \end{cases} \qquad \sigma_{g_1}{}^2 = b_c^2/12R^2 \tag{5.7}$$

ii) flat specimen aberration

$$g_2(\delta) = \begin{cases} \text{const} \times (-\delta)^{-1/2} \text{ for } s \le d \le 0 \\ 0 \text{ for } -s \ge \delta \ge 0 \end{cases} \qquad \sigma_{g_2}{}^2 = \frac{\gamma^4 \cot^2\theta}{0.45} \qquad (5.8)$$

where R is the goniometer radius and $s_2 = 1/2\ \alpha^2 \cot\theta$ may be regarded as the width of the instrument function g_2.

From equations (5.7) and (5.8), we can see that if the widths of the two functions g_1 and g_2 are equal ($s_1 - s_2 = b_c/R$), the corresponding variances are also approximately equal ($\sigma_{g_1}{}^2 = \sigma_{g_2}{}^2$).

The integrated intensity of the diffraction peak is a linear function of the beam divergence α (Parrish et al. 1966) and the receiving slit width b_c (Zevin 1981). Thus, the integrated intensity may be given by:

$$I = K\alpha\, b_c \qquad (5.9)$$

where K is a constant that does not depend on the variables α and b_c. By application of equation (5.6), we obtain the relationship between the two instrument parameters:

$$(\alpha)^o = 66.8 \sqrt{(b_c/R)\tan\theta} \qquad (5.10)$$

The irradiated length of the specimen is given by:

$$l_p = 1.65 \sqrt{(b_c/R)}\ \text{cosec}\,\theta \qquad (5.11)$$

These relationships are shown graphically in Fig. 5.3 for R = 173 mm. Due to the strong dependence of $\sigma_g{}^2$ on the scattering angle [equation (5.8)], the ratios α/b_c and l_p/b_c are also dependent on θ.

There are two major techniques employed in X-ray diffractometry: constant beam divergence and constant irradiated length. If the latter is used, then the optimal value of l_p for a given receiving slit width is almost constant for the entire scattering range $20° < 2\theta < 120°$. The optimal value of l_p is approximately 10-12 mm for $b_c = 0.2$ mm. In the range $2\theta < 20°$, this value changes sharply, and only approximate optimal values can be obtained, e.g., $l_p \approx 20$ mm for $b_c = 0.2$ mm. In this respect, the constant divergence technique is less versatile: only a step-wise change of α may be proposed. An idea of the length of the steps may be obtained from Fig. 5.4. For example, if the slit width $b_c = 0.2$ mm is used, the best approach to obtaining the optimal divergence α involves the following stepping change of α: α = 0.5° in the range $10° \le 2\theta \le 20°$, α = 1° in the range $20° \le 2\theta \le 60°$, and α = 2° in the range $60° \le 2\theta \le 100°$. In the range $2\theta < 10°$, the possibility of choosing proper values for the

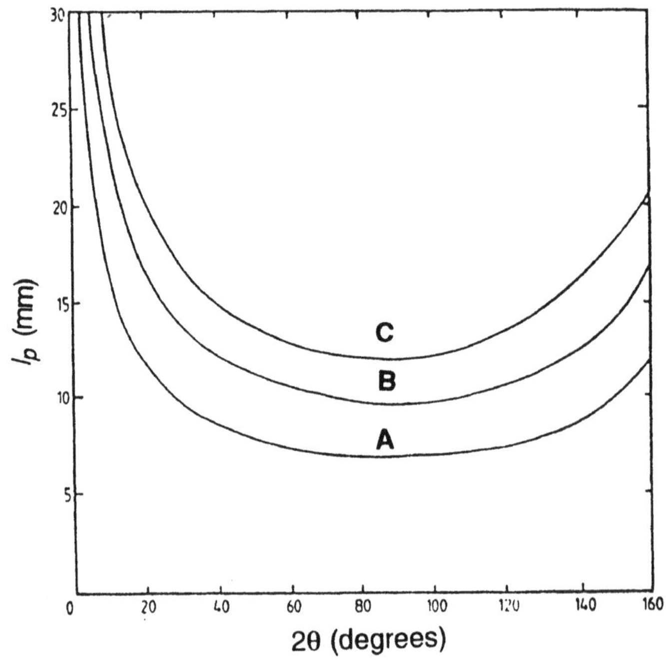

Figure 5.3. Optimal value of the irradiated length of a specimen l_p for a receiving slit width of A = 0.1 mm, B = 0.2 mm, C = 0.3 mm.

angular divergence α depends on the maximum length of the specimen in the focusing plane. If, for example, the maximum value of l_p is 20 mm, the application of equation (5.10) in the range $2\theta < 10°$ is hardly possible.

Additional examples of optimal instrument parameter ratios are to be found in the works of Zevin et al. (1963) and Zevin and Kheiker (1963a). The optimal ratio of the beam divergence (in the focusing plane) α to the divergence of the Soller slits δ is illustrated in Fig. 5.5. In the wide 2θ area, including the low-angle range, which is especially prone to instrumental aberrations, the ratio α/δ is almost constant and is equal to ~ 0.85. If, for example, Soller slit divergence $\delta \approx 2.3°$ (Philips diffractometer), then the optimal horizontal divergence α must be close to 2°. Unfortunately, implementation of this rule at low 2θ angles is difficult due to a sharp increase in the length of the irradiated sample l_p.

5.2.5 *Detectors and monochromators*

There is extensive literature on the construction and performance of X-ray detectors used in X-ray diffractometry (Klug and Alexander

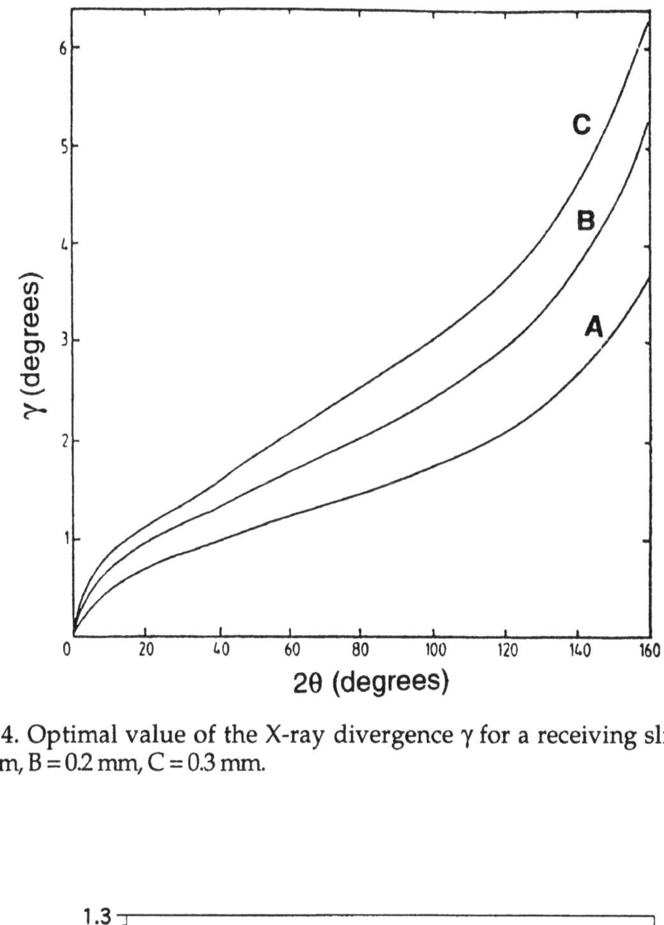

Figure 5.4. Optimal value of the X-ray divergence γ for a receiving slit width of A = 0.1 mm, B = 0.2 mm, C = 0.3 mm.

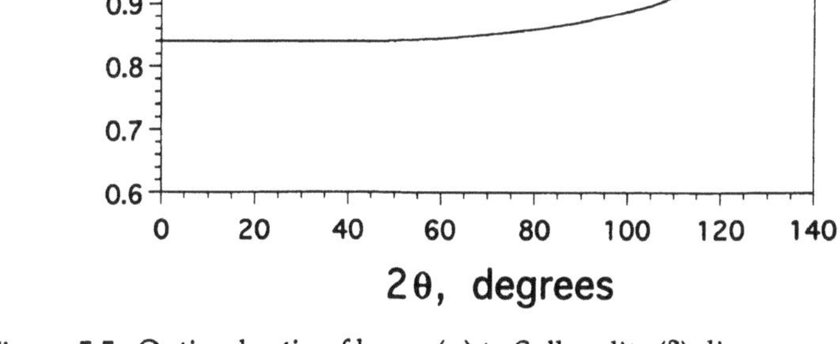

Figure 5.5. Optimal ratio of beam (α) to Soller slits (δ) divergences.

1954, 1974; Dowling et al. 1956; Kheiker and Zevin 1963). Here, we will discuss only the spectral characteristics of the various detectors, or, in other words, the ability of the detector to separate monochromatic radiation. Our aim at this stage of QXRD is the acquisition of a diffraction pattern of monochromatic, usually K_{α_1}, radiation. The radiation emitted by the sample that has been exposed to the X-ray source comprises the desired Raleigh scattering of K_{α_1} radiation as well as undesirable radiation composed of:

- scattered white radiation;
- fluorescence of sample elements excited by the incident beam;
- scattered K_β radiation; and
- scattered K_{α_2} radiation.

All these types of side radiation interfere with the basic diffraction pattern by increasing the background level and causing deterioration of the angular resolution.

Let us now consider the ability of X-ray detectors to screen out unwanted radiation. This ability is characterized by the energy resolution of the detector, i.e.:

$$W = \frac{\Delta E}{E_o} \tag{5.12}$$

where E_o is the energy for which the counting rate is maximal and ΔE is the FWHM of the counting rate as a function of the energy. The energy is determined in terms of the pulse height, as follows. If we suppose that the pulse height distribution has a Gaussian shape and that the energy resolutions are approximately equal for radiations with close energies, then in practice pulse height distributions for photons with energies $(E_o)_1$ and $(E_o)_2$ will not overlap if

$$(E_o)_1 - (E_o)_2 \geq 2\Delta E \tag{5.13}$$

If this condition is met, $(E_o)_1$ and $(E_o)_2$ can be separated. Typical values for energy resolution for CuK_α radiation are ~ 50% for a scintillation counter, 15% for a proportional counter, and ~2.5% for a solid-state [Si(Li)] detector. The corresponding energy bands ΔE are equal to ~4 keV, 1.2 keV, and 0.2 keV for these three types of detector. Photon energies for some radiations that can supposedly be separated from CuK_α are given in Table 5.1.

Applying condition (5.13), we find that scintillation and proportional detectors are suitable for separating white radiation but are generally unable to separate Fe fluorescence (Kheiker et al. 1962). The resolution ability of a Si or Ge detector, which is available in monochromatic diffractometers with a $2\theta/\theta$ setting (see Section 3.1), is good enough to

Table 5.1. Photon energies for various types of radiation

Radiation	E, keV	$E_{Cu} - E_i$*
CuK_α	8.047	
White radiation (maximum)	25–30	~20
FeK_α	6.403	1.64
FeK_β	7.057	0.99
CuK_β	8.904	0.857
CuK_{α_2}	8.027	0.020

*E_{Cu} is the photon energy for CuK_{α_1}, E_i are the photon energies for the corresponding radiations in this Table.

separate both close fluorescence and CuK_β, thus obviating the necessity for a β filter. The Si or Ge detector is, however, rather expensive and complicated in terms of maintenance. A very good substitute is to be found in a combination of a proportional or scintillation detector with a diffracted beam bent graphite monochromator. We can characterize the spectral quality of monochromator by the same concept of energy resolution that we used to characterize detectors. If $\Delta\omega$ is FWHM of the rocking curve of the monochromator crystal, then its energy resolution is given by:

$$W = (\cot \theta_m)\omega \tag{5.14}$$

where θ_m is the Bragg angle of the monochromator.

For a pyrolytic graphite crystal, ω is rather large, up to 0.4°-0.5°. If $2\theta_m \approx 26.6°$ (CuK_α), then the energy resolution $W \approx 3.3\%$ and the corresponding energy band is 0.26 keV. Thus, a graphite monochromator will separate any fluorescence radiation and CuK_β but not CuK_{α_2}. It must also be noted that the monochromator reduces the axial divergence of the reflected beam because it does not reflect oblique rays. Although the ability of the graphite monochromator to reduce axial divergence is slightly inferior to that of the 2.3° divergence Soller slit, the latter can nevertheless be removed from the diffracted beam if a monochromator is installed. In the case of the graphite monochromator, the intensity gain due to removal of the β filter and one set of Soller slits is greater than the intensity deficiency due to reflection by the monochromator crystal. Thus, a geometric arrangement that includes a reflected beam graphite monochromator facilitates a decrease in the level of background radiation combined with a gain in diffracted intensity.

Separation of the K_{α_2} component can be achieved only by a monochromator crystal with a narrow rocking curve. Usually, good-quality quartz crystals are used for this purpose. As a rule, reflection by this crystal is accompanied by appreciable intensity losses on the one hand and by an improvement in the shape of the diffraction peak shape on the other hand.

5.2.6 Fixed divergence versus fixed irradiated area geometries

In a conventional diffractometer with fixed beam divergence in the focusing plane (Fig. 3.3), the irradiated length of the sample is approximately:

$$l_p \approx R\alpha / \sin\theta \tag{5.15}$$

where we suppose that $\tan(\alpha/2) \approx \alpha/2$.

There are certain disadvantages inherent in this geometry. If the divergence α is chosen on the condition that $l_p \leq l_o$ at low scattering angles (l_o being the sample length), then at medium or high scattering angles $l_p << l_o$; this shows that the available sample volume has been neglected, which results in a corresponding decrease of diffracted intensity. On the other hand, if $l_p \approx l_o$ at medium angles, l_p might greatly exceed l_o at low angles, thus increasing the level of background radiation due to scattering by the sample holder and other parts of the sample assembly. The obvious advantage of this scheme is the angle-independent absorption correction, which, of course, holds true on the premise that the sample is infinitely thick and intercepts the primary beam.

Jenkins and Paolini (1974) proposed a variable divergence slit assembly, which maintains the irradiated sample length l_p constant throughout the whole diffractometric run, thus eliminating the disadvantages of fixed divergence geometry. This assembly has become very popular and can be found in most contemporary diffractometers. The expected ratio of intensities for the two geometries—fixed divergence (FD) and variable divergence (VD)—is obvious:

$$I_{FD}/I_{VD} = 2\alpha R / (l_p \sin\theta) \tag{5.16}$$

where l_p is the constant irradiated length in variable divergence geometry. If we compare the ratio of intensities for the diffraction peaks, for example, RIRs, then:

$$\frac{(RIR)_{FD}}{(RIR)_{VD}} = \frac{\sin\theta_R}{\sin\theta_j} \tag{5.17}$$

where $2\theta_R$ is the scattering angle for the reference (corundum) peak and $2\theta_j$ is the scattering angle for the analytical peaks of phase j. Equation (5.17) is very important because tabulated RIR values are usually obtained with fixed divergence geometry (Davis and Smith 1988) and must be converted, according to this equation, if variable divergence geometry is used in analytical practice.

Jenkins and Schreiner (1989), in a survey of a recent round-robin experiment undertaken by the JCPDS-ICDD, detected large systematic deviations of intensity ratios predicted by equation (5.17). The explanation of the observed discrepancy suggested by Bowden and Ryan (1991) lies in the probable variation of the diffracted beam as a result of the use of different antiscatter slits. In principle, the aperture of the antiscatter slit must match the aperture of the divergence slit (Fig. 3.3). Thus, in a variable divergence assembly the antiscatter slit must also be variable. If not, the divergence of the antiscatter slit must correspond to the maximum divergence of the variable divergence slit, say 4°.

5.3 Specimen preparation

Selection of a particular preparation technique depends on specimen factors, such as grain size and morphology of the particles, the amount of material available, and the amount of time required for the preparation. The latter factor is important for large-scale analysis, for which the time spent on specimen preparation may be important or even crucial. The optimal preparation techniques must be:

1. reproducible (i.e., ensuring minimal variability between replications);
2. unbiased (i.e., free of any systematic variations connected with specimen preparation, for example, variations in the degree of preferred orientation with phase abundance); and
3. simple and quick.

A large variety of preparation techniques has been suggested for QXRD. Each is suitable for a particular class of material, but there is no universal technique applicable to all materials. A critical discussion of the major specimen preparation techniques is given below [see also the review of Bish and Reynolds (1989)]. The discussion does not include a number of special and important specimen-handling techniques or "tricks" that may be found in the review of Smith and Barrett (1979).

Two kinds of samples are used in QXRD—powder samples, which are undoubtedly in the majority, and bulk samples, which cannot or should not be powdered. The latter are ductile metals, polymers, ceramics, or rocks, for which the composition of the surface layer or depth profiling are essential.

5.3.1 *Preparation of bulk specimens*

Bulk materials in the form of flat plates are well suited to X-ray examination. A sample of suitable dimensions is cut from the plate and inserted into the sample holder. Coatings on the flat surfaces of metals or other materials may be studied in the same way. Only in limited cases, when, for example, a strained coating must be studied in a relaxed form, is the coating removed from the substrate and examined in powder form. If the sample is a thin coating on a single crystal wafer, we must be aware that in this case the substrate acts as an analyzer crystal and might produce a spectrum of all the impurities of the anode in the ppm range. This complication of the pattern may be avoided if a diffracted beam monochromator is used or if the wafer is tilted by 1°-2° to slightly violate the $\theta/2\theta$ alignment (Zevin and Zevin 1987).

On many occasions, the sample to be examined is a small, irregularly shaped laboratory alloy casting. If a metallographic polished section is prepared from a bulk sample, then there is a ready flat surface for X-ray diffraction study. It is preferable to remove the sample from its mount because the mount usually gives intense scattering if the primary beam is not intercepted by the sample surface. If a metallographic section is not available, a flat surface must be specially cut.

The next step is mounting the sample and leveling the sample surface to the reference diffractometer plane tangent to the focusing cylinder. The simplest case is a flat plate with a recessed cavity, deep enough to accommodate the specimen. The specimen is held in the cavity by a piece of plasteline, and in the initial position the sample surface must project above the plane of the sample holder. A glass plate is put on the specimen from above, and its surface is made level with the plane of the sample holder by applying light pressure. The plate with the sample in the cavity is inserted into the goniometer sample assembly, as a conventional sample holder would be. For a small sample, the same operations of fixing and leveling are carried out more precisely by means of various bracket devices with fixed reference planes that are usually applied by major diffractometer manufacturers.

If the dimensions of the sample exceed those suitable for the diffractometer, specimens are cut from the bulk sample with a mechanical tool or spark cutter. The length and width of the specimen must be large enough to intercept the primary beam but small enough to fit into the specimen chamber. The ideal dimensions for a specimen plate are 20 x 30 mm x 2-3 mm (thickness). The irradiated surface of the bulk specimen must be flat. The necessity for additional finishing depends on surface roughness, since large surface irregularities cause peak shifts and peak broadening. Surface roughness can be estimated from equation (5.19)[1],

---

[1] The derivation of equation (5.19) is given in Section 5.3.5 in the description of side packing.

assuming $\Delta 2\theta$ is the additional broadening and S the depth of a scratch. If the FWHM of the peak is no less than 0.12°-0.15°, a surface roughness of 0.02-0.04 mm is acceptable.

For ductile metals, there is an additional consideration: the cutting of such metals produces a thin layer of plastically deformed metal on the cut surface. Due to the low penetration of X-rays, this layer might contribute considerably to the diffraction intensity, causing very significant peak broadening. We have even found this effect in the spark cutting of Ti-based alloys. The thin plastically deformed layer can be removed by chemical or electrochemical polishing rather than by etching, which is not uniform.

Bulk metallic samples have to be checked for grain size. It is not rare to find some grains as large as a millimeter, which makes the sample completely unsuitable for QXRD. One of two X-ray techniques may be used for investigating grain size: slow sample rotation (spinning) in the plane of the sample or a photographic technique that reveals coarse grains as spottiness of the diffraction line.

If a brittle material such as a rock sample or a ceramic material is to be studied as a bulk sample (as opposed to a powder), a flat specimen 3 mm thick should be cut from the bulk sample, usually with a diamond saw. If the material is not sufficiently strong, it can be glued to a glass plate and reduced to a thickness of 1–2 mm by the conventional technique used in petrography. Thin petrographic sections may also be suitable samples for diffraction study and may even satisfy the condition for infinite thickness. Of course, thin sections should be run before covering with a glass slide.

In general, it should be remembered that bulk samples may be predisposed to preferred orientation, which cannot be eliminated by any sophisticated sample preparation technique, as is possible for powders. Thus, a check for preferred orientation must be performed in QXRD of bulk samples.

It may be necessary to handle the material to be analyzed in the form of a powder, either to obtain the required grain size or to prevent preferred orientation, or simply because a representative sample may be more easily prepared from a powder than from the bulk material. For brittle materials, it is preferable to grind them into fine powders and then to prepare powder samples for QXRD.

5.3.2 *Sampling of powders*

A few grams of fine-grained material are usually needed for specimen preparation. It is, of course, impractical to grind the whole sample to optimal fineness, since its mass is often much larger than the quantity required for analysis. Thus, a certain sampling procedure should be carried out. Sampling is a general analytical problem that is well known in analytical chemistry, X-ray spectroscopy and many other analytical techniques. The major steps of sampling include comminution of the initial coarse-grained, and probably nonuniform, material in series of grinding apparatuses, such as jaw-crushers, rod mills, and ball mills. The small quantity (a few grams) of the material needed for further fine grinding is obtained by quartering or mechanical splitting of the initial coarse-grained material. Quartering is performed as follows. The powdered specimen is poured onto a flat surface to form a cone. The cone is divided into four quarters by means of a straight plate, and two opposing quarters are rejected. The remaining two quarters are mixed together and formed into another cone. This procedure is repeated until the desired specimen size is obtained. When mechanical splitting is used, the initial powder sample is first separated into two or more identical parts, and then the desired mass of material is obtained by repeated splitting. The splitting devices and the procedure itself are standardized by the American Society for Testing and Materials.

5.3.3 *Fine grinding of powders*

5.3.3.1 General remarks

We will discuss here the final grinding of a small amount of material (usually less than 10 g) needed for accurate intensity measurements. Let us start by defining the various bodies found in a powder:

- A *grain* is a mechanically free particle present in the powder.
- A *crystallite* is a continuous single phase that occupies a part or all of the grain volume.
- A *coherent domain* is a crystallite with a perfect or almost perfect crystalline structure that occupies part or all of the crystalline volume.

There are three major size-related phenomena that affect intensity measurements in powder diffraction:

- Reduction of the intensity due to primary extinction;
- Statistical fluctuations of the intensity from sample to sample due to variations in the number of reflecting crystallites; and
- Systematic variations in the intensity due to microabsorption.

The visible effect of grinding is the reduction of grain size, but we can easily appreciate that this operation need not necessarily affect crystallite size or coherent length. Indeed, crystallite size will be effectively reduced only for single-phase grains. For example, a grain of sedimentary phosphate (from the Negev, Israel) contains tiny (up to 1000 Å) fluorite-carbonate-apatite crystallites embedded in a calcite matrix. Due to the small crystallite size, diffraction by apatite does not exhibit extinction and crystallite-statistics effects. Slight peak broadening is a good indication of the small (less than 1 μm) crystallite or coherent domain size. It is a good rule to compare the diffraction peak profile of the analyte with a well-crystallized material, such as silicon or quartz.

Microabsorption is probably a major factor affecting accuracy in the acquisition of intensity data, and the best way to eliminate microabsorption effects is size reduction. The crucial question is to what extent we have to grind the sample. In many cases, the sample must be reduced to 1 μm in order to satisfy Brindley's condition for a fine powder. This task is, however, impractical with currently available grinding equipment, since prolonged and intense grinding causes severe damage to the crystal. A reasonable and attainable upper grain-size limit seems to be 10 μm if microabsorption effects are expected or the sample contains large perfect crystallites. If neither of these conditions exists, less drastic reduction to 20-30 μm may be satisfactory, particularly for easily damaged organic crystals.

5.3.3.2 Grinding equipment

Agate mortars and pestles can be found in almost all powder diffraction laboratories, and hand grinding are often used to prepare material for diffraction examination. However, too much time and effort is usually required to reduce the powder by hand grinding to the fineness required for quantitative measurements (10 μm and less). Mechanical mortars and pestles imitating the hand-grinding operation are thus available. In two cases, however, hand grinding in a mortar remains indispensable: grinding of a small amount of material and grinding of very soft (organic) materials. In the latter case, gentle grinding is essential, and most mechanical devices severely damage these crystals.

In most mechanical grinding devices, the required grinding action is produced by freely moving milling pieces—balls or rods. A ball mill is the machine conventionally used to reduce large amounts of coarse materials to fine powders, for example, in the cement industry. The ball mill consists of a steel or ceramic cylinder filled to ~50% of its capacity with steel or ceramic balls. The charge fills the space between the balls. The device rotates at a speed of 70-80 rpm, and size reduction is achieved by impact and shear between the balls. However, if this equipment is reduced to laboratory size, its efficiency drops as a result of the inevitable drastic reduction in the mass of the balls and the dimensions of the mill. Thus, size reduction of a powder to 10 μm may take as long as 24 h in a laboratory ball mill. Grinding time may be reduced drastically by the use of a vibratory mill, in which the movement of the milling pieces is facilitated by vibration. The energy attained by a milling piece—a ball or a rod—is proportional to mf^2, where f is the frequency of vibration and m is the mass of the milling piece. The energy of such a milling piece in a vibratory mill is thus likely to greatly exceed that of a falling ball in a ball mill.

A disk mill comprises a cylindrical chamber in which is placed a second cylinder—the milling piece—of slightly smaller dimensions. The charge of material fills the space between the milling piece and the chamber. A variety of disk mills with chambers of various dimensions and materials are available on the market today.

Depending on the type of mill, reduction of the coarse powder (0.1-0.5 mm) to a fine powder (below 10 μm) may take up to 20 min. However, time efficiency is only one characteristic of the mill. No less important is the resultant particle-size distribution of the reduced powder. A narrow distribution is desirable, because a significant fraction of very small particles (below 1 μm) leads to peak broadening or, in other words, to a significant transfer of intensity to the peak tails. The outcome is overestimation of the background and a systematic reduction of the integrated intensity. The same result may be expected from a strained outer layer of particles, which constitutes a progressively greater fraction of small particles. Unreduced or coarse particles will contribute to extinction or nonreproducibility of the intensity.

O'Connor and Chang (1986) compared the action of two types of vibratory mill—disk type and rod type—and found wet milling in a McCrone rod mill to be superior (5 to 10 min). There was a marked difference in particle-size distribution produced by dry milling in a disk mill and wet milling in a McCrone rod mill. In a comparison of a number of mills, Battaglia et al. (1990) also found distinct differences in grinding action. Maximal intensity reduction caused by supposed *amorphization* of the outer particle layers was found for powder ground in a jet mill or in a mortar by hand. Losses were lower for two types of vibratory mill—the vial mill and the ring mill.

5.3.3.3 Grinding media

Grinding may be carried out in the dry state or in slurry (when any liquid is added to the chamber of the mill). Wet grinding is usually superior for the following reasons. First, particle-size distribution is narrowed in wet grinding. It seems that this uniformity of particle size is achieved as a result of free movement of the particles in the slurry: no material escapes grinding action by concentrating on the lid or the bottom of the chamber. Second, the crystals are less damaged by wet grinding than by dry grinding, perhaps due to smaller shear strains in the thin outer layer of the particles. Third, though less important, wet grinding is faster than dry grinding. Wet grinding, however, involves an additional technological step—recovery of the powder from the slurry. Nonetheless, since wet grinding is superior, it is highly recommended. The availability of the wet grinding option must thus be one of the criteria considered when choosing grinding equipment.

The liquid used in grinding must be inert to the material to be ground and must have a low boiling point. The latter is essential in the recovery of the powder from the slurry, since the residual liquid is removed by gentle heating. Commonly used liquids in grinding technology are low-boiling-point alcohols (methanol, ethanol, propanol, iso-amyl alcohol), cyclohexane, acetone, and water. Chambers and milling pieces are manufactured from a variety of hard materials: tungsten carbide, agate, corundum, mullite, steel, and porcelain. Tungsten carbide tools are excellent but rather expensive, since both the hardness and the specific gravity of tungsten carbide ensure high impact energy in vibratory mills. Softer materials, such as porcelain or steel, may contaminate the material (although contamination is not such a critical factor in powder diffraction, which has a lower sensitivity than some spectroscopic methods that have detection limits in the ppm or ppb range).

Fibrous or plate-like materials (e.g., chrysotile, asbestos, and mica) pose a special problem in fine grinding. Due to strong bonds in the direction of the fiber or parallel to the plate, conventional grinding produces thin plates or fibers rather than a powder. McCrone Research Associates reported the effective grinding of biotite in a micronizing vibratory mill. The raw powder that passed through a 0.4-mm sieve was ground in propanol to a mean size of 5 μm in 10 min.

5.3.3.4 Control of grinding

Since grinding is such an important step in quantitative diffractometry, control is essential in the development of an optimal procedure. In this context, our aim is to determine the time needed to grind the powder to the required fineness, say grain size of less than 10 μm, in a given grinding machine. To this end, we require a reliable means of measuring particle size. Sieving, a technique commonly used for size classification, is not much help here, since the size range frequently needed—less than 10 μm—is beyond the practical sieving range. Even sieving through a "large" 325-mesh sieve (mesh size 44 μm) is time consuming and tedious. Sieves with smaller mesh sizes do exist, but special equipment is required for their use. A number of automated (but expensive) instruments are available for particle-size measurements in a grain-size range greater than 1 μm. These include such excellent instruments as the Sedigraph, which is based on particle sedimentation, and the Coulter counter, which is based on particle counting. With both of these instruments, particle-size distribution may be obtained in less than an hour. A simpler way of controlling the comminution process is measurement of the specific surface area by air permeability. This method is used to characterize the fineness of cement powder (Blain instrument, ASTM). Although the mean particle size can be evaluated only very approximately, the progress of grinding with time may be easily controlled. Finally, microscopic observations of the ground powder are, of course, very useful for following the grinding process.

Another way of controlling grinding is direct measurement of the diffracted intensity. If the initial nonground material has large coherent domains, significant nonreproducibility of intensity measurements due to poor crystallite statistics may be expected. Poor intensity data may be revealed by consequent intensity measurements of replicate specimens or by rotation of a single specimen around the sample normal. Thus, the coefficient of variation of such measurements may be used as a direct indication of the degree of comminution. Intensity fluctuations during rotation of the quartz sample are shown in Fig. 5.6. The coefficient of variation falls progressively with grinding time and increasing specific surface area of the powder (Table 5.2). Intensity measurements become reproducible within a few percent if the size of the coherent domain does not exceed several microns. Such domains are small enough to exhibit any effects of extinction. Thus, the procedure described above can be used to follow comminution when the effects of extinction and particle statistics are depressed.

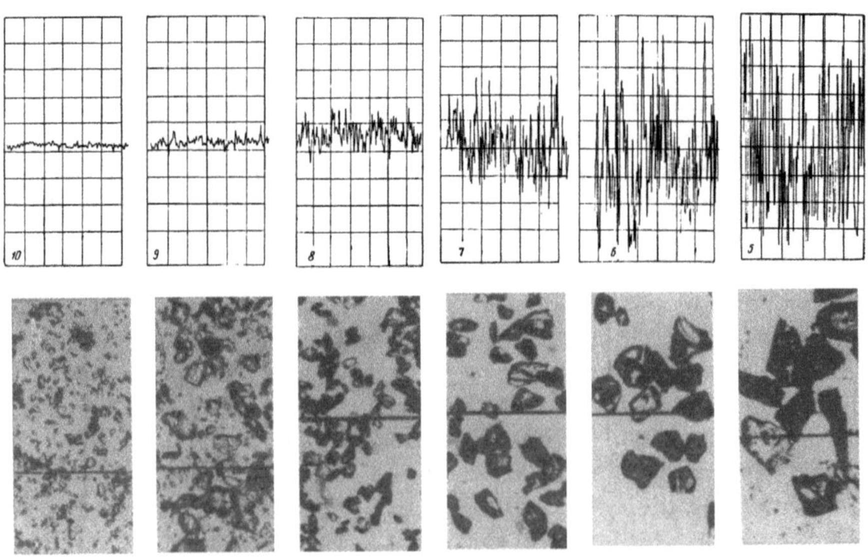

Figure 5.6. *Top*: Intensity vs. sample position during rotation at a fixed Bragg angle ($2\theta = 26.66°$) for crystal (grain) sizes of 6.3, 13.0, 21, 47, and 63 μm. *Bottom*: Corresponding photomicrographs.

Table 5.2. Fluctuations of intensity of the $10\bar{1}1$ quartz peak

	Grinding time, min			
	3	6	10	5
Specific surface area, cm^2/g^*	1200	1200	3100	4500
Coefficient of variation, %	12	8	5	3

*The specific surface area was measured by air permeability.

5.3.3.5 Grinding-related effects

Prolonged and intense grinding gradually destroys the perfect atomic or molecular arrangement in a crystal and may ultimately produce amorphous material, especially in organic crystals, which are built up of weakly bound molecules. Peak width increases gradually as crystallite size decreases and lattice strains accumulate. A similar process takes place in inorganic crystals with strong atomic bonds, but greater energy is needed for amorphization. This destruction of crystal structure starts from the particle surface, as is evidenced by strains. Although strained, the crystal will usually not exhibit decreasing integrated intensity of the diffraction peaks, but it will certainly exhibit peak broadening. As noted by Suortti and Jennings (1971), overestimation of the background in this case leads to an apparent decrease of the integrated intensity. The process of straining continues up to real amorphization. It can be expected that straining and amorphization will be stronger in fine particles with a larger ratio of surface area to particle volume. The reduction of intensity as particle size decreases is associated with the formation of an amorphous shell. Etching with hydrofluoric acid removes the amorphous material, resulting in a remarkable increase of peak intensities. The reduction of peak intensity for the coarser particles is associated with the extinction effect. The size range relatively free of both of these effects is 5-20 μm. Other workers have also reported a considerable (up to 10%) reduction of integrated intensities with grinding time for quartz (Altree-Williams et al. 1981). Although most work has been done on quartz, there is little doubt that the same effects will be found for other minerals and chemical compounds. Moreover, the expected amorphization and intensity losses may be greater for softer phases, especially if they coexist in a sample with harder phases. Two important conclusions may be drawn from these observations:

1. Grinding conditions must be constant for all batches of samples to be analyzed, including analytical standards.
2. Overgrinding must be avoided, and the requirement for grinding to a particle size below 8-10 μm must be carefully evaluated: an

examination of intensity variations versus grinding time, as described earlier, is a good check for optimal grinding.

5.3.4 *Powder aggregation and mixing*

5.3.4.1 Powder aggregation

Disaggreggation or comminution is not the only size-related operation involved in the preparation of powders. In some cases, the opposite operation, namely, *powder aggregation* is essential. In powder diffraction, this technique is used for platelike or fibrous particles, which are sometimes combined into spherical aggregates in order to avoid preferred orientation in specimen mounts. (Such aggregation is, in fact, known in ceramic technology—tiny particles are aggregated into spheres in order to improve filtration characteristics or the ability to fill a die.) The process employed is spray drying, in which a slurry with a low solid content is sprayed into a heated space. Spherical liquid droplets are evaporated in the air before they fall on a collecting surface, producing particle aggregates of almost spherical shape. Flocculation in the slurry must be prevented (by adding deflocculant) since floccules interfere with formation of spherical aggregates. The spherical particles then require further processing, such as collection and sample mounting. In order to ensure mechanical strength of the particles after drying, a binder is added to the slurry liquid.

Flörke and Saalfeld (1955) used this method as long ago as 1955 to produce spherical aggregates of clay particles. A suspension of 1 g of solid material in 10 ml of solution (2% cellulose acetate in acetone or 3%–4% of polystyrol in benzene) was sprayed into a glass box. Droplets congealed in midair, and solid spherical aggregates were collected on the bottom of the box. Later, Hughes and Bohor (1970) proposed a simple device for producing nearly spherical aggregates from clay particles less than 2 μm in size. A suspension of clay particles in a nebulizer was sprayed into a heated glass tube by means of compressed air. Particle aggregates were then collected in a paper bag connected to the end of the glass tube. The diffraction pattern obtained from the spherical aggregates (well-crystallized kaolinite) was in good agreement with the calculated pattern, a fact that confirms good randomization.

The spray-drying aggregation technique was further developed by Snyder and his co-workers (Smith et al. 1979a,b), who extended it to fibrous and rodlike materials and carried out a careful comparison of calculated and experimental patterns. The results obtained by Smith et al. (1979a) for siderite are especially impressive (Table 5.3). From Table 5.3, we can see that the diffraction pattern for the nonaggregated sample exhibited the characteristic strong orientation of the 104

Table 5.3. Diffraction pattern of siderite $FeCO_3$

d(Å)	h	k	l	Nonspray-dried powder	Calculated pattern	Spray-dried powder
3.5900	0	1	2	3	52	34
2.7910	1	0	4	100	100	100
2.3440	1	1	0	3	17	24
2.1320	1	1	3	41	16	17
1.9627	2	0	2	2	21	19
1.7950	0	2	4	2	9	8
1.7371	0	1	8	6	17	19
1.7296	1	1	6	1	27	27
1.5269	2	1	1	1	2	0
1.5048	1	2	2	6	11	0
1.4378	1	0	10	1	2	2
1.4251	2	1	4	4	7	8
1.3956	2	0	8	0	3	4
1.3806	1	1	9	0	2	2
1.3730	1	2	5	0	1	2
1.3533	3	0	0	0	7	10

Taken from Smith et al. (1979a).

rhombohedral plane and very poor agreement with calculated intensities. For the spherically aggregated powder, on the other hand, the measured and calculated intensities are in good agreement.

Calvert and his co-workers (Calvert and Sirianni, 1980, Calvert et al. 1983) proposed a technique for spherical aggregation in the liquid phase. The fine powder was suspended in a liquid, and a small amount of immiscible liquid, one that preferentially wets the solid, was added. Spherical aggregates were then produced by shaking the suspension. To produce spherical aggregates of platelike MoO_3, 1 g of powder was suspended in 25 ml of a 1:1 mixture of hexane and octane with 0.02 ml of MeOH. Water (1 ml) was added to the suspension, and spherical aggregates were obtained by shaking in a Spex blender.

It should be remembered that the spray-drying technique is influenced by operating conditions, such as air pressure, viscosity of the slurry, and chamber temperature, and is thus more complicated than many other sample preparation techniques.

5.3.4.2 Powder mixing

This operation is often encountered in sample preparation for QXRD. A number of techniques—internal standard, addition, and dilutions—involve the mixing of each sample with a certain amount of the internal standard, a neutral diluent, or the phase to be analyzed. The mixing operation must produce a powder mixture of uniform composition for volumes smaller than the volume irradiated by the X-rays. If the penetration depth of the X-rays is ~10 μm, then the irradiated volume is of the order of 10^{-3} cm^3. Usually, the overall volume of the mixed component is several cubic centimeters, and due to the obvious demand for uniformity, each 10^{-3} cm^3 fraction of the mixed material must have a composition identical to the macroscopic composition of the mixture.

A variety of factors affect the mixing of powders, the proportions of the components, the fineness of the powders, and the differences in the densities of the components. There is usually no problem in mixing relatively large fractions of internal standard (10%-20%) or diluent (30%-50%) with the materials to be analyzed. Two components of a mixture may be combined before final grinding in order to carry out grinding and mixing simultaneously. However, if the hardness of the sample material differs significantly from that of the internal standard or the diluent, unwanted differential grinding may occur. The softer material may become amorphous at a time at which the harder material may not yet have been sufficiently ground. In this case, it is advisable to mix the materials after the grinding procedure.

The spiking of the sample to be analyzed with a small amount (a few percent or less) of pure phase may also pose a problem. For extremely low spiking, a two-stage mixing regimen is recommended. First, the mixture containing 10%-20% of the pure phase is prepared. This mixture is then used for preparation of the final low-percentage mix. The actual mixing is carried out in a plastic vial with plastic balls added as mixing aids. Vibratory or rotary laboratory mills may also be used for mixing powders. The usual mixing time is 5-15 min. A dispersing liquid is used to facilitate particle movement and hence to accelerate and improve mixing.

In developing a procedure for mixtures for large-scale analysis, it is advisable to carry out intensity control by measurements of replicate samples. Suppose that we measure the ratio of the intensity of an analytical peak to that of an internal standard. If S_2^2 and S_1^2 represent the dispersion of the results obtained with and without reloading the sample holder, respectively, the ratio S_2^2/S_1^2 will give us some idea of the uniformity of mixing. Let us look at the example given in Table 5.4, in which Portland cement is mixed with 20% of an internal standard, CaF_2. The value of S_1^2 is equal to $1.3 \cdot 10^{-2}$, which is close to the counting

Table 5.4. Analysis of the mixing of Portland cement with an internal standard

	Time of mixing, min:					
	1	5	15	30	90	120
$S_2^2 \cdot 10^{-2}$	6.3	22	3.6	0.6	1.2	0.8
$F = S_2^2/S_1^2$	7	24	4	0.7	1.3	0.9

statistics dispersion of $0.9 \cdot 10^{-2}$. The value of S_2^2 is $0.9 \cdot 10^{-2}$, and the ratio $F = S_2^2/S_1^2$ decreases progressively with grinding time (Table 5.4). The uniformity of the mix as the mixing progresses is estimated in terms of the F criterion. Six replicates were measured for S_2^2, and a Poisson dispersion of $0.9 \cdot 10^{-2}$ was taken for S_1^2. The tabulated value of $F(5;\infty) = 3$ at the 0.01 significance level. Thus, the optimal mixing time in the case under discussion is 15-30 min.

5.3.5 *Powder mounts*

On rare occasions, such as in the analysis of dust collected on membrane filters, the powder is ready for a diffractometric run. In the overwhelming number of cases, however, the powder must be prepared in special form, i.e., a powder mount. The particular technique chosen for the powder mount depends on:

1. The actual diffraction mode, i.e., reflection, transmission, or Debye-Sherrer;
2. The amount of sample available (from ≤ 1 mg up to many grams);
3. The morphology of the particles, i.e., equidimensional, plate-like or fibrous; and
4. The enhancing or depressing of preferred orientation of the particles.

We can list the following requirements for a powder mount:

1. The powder mount must be sufficiently strong mechanically to maintain its shape during the diffractometric run.
2. A minimum of foreign materials, such as covers and substrates, must be exposed to the X-ray radiation in order to avoid a possible increase in the background radiation and the appearance of extraneous diffraction peaks.
3. The requirement for an infinitely thick sample must be maintained for the reflection mode, if sufficient sample material is available.
4. The thickness of the specimen in the transition mode must be close to optimal.

A flat specimen with a surface of several square centimeters is ideal for conventional diffractometers with Bragg-Brentano geometry.

We will now proceed to a discussion of mount techniques. The cavity mount technique is the standard preparation method in powder diffractometry. A rectangular or circular cavity in a flat plate is filled up with the powder to be analyzed, and the sample surface to be irradiated is made flat. After the sample holder is placed in the goniometer, the position of the surface of the sample to be analyzed is tangent to the focusing cylinder. The material of the sample holder must conform to a number of requirements:

1. It must, of course, be mechanically strong enough to prevent any deformation during the sample mounting.
2. It must exhibit minimal self-scattering (fluorescent or coherent), since in certain circumstances the sample holder itself is subjected to irradiation. This often happens at low scattering angles, when the width of the primary beam is greater than the mount opening as a result of an improperly chosen divergence slit.

Light elements are thus preferable, and aluminum is the material commonly used for sample holders. Sometimes several Al peaks might thus appear on the diffraction pattern. Glasses or noncrystalline plastics are, of course, free from this drawback, although these materials might increase the background if they are irradiated at low angles.

Three techniques are used for powder packing in a cavity—front packing, back packing, and side packing—as described below.

Front packing

In this option, the cavity in the sample holder is packed from the side to be irradiated by the X-rays, i.e., the front. A slight excess of powder is poured into the cavity and pressed lightly with a glass slide or a small metal plate. The excess of powder is gradually removed with a knife having a straight sharp blade edge until the surface of the powder is level with the reference plane of the sample holder. A glass slide or any other type of plate often used for this leveling operation is not satisfactory, because with such a tool not only the excess powder but also powder from the cavity might be removed, leaving deep grooves on the sample surface. Even the knife edge may sometimes leave a damaged surface if the powder is susceptible to lumping.

In vertical goniometers, adhesion of the particles is usually strong enough to preserve the integrity of the mount during a diffractometric run. However, in horizontal diffractometers, especially those with sample spinning, some binder is needed. We have found that wetting the loose powder with a few drops of ethyl alcohol in the first step of sample preparation prevents the powder from falling out.

The main drawback of front packing is that nonisometric particles (plates, fibers) will tend to orient parallel to the sample surface. Smith and Barrett (1979) advise that the surface of the prepared sample be retouched with a serrated disk or a knurled surface to make small grooves that disturb preferred orientation.

Back packing

In this procedure, the rectangular machined hole of the sample holder is put down on the supporting flat plate, which has been filled up with an excess of powder. The powder is pressed slightly to ensure filling of the cavity and some adhesion between the particles. The excess powder is removed with a sharp-edged instrument, and the cavity is closed with a suitable thin plate. In the Phillips sample holder a thin elastic steel foil clamping into two slots in the sample holder plate closes the cavity. Thereafter, the sample holder and its supporting plate are turned upside down, and the supporting plate is carefully removed, exposing the front sample face.

The surface of the back-packed sample is flat and does not have the large flaws often encountered in front packing. However, the particles are also susceptible to preferred orientation. This may be diminished if the supporting plate is not strictly flat but has some microrelief, such as that found in serrated or frosted glass or simply in cardboard. This relief is copied onto the sample surface, destroying, to some extent, the preferred orientation of the nonisometric particles. If a dry mount is not strong enough, the powder may be wetted before packing with alcohol or a dilute solution of a binder.

Side packing

In terms of reduction of the preferred orientation of particles, this technique appears to be superior to the two procedures described above. A cavity 1-2 mm deep in the sample plate is provided with a side opening. A glass plate is clamped to the front of the sample holder. Powder is sieved into the cavity through the side opening. Some tapping is needed to densify the mount. Then the plate covering the front of the sample holder is removed so as not to disturb the front surface of the mount. No pressure is applied in the direction normal to the sample surface, and thus orientation of nonisometric particles is retarded. This method was suggested by Swanson et al. (1966), and later Calvert et al. (1983) showed its efficiency in the reduction of preferred orientation of the particles. Byström-Asklund (1966) proposed a simple device for reproducible filling of the sample holder. In their device, the glass plate covering the front of the sample holder has a coarsely ground surface to prevent packing of platelike particles parallel to the sample surface. Powder is sieved into the cavity through the funnel. During filling, the device is placed on a vibrator table, thus making the densification of the powder reproducible. After filling, the side opening of the cavity is closed with a rectangular rod. (A similar device was used by Byström-Asklund to fill a circular sample holder for a sample spinner.)

The apparent density for the front- and back-packing techniques is ~60% of theoretical value, leaving approximately 40% of the mount volume for pores. For the side-packing mount, the porosity is greater, being ~50% of sample volume. Two effects result from the substantial porosity of cavity mounts:

1. Intensity reduction, and

2. Increasing X-ray penetration due to decreasing the linear absorption coefficient, which in turn leads to diffraction peak broadening and displacement.

Intensity reduction in porous samples is not problematic in QXRD: due to the usually good reproducibility of packing density, the degree of intensity reduction will be almost constant for all analyzed samples.

The effect of increased X-ray penetration is substantial for low-absorbing organic compounds or light elements, for which penetration depth is relatively high, even for bulk samples. Significant peak broadening and displacement are to be expected if the conditions for infinite thickness and low absorption are to be met. Increased broadening is obvious for the cavity-mounted infinitely thick sample.

If we want to relate the peak position to a particular surface in a low-absorbing sample, it will certainly not be the sample surface tangent to the focusing circle but rather a surface inside the sample. On the assumption that this effective surface corresponds to the half-penetration depth, Frevel (1978) constructed an adjustable sample holder in which the position of the sample may be fixed so that the effective reflecting surface is tangent to the focusing circle. Thus, no peak displacement is observed. This technique does not, however, prevent peak broadening.

The effect of beam penetration is reduced for mounts of most inorganic compounds. The depth of half-attenuation of the diffracted beam may be calculated by:

$$S_{1/2} = (\ln 0.5/2\mu_a) \sin \theta \tag{5.18}$$

Thus, the deviation of X-rays scattered by crystallites at that depth ($\Delta 2\theta$) vs. crystallites on the sample surface may be calculated by equation (5.19), where the depth S is defined in equation (5.18):

$$\Delta 2\theta = \frac{2S \cos \theta}{R} = \frac{\ln 2 \sin 2\theta}{2\mu_a R} \tag{5.19}$$

where μ_a is the linear absorption coefficient for the powder mount. The value $\Delta 2\theta$ may also be used as a measure of probable peak broadening. Let us take a relatively low absorbing powder, such as MgO, Al_2O_3, or SiO_2, with an absorption coefficient for the bulk material of ~100 cm^{-1}. Suppose that the apparent density of the powder mount is 50% of the theoretical density and $\mu_a \approx 50$ cm^{-1}. Then, the value of $\Delta 2\theta$ calculated by equation (5.19) for $2\theta = 30°$, 60°, and 80° will be 0.011°, 0.020°, and 0.023°, respectively. Although these values are rather small, the effect may be noticeable. The effects of aberrations in the diffraction of light materials and a comparison between bulk and thin samples are given by Kimmel (1987).

For higher-absorbing compounds, the effect of X-ray penetration will be negligible if there are no large flaws on the sample surface. As was noted earlier, such flaws sometimes appear in front-packing techniques.

Let us now look at the amount of powder required for a cavity mount. If the opening in the sample holder plate is 20 mm x 15 mm and the thickness of the plate is 1-2 mm, then the volume of powder is 0.3-0.6 cm^3, or the mass of the sample will be from 0.5 g to a few grams, depending on the density of the material. Note that only a few milligrams of this material are actually irradiated by the X-rays to give the diffraction pattern.

A comparison of the three cavity mount techniques—front, back, and side packing—shows that in most cases the two latter techniques are superior. The main reasons are the smaller susceptibility to preferred orientation and the better quality of the sample surface (fewer irregularities). However, there is one case—the availability of only a small amount of powder—when front packing has a clear advantage. The depth of the cavity for front packing may be as small as 0.1 mm deep, and the amount of powder needed for the usual 20 mm x 15 mm opening is as little as 3.10^{-2} cm^3 or 100 mg or less. For most inorganic compounds, a powder layer 0.1 mm thick may be regarded as infinitely thick. (Such a small amount of powder is hardly adequate for side or back packing.)

5.3.6 *Pressing*

This technique of powder compacting is very similar to back loading of a cavity mount, but it is carried out with substantial pressure. This ensures higher apparent density of the powder mount and a flat front side that is subjected to X-ray illumination. On rare occasions, the green strength of the pellet is sufficient for the pellet to be self-supporting. As a rule, the powder is pressed into a circular holder or a soft aluminum cup. The pressing technique is used conventionally in sample preparation for fluorescent X-ray spectroscopy. Since pressing exaggerates the preferred orientation of nonisometric particles, it is suitable largely for materials not exhibiting orientation, such as cements. The advantage of pressing is that it is a reproducible technique that gives a constant apparent density of the powder and a good surface finish. Sometimes, even a degree of preferred orientation is reproduced with acceptable deviation. In that case, pressing can even be used for oriented sample mounts, as will be discussed in Section 5.3.9.

The pressure may vary over a wide range, usually between 10 and 100 kg/cm^2. The apparent density and strength of the pellet increase with increasing pressure, although there is no apparent influence on intensity. Hidalgo and Renton (1970), however, observed that the diffracted intensity increased with increasing pressure up to a value of approximately 400 kg/cm^2, above which the intensity remained constant as the pressure increased.

A cylindrical die with an inner diameter of 25-30 cm is required for pressing. Any manual hydraulic press with a total loading up to 20-25 tons is suitable for the technique. The thickness of the pellet must be 3-4 mm, and usually 0.5-1 g of sample material is needed to prepare a single pellet. A flaw often encountered in the pressing of a dry powder is lamination (i.e., the pellet becomes separated into layers). Lamination may be prevented by decompressing the die slowly over a period of 2-3 min after it has been subjected to maximal pressure for several minutes. An abrupt drop of pressure must certainly be avoided. If this procedure fails to prevent lamination, wetting of the powder with water or ethyl alcohol may be helpful. The usual practice in X-ray spectroscopy is mixing the material to be analyzed with a binder, such as cellulose powder. This practice is undesirable in our case, not only because of the extra time and labor required for mixing each sample but also because of the parasitic coherent and incoherent scattering caused by the binder.

If the amount of material available is not sufficient to prepare a self-supporting pellet, usually 2-3 mm thick, a two-layered pellet should be prepared. The die is filled first with the material to be analyzed, which is compacted under the chosen pressure. Then, the piston is removed, and the die is filled with a supporting material, such as plastic clay (montmorillonite). Finally, pressure is applied once again to produce a two-layered pellet. The outer layer that consists of the material to be analyzed may be as thin as a few tenths of a millimeter.

5.3.7 Dusting of loose powder on a plate

Dusting of loose powder onto a plate is a very simple technique that is suitable for small (milligram range) amounts of material. Although a glass slide is often used as the supporting plate in this type of sample mount, a single-crystal "backgroundless" plate is preferable. The latter is cut from a single crystal (quartz or silicon) in a direction that is not parallel to any crystallographic plane. The crystal plate ensures a significant reduction of background due to reduced coherent scattering.

In the dusting technique, the ground powder is dusted on the support plate through a laboratory sieve. The mesh size of the sieve must be such that powder passes through easily when the sieve is tapped with a metal spatula. During the dusting, the sieve must be oscillated in the horizontal plane to ensure uniform coverage of the plate.

In many cases, adhesion of the particles to the plate surface is strong enough to prevent them from falling off if a vertical goniometer is used and the layer is thin. If this is not so, then the plate surface should be covered with a thin layer of adhesive such as silicon grease or glycerol.

It must be remembered, however, that such an adhesive might not only increase the background level, but it might sometimes also produce its own diffraction pattern.

The dusting technique is very good for identification purposes when a limited amount of material is available or when low-absorbing (organic) materials are being studied. In the latter case, beam penetration of the conventional cavity mount causes considerable peak broadening and displacement. However, for quantification, certain conditions for sample absorption must be met, as discussed below.

We will characterize the amount of dusted material in terms of the surface density, i.e., the mass of material per square centimeter of sample surface. The surface density needed to ensure that the diffracted intensity is no less than 99% of the intensity diffracted by an infinitely thick sample may be calculated from the following equation:

$$\rho_s = (2.3 \sin \theta)/\mu^* \qquad (5.20)$$

For aluminosilicates with $\mu^* \approx 30$ cm^2/g for CuK_α ρ_s= 20 mg/cm^2 for $2\theta = 30°$ and 38 mg/cm^2 for $2\theta = 60°$. If the apparent density of the layer is $\rho_a \approx 0.6$ g/cm^3, then the thickness of the sample that approximates to infinite thickness ($\rho_s \approx 38$ mg/cm^2) is close to:

$$t = \rho_s/\rho_a \approx 6.10^{-3} \text{cm}$$

The linear absorption coefficient of the layer is thus as little as $\mu_a = \rho_a \mu^* \approx 18$ cm^{-1}. Due to the low absorption and the relatively large thickness, considerable peak broadening and peak displacement can be expected for dusted samples close to the "infinitely" thick condition. The effect is known as *transparency broadening* (Kimmel 1987, Kimmel and Schreiner 1989). Thus, the preparation of an "infinitely thick" dusted sample is practical only for highly absorbing materials with a mass absorption coefficient greater than 100 cm^2/g.

Another approach to the quantification of diffraction measurements is the preparation of a very thin layer with negligible absorption. Kimmel and Schreiner (1989) showed good agreement between observed and calculated intensities for very thin layers of Al_2O_3 and tungsten. If this technique is employed, the mass of the sample material in the supporting plate (or rather the surface density in g/cm^2) must be known, and uniform distribution of the sample material over the supporting plate must be ensured.

In any intermediate case between an infinitely thick and an "infinitely" (in the sense described above) thin layer, the sample absorption must be known. This information is usually obtained by measurement of the attenuation either of a transmitted beam (if the sample support is transparent enough to X-rays) or of a beam reflected by a crystalline support plate. Details of these techniques will be discussed in subsequent sections.

5.3.8 Powder deposition on a membrane filter

A thin, uniform layer of powder may be deposited on a membrane filter by suction of the air from a dust chamber in which fine particles are suspended. Altree-Williams et al. (1977) used this technique to prepare filter samples with a known mass of deposited quartz. Quartz dust of breathable size was generated in the dust chamber, and samples were collected with a Casella model sampler including a pump and sampling head.

Davis (1986) has proposed a similar technique for general sample preparation for quantitative analysis. His suspension (dust) chamber consists of a 500-ml gas burette without the upper stopcock. A sample of a few tenths of a milligram in weight, together with fine glass beads (less than 1 mm diameter), is loaded into the lower part of the chamber through a funnel. A flow of air from the pressure part of a vacuum pump produces a fluidized bed of glass beads, which aids in disaggregation of the powder particles. The upper end of the dust chamber supports a filter cassette fitted with 4.25- or 4.7-cm filters, usually glass fiber filters (Whatman GF/C). The filter cassette is connected to the vacuum pump. Suspended particles of any particular diameter may be separated out by adjusting the air flow with a rotameter. Thus, all particles up to a certain size will reach the filter surface. The uniformity of the deposition was demonstrated by the surface density of montmorillonite deposited on a glass-fiber filter. (This result is especially convincing if the tendency of montmorillonite to aggregate is taken into account.)

Another very important feature of filter deposits is the reduction of preferred orientation of nonisometric particles, which will be discussed in Section 5.9. There are, however, several disadvantages to the technique. The upward velocity of a particle depends on the particle mass and size, and the particle-size distributions for the various minerals in a ground multiphase rock sample are not identical. Moreover, the ability of the powder grains to penetrate inside the filter material will also depend on the shape and size of the particles. Thus, even for short-time (several minutes) deposition, significant particle segregation may be expected. Partial penetration of the material to be analyzed into the filter body, although essential for depression of preferred orientation, complicates the absorption correction. The situation is less complicated if the mass absorption of the sample material is much greater than the mass absorption of the filter material. In this case, filter absorption can be ignored. However, when the two absorption coefficients are close, as in the deposition of siliceous materials on a glass-fiber filter, a special absorption correction is needed.

5.3.9 *Preparation of oriented mounts*

Oriented mounts are mostly prepared for clay minerals. Identification and quantification of clay minerals is based on the analysis of basal (001) reflections, and their enhancement by preferred orientation is thus essential. In fact, it is much easier to prepare an oriented clay mount than to depress preferred orientation. Clay minerals exist as very fine particles mixed with very fine rock fractions. The first step in sample preparation is thus the removal of coarse particles, usually greater than 2 μm. This procedure is carried out by sedimentation. Details are to be found in standard textbooks (Jackson 1956; Brown 1961).

The three major techniques used to prepare oriented mounts of clay minerals are sedimentation, smearing of a slurry, and filtration. In the sedimentation technique, a standard microscope slide is loaded by means of a pipette with a few milliliters of the clay mineral in water. The loaded slide is left undisturbed until the water evaporates off, a step that may take up to 24 h at ambient temperatures. The resulting clay mineral layer is dense and even, with good adhesion to the slide. However, sedimentation of clay minerals in the layer of suspension loaded on the slide is not uniform, because the coarser particles settle first. Gibbs (1967) found significant segregation of clay minerals in the sedimented layer, with the lower layers being enriched with the better crystallized kaolinite and the upper layer with montmorillonite. Due to significant absorption in the deposited layer, segregation results in the enhancement of montmorillonite reflections and the depression of kaolinite reflections. To prevent this from happening, Gibbs proposed the technique of smear mounting of clay minerals. Clays are suspended in a small amount of water to give a thick slurry. A drop of slurry is placed on a microscope slide, and the suspension is smeared over the surface of the slide by passing the edge of another slide over it in one stroke. This type of mount does not exhibit any segregation of minerals, although the reproducibility of the results might depend on the skill of the person preparing the smear.

The filtration technique seems to be less dependent on the human factor. A concentrated suspension is filtered onto a membrane filter by vacuum suction. While still wet, the filter is placed clay side down on a microscope slide, and the membrane is peeled off. Settling of the particles takes a matter of minutes by this method compared to hours by sedimentation, and no mineral segregation has been detected in the former technique.

In the quantification of clay minerals, the first question is the reproducibility of their orientation. When one of the conventional comparative techniques is used for quantitative analysis, the diffraction peaks of the clay minerals in the analyzed sample are compared with the diffraction peaks of pure standard clays or of a mixture of standard clays. In order to ensure meaningful results, we have to assume intersample constancy of preferred orientation. This means that the preferred orientation of any given clay mineral must be independent of the sample composition, although the degree of preferred orientation can be different for different clay minerals.

When calculated patterns are used for quantification, the diffraction peaks of various minerals within the same sample may be compared. In this case, preferred orientation must not vary between minerals, i.e., various minerals must exhibit the same degree of preferred orientation, although preferred orientation may be different for samples of different compositions.

Unfortunately, there is no clear proof of constancy of preferred orientation in one sense or another for any of the mount techniques used in clay diffractometry. Additional experimental work is needed in this field.

5.3.10 *Handling of reactive samples*

In some special cases, samples cannot be examined in air because they are susceptible to oxidation, hydration, or carbonization. The simplest way to handle these samples is to seal them in a sample holder. For example, in the determination of the hydration of cement, it is imperative that water exchange with the atmosphere be prevented. A standard cavity mount is filled with a cement-water mixture of the required composition and sealed with a piece of Mylar film, which is fixed in place with polyethylene glue. In order to follow the cement hydration, the sealed specimen is examined periodically over a period of several months.

If the sample components are unstable in air, e.g., if they are highly hygroscopic, as are some fertilizers, filling and sealing of the sample holder must be carried out in a dry box.

For reactive materials, the sealing windows are made of a thin beryllium sheet, a mica sheet, or Mylar or any other type of thin plastic film. The window itself, although positioned ideally for X-ray diffraction, produces several unwanted effects. It absorbs both the primary and the diffracted beams, especially at low scattering angles, and it contributes its own scattering to the sample pattern. If the window material is amorphous, it will increase the background or if it is crystalline, it will complicate the diffraction pattern with additional peaks.

Another way to prevent sample interaction with air components is to pass a protective or inert gas, such as nitrogen, hydrogen, or helium or mixtures of these gases, through the sample chamber or to use the evacuated chamber supplied by many diffractometer manufacturers. Further details on the handling of reactive samples can be found in the review of Smith and Barrett (1979).

5.3.11 Handling of low-mass samples

A sample that does not meet the conditions of "infinite" thickness, defined in Section 4.13, is considered to be a low-mass sample. Depending on the material absorption, the upper range of a low-mass sample is 2-20 mg (Table 4.23). The lower range is defined by the mass detection limit of the diffraction system. Diffractometric studies of samples with masses of 1 μg and less have been reported (Baldock and Parker 1973; Zevin and Zevin 1987); this indicates that the diffractometric technique is as practical for microgram samples as the photographic technique.

A uniform layer of milligram-range samples can be produced by conventional front filling of a shallow cavity. To fill a cavity 1 cm in diameter and 0.01 cm deep, we need $8 \cdot 10^{-3}$ cm^3 or 15-20 mg of sample material. Frevel and Roth (1982) have described in detail the procedure and tools needed to fill a shallow cavity. To keep the powder in the cavity, a thin layer of Vaseline is allowed to form on the cavity bottom by evaporating a drop of 10% solution of Vaseline in cyclohexane. The sample and the internal standard are weighed in a small weighing bottle. The mixture is stirred with a tungsten needle or a glass or steel rod, and a rubber adapter holds the weighing bottle in place during the mixing operation. Vigorous comminution is achieved in a micropulverizer containing a ball and fitted to a vibrator. The mixed powder is transferred with a spatula to an Al sample holder, and the excess of powder is carefully removed with a clean, sharp razor blade.

The same tools are used to handle samples that are less than 1 mg in weight. However, this amount of material is not sufficient to fill a cavity, and it has to be spread over a flat support. The optimal support is provided by single crystals of, for example, quartz or silicon, cut in such a way as to prevent any reflections. In spreading the powder over the support, we have to minimize the scattering by the support. The irradiated volume of the support might exceed the mass of a minute sample many times and thus enhance the background, even if the support is a single-crystal plate.

Zevin and Zevin (1987) have suggested that the optimal area S occupied by the sample material may be calculated from the expression:

$$S \sin \theta = 10\, G\mu^* \tag{5.21}$$

where G is the sample mass and μ^* is the mass absorption coefficient. Increasing S above the value given by expression (5.21) will slightly increase the diffracted intensity, but the background will increase in proportion to the growth of S. Optimal spreading of the sample over a small area S requires corresponding collimation of the primary beam. Collimation is easily achieved if the goniometer is equipped with a variable divergence slit, but not all commercial diffractometers have facilities for proper collimation in the axial direction.

Minute amounts of powder can hardly be distributed evenly, as is achieved by filling a small recess in a flat plate. Even if the sample mass is so small (less than the second critical mass) that no absorption correction is needed, one cannot be sure that these conditions are met in all parts of an unevenly distributed powder. Thus, for quantification of samples less than 1 mg in weight, it is highly advisable to use internal standards.

In extreme cases of low masses, crushing and weighing of the sample and the internal standard as well as mixing are carried out on a support plate. The amount of internal standard can be increased up to 50% of the mixture, thus facilitating handling of minute samples.

5.4 Analytical standards

In order to quantify the intensity diffracted by an analyte, the phase is scaled (directly or indirectly) against an analytical standard. It is apparently assumed that the analyte and the analytical standard are identical in composition and crystal structure, i.e., that they are identical not only in some "average" structure but also in the character and concentration of lattice defects. In the standardless technique discussed in Section 4.10, scaling is achieved in a set of properly chosen samples. In another technique, based on Rietveld structure refinement (Section 4.9.4), the crystal structure of the calculated standard is adjusted statistically to the structure of the real phase. In most of QXRD techniques, the analyte and the analytical standard (real pure phases or calculated ones) are different species, principally in terms of appreciable differences in chemical composition and crystal structure.

5.4.1. *Effects of solid solution on diffracted intensity*

The effect of solid solution on the diffracted intensity is due to differences in atomic scattering by cations. If the atomic scattering factors of the substituents are close, then intensity variations are only marginal, and, conversely, intensity variations are very significant if the atomic scattering factors are different. Relative intensity changes can be revealed by computer simulation of powder patterns with variable occupational parameters of certain atoms.

Structural positions occupied by foreign atoms are centers of local lattice distortion. The degree of distortion is characterized by the mean square atomic displacement $<U^2>$, which is defined as:

$$<U^2> = j\,X\,(\Delta R)^2$$

where X is the concentration of foreign atoms, ΔR is the difference in atomic (ionic) radius between matrix and solute atoms, and j is a numeric coefficient. Static atomic displacements cause a reduction in diffraction peak intensity and an increase in diffuse scattering. In terms of mathematical formalism, this effect may be described in the same way as the well-developed theory of X-ray scattering by a thermally vibrating lattice. The intensity of diffraction is expressed as:

$$I = I_o \exp(-2M) \tag{5.22}$$

where $M = 8\pi^2 \dfrac{\sin^2\theta}{\lambda^2} < U^2 >$.

Low-angle peaks are less sensitive to this effect, and this fact should be taken into account when choosing an analytical peak. For the mean square displacement (Baram and Zevin 1990) and CuK_α radiation, the reduction in intensity does not exceed a few percent for scattering angles below 30°-40°.

On some occasions, foreign atoms stabilize certain polymorphic modifications of the analyzed phase. A striking example is the stabilization of the tetragonal form in doped zirconia. Sometimes the structural features of polymorphs are rather similar, ensuring similar diffraction patterns, usually with close integrated intensities of the diffraction peaks, but different intensities in the peak maximum. Tricalcium silicate $3CaO.SiO_2$, an important phase of Portland cement, is an example of this phenomenon: the thermodynamically stable form at room temperature is the triclinic form, which transforms to more symmetrical monoclinic and trigonal forms at higher temperatures. Doping with 1%-2% of MgO or Al_2O_3, which are always available in the clinker mixture, stabilizes the high-temperature monoclinic polymorph. Certain reflections, which are separated in the triclinic form, will overlap with those in the monoclinic form due to symmetry conditions. The maximum intensity is greatly affected by this phase transition, but the integrated intensity remains almost constant (Table 5.5). Similar observations were made by Zevin and Zav'yolova (1974) for the $2MgO.Al_2O_3.5SiO_2$ cordierite/indialite phase transition and by McCarthy et al. (1981) for a series of solid solutions of

Table 5.5. Intensity of the (620) peak of tricalcium silicate

MgO doping, %	Intensity, arbitrary units	
	Maximum	Integrated
0	100	100
1	148	102
2	171	97

Cr_2O_3/Al_2O_3. Of course, integrated intensities are preferable in these cases for QXRD analysis.

Finally, deviation from the stoichiometric composition may have an appreciable influence on the absorption coefficient of the analyzed phase μ_j^*. In a number of the examples described in Chapter 4, this quantity must be known in order to calculate the calibration constants. It is clear from this short discussion that chemical variations of the analyzed phase may have a tremendous effect on the diffracted intensities. If real analytical standards are used, their composition must match the composition of the analyte. If calculated data are used for scaling in an analytical procedure, then the proper chemical composition and occupational and spatial parameters must be used in the calculations. There are several ways of acquiring the chemical compositions of real phases. The most universal method is X-ray fluorescence analysis, which gives satisfactory chemical data for our purposes. An alternative method is the determination of lattice constants, if they are affected by chemical composition (e.g., the feldspar series of Smith 1974).

5.4.2 *Linear and planar structural imperfections*

Linear and planar structural imperfections, or disorder of displacement, are as important in their influence on diffracted intensity as on the substitutional disorder discussed in the previous section. Let us look, for example, at the diffraction patterns of barium feldspar (celsian) obtained by crystallization at different temperatures (Fig. 5.7, Zevin 1992). Diffraction peaks with odd *l* indices become broad and almost absent in celsian crystallized at low temperatures. The reason for this is planar disorder consisting of random displacements of one part of the crystal in the [001] direction by c/2 occurring in the (010) plane. If the probability of this fault approaches ~0.2, then peaks with odd *l* indexes almost merge with the background. At the same time, diffraction peaks with even *l* indexes are relatively unaffected by this disorder.

Numerous examples of the tremendous influence of planar defects on intensities are to be found in phyllosilicates. In disordered kaolinite,

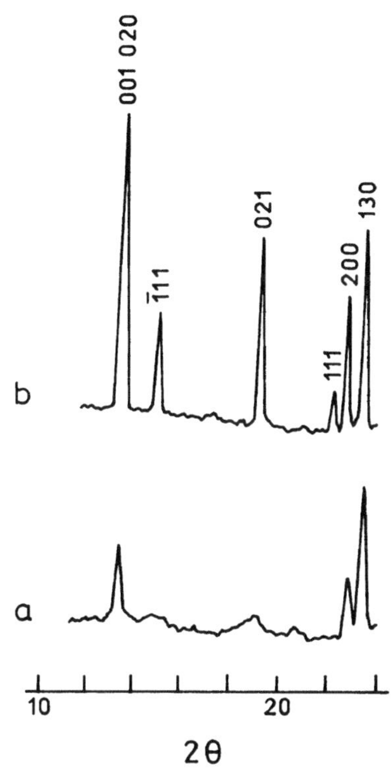

Figure 5.7. Diffraction patterns of celsian glass ceramics annealed for 2 h at 1350°C (top) and 1050°C (bottom).

all reflections with $k \neq 3$ have reduced intensity or disappear due to random plane displacement in the [010] direction by b/3. Smeclites with a high concentration of stacking faults in the (001) plane reveal (hk) bands of two-dimensional plane diffraction instead of regular (hkl) reflections. At the same time, if existing faults do not change spacings between basal planes, (001) reflections remain relatively intact.

Linear defects (dislocations) create a strain field in the surrounding medium, resulting in diffraction peak broadening at high dislocation densities. This kind of peak broadening increases with increasing tan θ. Deformation and twin faults, commonly found in deformed metals, also affect peak width and peak position (Warren 1969). Although these defects do not affect the integrated intensity, the tails of broad diffraction peaks are spatially lost in the background, making the apparent integrated intensity lower than that for a well-crystallized phase. A similar effect can be expected for phases with very small (less than 1000 Å) coherent domains, which also exhibit broad diffraction peaks.

Very often, the level of structural defects affecting the intensity diffracted by a crystalline substance is qualitatively defined as the degree of crystallinity. An impressive example was given by McCarthy et al. (1981), who measured the integrated intensity of the 222 reflection of Y_2O_3 produced by calcination of $Y(NO_3)_3$ at various temperatures. The ratio of the Y_2O_3 peak to the 113 corundum peak exhibited an almost twofold increase from low to high calcination temperatures. No evidence of an amorphous phase was found. The variation in intensity was attributed to the variation in the degree of crystallinity.

We should remember here that the term crystallinity has two different meanings:

1. the abundance of a crystalline (ordered) phase in two-phase materials also containing a disordered (amorphous) phase, as in polymers or glass ceramics; and
2. the meaning used above in the definition of perfection of a crystalline phase.

The first meaning is clear, although even in polymers a continuous transition may be found from an ordered to a disordered state through various paracrystalline states. The second meaning is rather vague: usually, poor crystallinity is also associated with morphological features such as small crystallites, giving broad diffraction peaks.

A comparison of the peak shapes of the analyte and the prospective analytical standard is a necessary step in developing a QXRD technique. A simple way to do so, giving immediate estimation of peak broadening, is the comparison of the α_1-α_2 resolution in corresponding peaks of the analyte and the standard. If the peak widths and chemical composition of the standard and the analyte are similar, one should not expect serious trouble to result from variations in structure perfection of the analyte and the analytical standard. However, this condition can be considered solely as an indication and not as the definitive requirement for choosing a particular material as an analytical standard. Very often, morphological characteristics affecting orientation of grains and sample absorption are no less important.

5.4.3 *Particle morphology*

There are numerous reports in the literature of astonishing differences in peak intensities for minerals from different geologic exposures, as is illustrated in a table of RIRs given by Davis (1988) (Table 5.6). Although Davis (1988) does not give a comprehensive analysis of the dependence of peak shapes versus Bragg angles on RIRs, it is unlikely that tremendous intensity variations are the result of structure imperfections in such relatively simple structures as calcite, hematite, or gypsum. A more probable explanation lies in the size and morphology of the crystals. Grinding of coarse-crystallized calcite, hematite, or selenite produces mostly single-crystal grains with well-

Table 5.6. Observed variability of RIRs for some naturally occurring compounds*

	hkl	RIR
Quartz (fine)	101	0.50
Quartz (pegmatitic)	101	1.47
Calcite (fibrous, hydrothermal)	014	0.09
Calcite (coarsely crystalline)	014	0.24
Calcite (crystals of dog-tooth spar)	014	0.48
Calcite (limestone)	014	0.18
Hematite (specular)	014	0.33
Hematite (earthy)	014	0.12
Gypsum (massive. Spearfish m., Black Hills)	020	0.07
Gypsum var. selenite (Wasatch from. E. Wyoming)	020	0.43

*Relative to CaF_2-111; CuK_α radiation.
Taken from Davis (1988).

developed clear planes ($10\bar{1}4$ for calcite and 020 for gypsum). If the grinding does not produce fine grains, the intensity will show severe deviation from randomly oriented powder diffraction. It seems that orientation effects are also responsible, at least in part, for the observed differences in the intensity of basal reflections between well- and poorly crystallized clays. Under certain conditions, the coarse particles of well-crystallized species usually lie in a particular orientation and thus produce stronger basal reflections. Grain size in itself may be very important for heavy absorbers, such as hematite, in CuK_α radiation due to microabsorption effects.

The effect of morphological differences on X-ray diffraction can be reduced if two materials exhibiting such grain size differences are finely ground, since small grains are less inclined to preferred orientation. This effect can also be diminished if the whole pattern, rather than a single peak, is used in the analytical scaling of the phase to be analyzed.

5.4.4 *Selection of analytical standards*

Errors in analytical scaling seem to be the main factor reducing the accuracy of QXRD techniques. There are two principal methods for the selection of analytical standards, which are especially intended to overcome the problems associated with the selection. One is the standardless technique, discussed in Section 4.10, in which a properly selected set of multiphase samples possesses self-accommodating scaling ability. The second is a combination of analytical scaling with structure refinement, in which the similarity between the analytical standard and the analyzed phase is obtained by structure variation and consequent best fit of observed and calculated intensity data. Despite the undoubted merits of these techniques, they are not universal, and there are an overwhelming number of conventional techniques based on real or calculated analytical standards. Although the importance of proper selection of analytical standards is appreciated by many workers in the field, literature on the subject is very scarce.

Many pure minerals from different deposit sites and an enormous number of pure chemicals are available commercially, and thus a mineral or compound with chemical composition similar to the material to be analyzed can usually be found. However, the crystal identity of a pure compound with the analyte may not be taken for granted. It is also necessary for the analytical standard to have the same type and level of lattice defects and the same crystal morphology as the analyte. For natural compounds (minerals), the composition, defect structure and crystal morphology depend on the geologic history, i.e., pressure, temperature, strain level, and composition of solutions. Thus, in many cases it is imperative to extract minerals used as analytical standards from the same deposits as the samples to be analyzed or from another geographical area with similar geologic conditions, as was discussed by Davis (1988). If the mineral to be used as the standard constitutes part of a macroscopic formation, separation from its surroundings does not present any problem.

Separation of any mineral from fine-grained rock is based on its physical properties, such as hardness, density, magnetic properties, and chemical properties. If the rock contains a soft mineral in a hard matrix, then grinding will produce a fine fraction rich in the soft mineral, which can be separated out by sieving or any other method of size separation. By repeating the grinding procedure a number of times, almost complete separation of minerals can be obtained. This technique has been used successfully to separate hard tridimite from calcite in Israeli porcellanite. Pure tridimite was then used as the analytical standard in the quantification of various porcellanites present in the area. In another case, almost pure carbofluorapatite (francolite) was obtained by dissolution of the major diluent mineral—calcite—in a solution of citric acid (Zevin et al. 1988).

Another commonly used technique is that of *gravity separation*, which is based on the density of minerals. In this technique, finely ground powders are sedimented in various liquids. Gibbs (1967), for example, developed an elaborate technique for the separation of analytical standards from multiphase clays. The first step was the division of the material into narrow particle size fractions, <0.2, 0.2–0.5, 0.5–1.0, 1–2, 2–6, 6–12 and >12 μm, by means of centrifugation for <2-μm fractions and free precipitation for >2-μm fractions (Jackson 1956). In order to facilitate size separation, the material was subjected to preliminary chemical treatments to remove organic matter and cementing ferrous-oxide particle coatings. Montmorillonite was concentrated in the <0.5-μm fractions. Final purification was achieved by gravity separation based on minute differences in densities, i.e., 2.6-2.8 g/cm^3 for kaolinite, 2.8-3.0 g/cm^3 for mica, and 3.05-3.3 g/cm^3 for chlorite. Further details on mineral separation techniques may be found in special texts on preparatory techniques in soil science and mineralogy (Jackson 1956). It must be noted, however, that in many cases it is difficult to achieve full separation of minerals by such readily available techniques as size or gravity separation. The latter is not suitable for minerals with similar densities, such as quartz, plagioclases, calcite, and bassanite.

We must be very careful in the application of chemical treatments in separation procedures, since they may cause precipitation of an unwanted new phase as a result of a chemical reaction with a mineral that we want to attack and remove. The development of special techniques, such as flotation, may be too laborious and time-consuming to be implemented as a stage of analysis. On the other hand, partial separation of minerals can be achieved by relatively simple means. Fukasawa et al. (1980) separated airborne particulates into five gravity fractions by centrifuge-aided precipitation in carbon tetrachloride/methylene iodide mixtures. Separation was better for heavy (magnetite and hematite) and light (magnesium sulfate hydrate) minerals than for quartz, gypsum, and plagioclases. Nevertheless, each precipitated fraction, although not a mono-mineral species, was enriched with one or more minerals. Thus, these precipitates constitute an ideal sample set for application of the standardless technique discussed in Section 4.10.

In the analysis of an anthropogenic material, the history of the material is usually known and must be taken into account in the preparation of analytical standards. This includes the chemistry of precursors, time and temperature of heat treatment, the surrounding atmosphere, and so on. A good rule of thumb is to compare the diffraction characteristics of the potential analytical standard and the analyzed phase, including peak breadth, α_1–α_2 separation, and peak position. If these characteristics are identical in both phases, then greater confidence can be placed in the chosen standards.

Finally, we have to look at the analysis of retained austenite in steels: only certified standards for QXRD, i.e., standard reference materials SRM 485, SRM 486, SRM 487, and SRM 488, may be used (Hicho and Eaton 1983). These standards were obtained by sintering a mixture of austenitic and ferritic steel powders. A similar set of cements with known, although not certified, mineralogical compositions are available as standards in the analysis of cement.

5.5 Intensity measurements

5.5.1 *Counting statistics*

The intensity of scattered X-rays in QXRD is expressed as a counting rate:

$$I = N/T \tag{5.23}$$

where N is the number of counts (X-ray photons) accumulated in time T.

In order to keep to the strict definition of intensity given in Section 2.2, we should divide the right-hand side of expression (5.23) by the receiving slit area to obtain the energy flow per square centimeter per second. However, the receiving slit is usually kept constant during a diffractometric run, enabling comparison of various diffraction peaks and scaling against calculated values given in equation (5.23). Counting time is a parameter that, by its very nature, must be allowed to vary; it is usually longer for weak diffraction peaks. In this way, equal precision is achieved for weak and strong peaks.

The dispersion (variance) of I is defined in the usual way as:

$$\sigma_I^2 = \left(\frac{\partial I}{\partial N}\right)^2 \sigma_N^2 + \left(\frac{\partial I}{\partial T}\right)^2 \sigma_T^2 = \frac{\sigma_N^2}{T^2} + I^2 \frac{\sigma_T^2}{T^2} \tag{5.24}$$

where σ_N^2 and σ_T^2 are the dispersion (variance) of counting and time measurements. The number of counts N is a random variable distributed according to Poisson's law. The probability P_N of accumulating N counts during time T is given by:

$$P_n = (1/N!)\,(T\,\langle I\rangle)^N \exp(-T\langle I\rangle) \tag{5.25}$$

where $\langle I\rangle$ is the mean counting rate.

The properties of this distribution are given below.

Mean value of N:

$$\langle N\rangle = T\,\langle I\rangle \tag{5.26}$$

Dispersion (variance) of N:

$$\sigma_N^2 = \langle N\rangle \quad \text{and} \quad \varepsilon_N = \sigma_N/\langle N\rangle = (\langle N\rangle)^{-1/2} \tag{5.27}$$

where ε is the relative mean square deviation or coefficient of variation.

The contribution of time errors to the dispersion of the counting rate σ_I^2 will be negligible if:

$$\sigma_T^2 << \sigma_N^2/I^2 \quad \text{or} \quad \sigma_T^2 << \sigma_T^2/<I> \qquad (5.28)$$

This is usually the case even for short counting times (1 sec) and high counting rates (10^4 counts/sec).

If condition (5.28) holds, the standard deviation of the counting rate is given by:

$$\sigma_I = \sigma_N / T = \sqrt{(<I>/T} \qquad (5.29)$$

and the coefficient of variation by:

$$\varepsilon_I = \varepsilon_N = (<I>T)^{-1/2} \qquad (5.30)$$

The probabilities of accounting for the deviations $\Delta I = |I - <I>|$ greater than σ_I, $2\sigma_I$, and $3\sigma_I$ are equal to 31.7%, 4.5%, and 0.3%, respectively.

Numerous tests have confirmed the validity of equations (5.27) and (5.29); the latest was carried out in an intensity round robin (Jenkins and Shreiner 1989). The standard deviation was calculated for 300 background points on the digitalized Al_2O_3 pattern submitted by 12 users. The ratio of the observed standard deviation σ_I to that calculated by equation (5.29) was expected to be 1. In fact, it was rather close to 1, ranging from 0.88 to 1.22.

5.5.2 *Difference and sum of intensities—counting strategy*

In conventional diffractometric practice, the background intensity is subtracted from the total (peak and background intensity). Summation of step-scanned intensities is conventional practice in obtaining the integrated intensity or a diffraction peak. Suppose that peak intensity is given by:

$$I_p = I - I_b \qquad (5.31)$$

where I is the total intensity (peak and background) and I_b is the background intensity. If T_1 and T_2 are the times spent in counting of I and I_b, then according to the general rule for the propagation of errors the standard deviation σ_p and the coefficient of variation ε_p are given by:

$$\sigma_p = \sqrt{I/T_1 + I_b/T_2}\,; \quad \varepsilon_p = \sqrt{I/T_1 + I_b/T_2}\ / \ (I - I_b) \qquad (5.32)$$

The total time spent on counting, $T = T_1 + T_2$, can be divided between measurements of I and I_b in order to minimize σ_p. This is the conventional problem of a conditioned extremum, which may be solved by the method of the Lagrange multiplier (Mack and Spielberg 1958, Kheiker and Zevin 1963). The total time must be distributed according to the following relationships:

$$T_1/T_2 = \left(I/I_b\right)^{1/2};\; T_1 = T\, I^{1/2}/\left(I^{1/2} + I_b^{1/2}\right);\; T_2 = T\, I_b^{1/2}/\left(I^{1/2} + I_b^{1/2}\right) \quad (5.33)$$

The expression for the minimal error is obtained by substituting the optimal values of T_1 and T_2 from expressions (5.33) into equation (5.32):

$$(\sigma_p)_{min} = \left(I^{1/2} + I_b^{1/2}\right) / T^{1/2} \quad (5.34)$$

The relative background level $K_b = I_b/I_p$ and

$$(\sigma_p)_{min} = \left(I_p/T\right)^{1/2} \left(K_b^{1/2} + (1 + K_b)^{1/2}\right) \quad (5.35)$$

$$(\varepsilon_p)_{min} = \left(I_p/T\right)^{-1/2} \left(K_b^{1/2} + (1 + K_b)^{1/2}\right) \quad (5.36)$$

Let us compare the minimal error defined by equations (5.36) with the errors obtained in two widely used counting strategies: the preset time mode, in which the total time is equally divided between I and I_b, and the preset count mode, in which equal numbers of counts are accumulated for I and I_b. In the former mode, $T_1 = T_2 = T/2$ and, according to equation (5.32), errors in the preset time mode are given by:

$$(\sigma_p)_{t\,=\,const} = \left(2I_p/T\right)^{1/2} (1 + K_b) \quad (5.37)$$

$$(\varepsilon_p)_{t\,=\,const} = \left(\sqrt{2}I_p/T\right)^{-1/2} (1 + K_b) \quad (5.38)$$

In the preset count mode, $N = I\, T_1 = I_b\, T_2$ and $T_1 = I_b/(I + I_b)$; $T_2 = I/(I + I_b)$. Errors in the preset count mode are thus given by:

$$(\sigma_p)_{N\,=\,const} = (I_p/T)^{1/2} \left(\frac{(1 + 2K_b)\,(1 + 2K_b + 2K_b^2)}{K_b\left(1 + K_b\right)} \right)^{1/2} \quad (5.39)$$

$$(\varepsilon_p)_{N\,=\,const} = (I_p/T)^{-1/2} \left(\frac{(1 + 2K_b)\,(1 + 2K_b + 2K_b^2)}{K_b\left(1 + K_b\right)} \right)^{1/2} \quad (5.40)$$

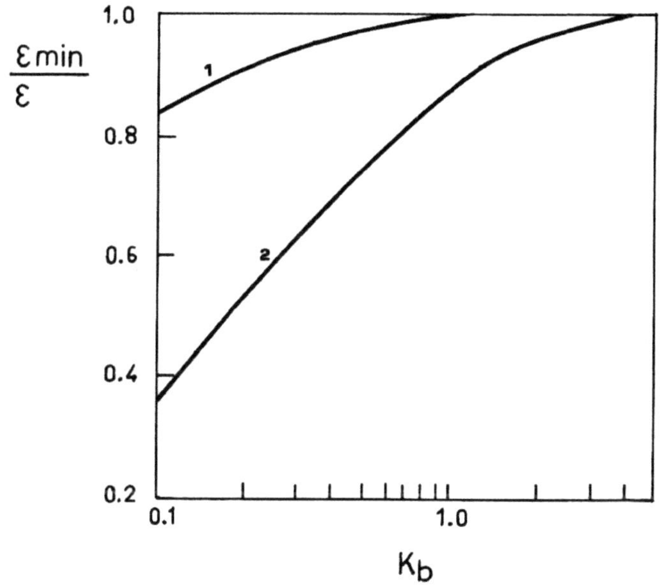

Figure 5.8. Minimum variation $\varepsilon_{min}/\varepsilon$ of the intensity vs. K_b for the preset time mode (line 1) and the preset count mode (line 2).

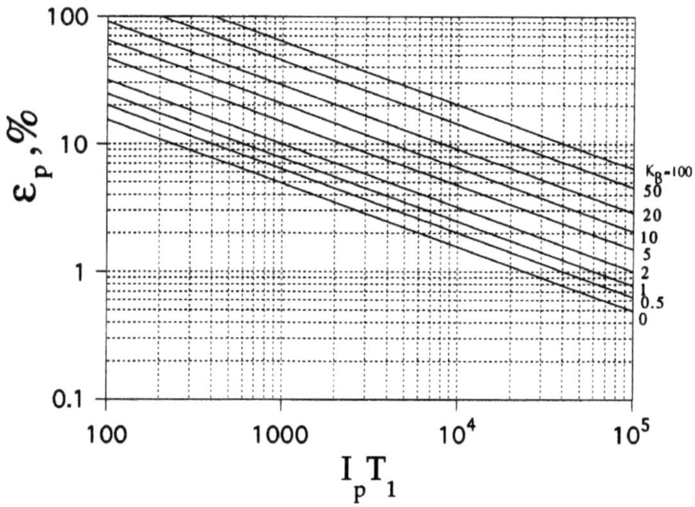

Figure 5.9. Counting statistic error ε_p vs. accumulated counts (I_pT_1) for different background levels K_b.

We calculated the ratio $(\varepsilon_p)_{min}/(\varepsilon_p)_{t = const}$ and $(\varepsilon_p)_{min}/(\varepsilon_p)_{N=const}$ for various background levels K_b (Fig. 5.8). Errors in the preset (constant) time mode were very close to minimal values, corresponding to the optimal partition of time. At low background levels ($K_b < 0.1$), optimal distribution of time is advantageous. The constant count mode deviates significantly from the optimal time partition and is therefore not recommended. For a quick estimation of the counting statistics error, values for $(\varepsilon_p)_{t = const}$ are given in Fig. 5.9 as a function of accumulated peak counts I_pT_1 and relative background level K_b.

5.5.3 *Ratio of intensities*

A ratio between two intensities must be determined in almost all analytical techniques, in particular, in internal standard methods, and in the evaluation of RIRs. If the ratio of intensities $S_{12} = I_1/I_2$ and that the counting times for I_1 and I_2 are T_1 and T_2, then the standard deviation is given by:

$$\sigma_s = S\sqrt{(I_1 T_1)^{-1} + (I_2 T_2)^{-1}} \tag{5.41}$$

and the coefficient of variation by:

$$\varepsilon_s = \sigma_s/S = \sqrt{(I_1 T_1)^{-1} + (I_2 T_2)^{-1}} \tag{5.42}$$

The total time $T = T_1 + T_2$ can be partitioned optimally between measurements I_1 and I_2 to ensure a minimal value of ε_s. In this case (Zevin and Zav'yalova 1974):

$$T_1/T_2 \sqrt{I_2 I_1)}\ ;\ \ T_1 = \frac{\sqrt{I_2}}{\sqrt{I_1 - I_2}}\,T;\ \ T_2 = \frac{\sqrt{I_1}}{\sqrt{I_1 - I_2}} \tag{5.43}$$

and the minimal relative error is:

$$(\varepsilon_s)_{min} = \frac{1}{\sqrt{T}}\,\frac{\sqrt{I_1} + \sqrt{I_2}}{\sqrt{I_1 - I_2}} \tag{5.44}$$

For the constant time and constant count modes, the relative errors are equal in this case:

$$(\varepsilon_s)_{t\,=\,const} = (\varepsilon_s)_{N\,=\,const} = \frac{\sqrt{2}\,\sqrt{I_1 + I_2}}{\sqrt{T}\sqrt{I_1 + I_2}} \tag{5.45}$$

When the ratio of $(\varepsilon_s)_{min}/(\varepsilon_s)_{t=const}$ is plotted against the intensity ratio I_1/I_2 (Fig. 5.10), the advantage of optimal time partition is

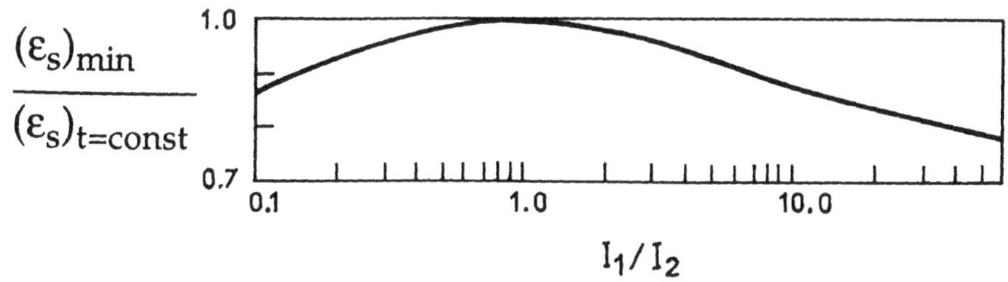

Figure 5.10. Variation of the ratio $(\varepsilon_s)_{min}/(\varepsilon_s)_{t=const}$ vs. the intensity ratio I_1/I_2.

evident only for very small (S < 0.1) or very large (S > 10) intensity ratios. In the intermediate situations, both t = const or N = const counting modes are viable.

5.5.4 *Integrated intensity*

The integrated intensity of a diffraction peak is defined as:

$$I_{in} = \frac{1}{\Delta 2\theta} \int_{2\theta_1}^{2\theta_2} I(2\theta)\ d(2\theta) - \frac{1}{\Delta 2\theta} \int_{2\theta_1}^{2\theta_2} I_b(2\theta)\ d(2\theta) \tag{5.46}$$

where $2\theta_1$ and $2\theta_2$ are peak boundaries; $\Delta 2\theta = 2\theta_2 - 2\theta_1$; $I(2\theta)$ is the total (peak and background) intensity at running angle 2θ; and $I_b(2\theta)$ is the background intensity. In the step-scanning mode with a constant step size:

$$I_{in} = (1/m) \sum_{i=1}^{m} N_i/t_i - (1/m) \sum_{i=1}^{m} I_{ib} \tag{5.47}$$

where $m = \Delta 2\theta/\delta 2\theta$ is the number of steps within peak boundaries; $\delta 2\theta$ is the step size, N_i are the counts accumulated in step i; t_i is the counting time for step i; and I_{ib} is the background intensity for step i. If we assume linear variation of the background within peak boundaries, then $\sum_{i=1}^{m} I_{ib} = mI_b$, where I_b is the background intensity corresponding to the middle of the $2\theta_2 - 2\theta_1$ range.

In the constant time mode, counting time per step $t_i = t = \text{const}$, and

$$I_{in} = I_\Sigma - I_t \tag{5.48}$$

where $I_\Sigma = N_\Sigma/T_1$; $N_\Sigma = \sum_{i=1}^{m} N_i$ net counts accumulated in step scanning; and $T_1 = mt$ is the total time spent.

Equation (5.47) is also retained in the constant count mode, but in this case: $I_\Sigma = (N/m)\sum_{i=1}^{m} t_i^{-1}$, where N is a preset number of counts per step and t_i is the counting time for step i.

Continuous mode scanning is similar to step scanning with very small steps $d(2\theta) = \eta dt$, where η is the scanning speed. In this case, I_Σ from the general expression (5.47) is equal to $I_\Sigma = N_\Sigma/T_1$, where N_Σ are the net accumulated counts, $T = (\Delta 2\theta)/n$ is the total scanning time, and $\Delta 2\theta$ is the scanning range.

The background within peak boundaries is usually supposed to vary linearly with the scattering angle. In this case, the background intensity I_b from equation (5.47) is defined in the middle of the counting range $\Delta 2\theta$, i.e., at the scattering angle $2\theta_b = (2\theta_1 + 2\theta_2)/2$. If the background intensity I_{b3} and I_{b4} is actually measured at two angles $2\theta_3$ and $2\theta_4$ (not necessarily $2\theta_1$ and $2\theta_2$), which are located symmetrically to $2\theta_b$, then:

$$I_b = (I_{b3} + I_{b4})/2 \tag{5.49}$$

where $I_{b3} = 2N_{b3}/T_2$; $I_{b4} = 2N_{b4}/T_2$; N_{b3} and N_{b4} are the net counts at scattering angles $2\theta_3$ and $2\theta_4$, and T_2 is the total background counting time, equally divided between two background points.

The integrated intensity defined in equations (5.46) and (5.47) has the advantage of being independent of such experimental factors as step size, counting time, and counting strategy. Thus, the proper intensity ratios will be retained independently of the time spent on counting (usually longer for weak peaks and shorter for strong peaks). Of course, counting statistics errors will depend on counting time, and equations (5.32), (5.35), and (5.39) derived for the peak intensity are also applicable here, with I replaced by I_Σ. The optimal distribution of the counting time is also appropriate in measurements of the integrated intensity, although in this case there are two possibilities for time partition:

1. Partition of the total time T between measurements of I_Σ and I_b (T_1 and T_2, respectively), as was done in the previous section for peak intensity; and
2. Partition of time T_1 between m steps in the step-scanning mode.

However, in both cases, the constant time mode gives a very good approximation to the optimal time distribution, and the gain acquired by the latter can hardly justify the obvious complication of the counting strategy imposed by optimal time partition.

Another parameter of some importance in the measurement of integrated intensity is the width of the counting range $\Delta 2\theta$. Intuitively, we can feel that there must be an optimal range in which the counting error is minimized. Indeed, extending the counting range will include more background counts in N_Σ [equation (5.46)]. On the other hand, excessive reduction of the counting range simply reduces "useful" diffraction counting. Let us assume a constant time mode with $T_1 = T_2 = mt$, where m is the number of steps, and t is the counting time per step. Then, the relative error in the integrated intensity is given by:

$$\varepsilon_{in} = (\sigma_{in}/I_{in}) = \sqrt{N_\Sigma + N_b}\,/(N_\Sigma + N_b) = \sqrt{N_p + 2N_b}\,/N_p \qquad (5.50)$$

where N_Σ are the total accumulated counts within the counting range; N_b are the background counts within the same range, and $N_p = N_\Sigma - N_b$ are the net diffraction peak counts. We approximate the peak profile by the function F(x) with $x = 2\theta - 2\theta_m$, where $2\theta_m$ is the scattering angle corresponding to the peak maximum and we assume normalization in which F(0) = 1. If the peak intensity is I_p, then the net diffraction peak counts $N_p = \sum_i I_p t F_i$, where F_i is the value of F(x) at step i. As usual, we introduce the relative background level $K_b = I_p/I_b$ and background counting $N_b = K\,I_p\,mt$, where m is the number of steps. Substituting these expressions for N_p and N_b into equation (5.50), we obtain:

$$\varepsilon_{in} = \frac{\left(\sum_{i=1}^{m} F_i + 2mK_b\right)^{1/2}}{(I_p t)^{1/2} \sum_{i=1}^{m} F_i} = \frac{\left(\int_{-x}^{x} F(x)\,dx + u_x K_b\right)^{1/2}}{(I_p t)^{1/2} \int_{-x}^{x} F(x)\,dx} \qquad (5.51)$$

where 2x is the counting range centered around x = 0.

We calculated ε_{in} as a function of the counting range for three different functions F(x), Gaussian, Lorentzian, and Voigt functions (defined in Section 5.8), the latter being a convolution of the Gaussian and Lorentzian functions in a 1:1 proportion. Results for a relatively high background level are shown in Fig. 5.11, in which the counting range (FWHM) is expressed in w units. The curves for all three profile functions exhibit a minimal error for a counting range close to 2w. However, for a low-decaying peak intensity, expressed by a Lorentzian or a 1:1 Voigt function, the value of F(w) is equal to 0.15-0.20, i.e., it is still very significant. Moreover, the derivative dF/dx at x = w is also significant, increasing the intensity error due to probable small peak shifts. On the other hand, increasing the counting range up to $\Delta 2\theta$ = 3-4w does not appreciably increase the intensity error for slow-

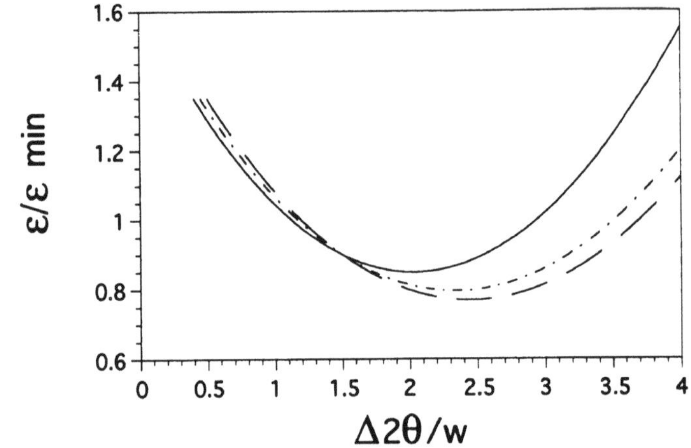

Figure 5.11. Plot of $\varepsilon/\varepsilon_{min}$ vs. $\Delta(2\theta)/w$. ——— Gaussian; – – – Lorentzian; · · · · · · · · Voigtian.

decaying peaks. Thus, a counting range closer to 3w is optimal for fast-dropping (Gaussian) profiles and close to 4-5w for slow-decaying (Lorentzian) profiles.

In this discussion, we assumed that the diffraction peak is free standing, i.e., that there is no overlapping with neighboring peaks. If this is not the case, the counting range must be chosen on the basis of minimal contamination of N_Σ by foreign peak counts.

5.6 Definition and subtraction of background

The signal recorded by a diffractometer can be considered to be composed of Bragg reflections, background, and noise. Our goal is the determination of the intensity of the Bragg reflections, purified of the noise and background components. The angle dependence of the signal components may be arranged in the following order of intensity/angle slope: background > diffraction peak > noise [with the background being the slowest (smallest slope) and noise the fastest (highest slope) component]. Separation of the diffraction peak from the interfering factors is, in fact, based on differences in the intensity/angle slopes. We do not have any means of determining the background directly under the peak profile. Instead, background is measured in places in the diffraction pattern that are free of Bragg reflections and then interpolated between these places to produce a background line for the diffraction peaks. In carrying out this procedure, we assume a slow change of background intensity with 2θ, which can be approximated by a linear or parabolic regression. In one special case of QXRD, namely, the analysis of an amorphous component (see Section 4.12), scattering by the amorphous phase is also a weak function of the scattering angle. In this case, separation of the background demands special effort, including theoretical calculations.

5.6.1 Definition and subtraction of the background for a single diffraction peak

If an analytical peak (or cluster of peaks) is well separated from neighboring peaks or clusters, then two suitably remote portions of the pattern on each side of the peak are taken for background counts (points $2\theta_3$ and $2\theta_4$). It is usually assumed that the background under a peak varies in a linear manner. If a step-scanning mode is used for data acquisition, then the background points on each side of the peak are fitted to a linear regression curve, which provides the background counts for each peak profile point. There are two questions that need to be answered in this approach:

1. What is the probable extension of the angular range with linear variation of the background?
2. How far from the peak maximum or centroid do the peak tails decay into the background?

The background varies slightly with scattering angle, and usually the 5°-7° 2θ range can be fitted to a linear function. An exception is the case in which a diffraction peak is superimposed on the hump of amorphous scattering and the slopes of the hump have different signs on the two sides of the peak. In this case, a parabolic approximation of the background under the peak gives satisfactory results.

The second question deals with the location of the two points $2\theta_3$ and $2\theta_4$ assigned to background counting. The answer depends on the rate of peak profile decay, which can be quite different from case to case.

Let us now take a look at the three peak profile functions used for estimation of peak counting range in the previous section, arranged in order of decreasing decaying rate: Gaussian < 1:1 Voigt < Lorentzian (a definition of these functions is given in Section 5.8). Table 5.7 gives the distance (in FWHM, w) from the peak maximum to the point of intensity decay to 1% and 0.5% of its maximal value.

Table 5.7. Rate of decay of peak profiles

	Decay distance in FWHM, w	
Peak profile	1% terminal intensity	0.5% terminal intensity
Gaussian	1.28	1.38
Lorentzian	5.0	7.0

For a low-decaying Lorentzian profile, background counting must be started at least 7w from the peak maximum in order to approach the true background level. This does not mean, however, that peak counting must be performed within the same limits, say ± 7w. Some peak truncation, as discussed in Section 5.5, is essential to maintain the proper balance between accuracy and precision in intensity measurements. A numerical method for the determination of true background has been suggested by Enzo and Parrish (1985).

The following conditions are recommended for the determination of the background for most cases: intensity fluctuation in background counting within ~5%, step size ~0.1w, and angular range ~4w on each side of the peak.

5.6.2 *Full-pattern approach*

In advanced QXRD techniques, the full pattern is used for phase quantification. The background must be fitted automatically to the full pattern and not to a single peak, as was discussed in the previous section. In this technique, the approach to the definition of the background is quite different, depending on the method used for data processing. If any kind of profile fitting (with or without a structural model) is used in quantification, then the background appears as a product of the best fit experimental and calculated patterns. This technique will be discussed in Section 5.8. In a different technique, the pattern of an analyzed mixture is presented as a scaled sum of the patterns of the individual phases (standards). In this latter case, which is discussed in this section, the intensity data must be stripped from each experimental pattern, using any empirical technique. There are several such techniques for background fitting, all of which are based on a preliminary search for pattern sections free of diffraction peaks; these sections can be assigned as background sections. As soon as these sections are established, a first- or second-degree polynomial is fitted to the background. In addition to the simplest "manual" searching for background sections, there are two major approaches in computer-aided searching: an iterative procedure and a second-derivative technique.

In the iterative technique proposed by Goehner (1978), a linear regression line is fitted to the first segment of the pattern containing 200 raw data points for intensity. Then, all the points that lie above this line by one standard deviation are rejected, and a new fit is made. This procedure is repeated until convergence. The final line, which is supposed to be the true background, is subtracted from the first 150 raw data points. The next segment of 200 points is then chosen, starting from point 151, i.e., giving a 50-point overlap with the previous section. The same procedure is carried out for the second section and so on up to the end of the pattern. The number of points in a section must be small enough to follow the probable background curvature but large enough to include true background areas. This technique has proved to be very good for background fitting for patterns superimposed on the hump of amorphous scattering. However, it can be problematic if the diffraction pattern contains segments of overlapping peaks of a length greater than the basic moving section of the raw data points.

A somewhat similar iterative procedure was used in the peak-hunting program developed by Sonnefeld and Visser (1975). In this method, every 20th data point is used as the first approximation to the background. The intensity at each point of the pattern is replaced by the mean of its neighbors when the neighbors' mean has a lower value or is retained for further iterations when the neighbors' mean has a higher value. After several dozen repetitions of this procedure, it converges to a linear background. This technique seems to be less sensitive to extended sections of overlapping peaks, although it might be inferior to the Goehner algorithm in fitting a curved background.

A second derivative method for locating a diffraction peak also provides a way of separating peak and background sections. If the background can be approximated by a second-order polynomial, then on a second derivative plot, the background sections will appear as straight lines with a zero slope and superimposed noise (Pyrros and Hubbard 1983). As soon as the background sections are identified, the second-degree polynomial is fitted to the corresponding intensity points in the raw data pattern. Observed points that are more than 2σ above the calculated background are discarded, and the background polynomial is recalculated. Thereafter, the calculated background is subtracted from the raw data intensities. This technique has a sound mathematical basis and appears to be superior for identification of background sections. However, derivation of the raw data itself increases the noise, thus masking the low-intensity peaks, which may be erroneously included in background sections.

We now come to a novel full-pattern approach in QXRD, in which quantification is achieved by scaling of computer-simulated patterns of analyzed phases, or a combination of Rietveld structure refinement and quantitative analysis. The background is either given by the operator or refined. The refined data are obtained from "refinable" background functions [see also Section 5.8.1, equations (5.75) and (5.76)].

5.6.3 *Subtraction of incoherent scattering*

This problem is of particular relevance in QXRD of partially crystallized organic polymers for two reasons: the high intensity of incoherent scattering for low-atomic-weight elements, such as carbon and oxygen, and the necessity of measuring the overall coherent scattering by the sample material or the coherent scattering by the amorphous component. Removal of incoherent scattering is based on the assumption that, due to intrinsic lattice imperfections common in these materials, the diffracted intensity in reciprocal space is dissipated between the lattice points at large reciprocal vector values. In this case, the total scattered intensity per atom is expressed as:

$$I_{tot} = \langle f^2 \rangle + \langle I_{inc} \rangle \tag{5.52}$$

where $\langle f^2 \rangle = \sum_i n_i f_i^2 / \sum_i n_i$ is the mean coherent scattering per atom; f_i is the atomic factor for atom i, n_i is the number of atoms i in the chemical formula; and $\langle I_{inc} \rangle = \sum_i n_i (I_i)_{inc} / \sum_i n_i$ is the mean incoherent scattering per atom, in electronic units.

The data for f_i and $(I_i)_{inc}$ are available from International Tables for X-ray Crystallography as a function of $\sin\theta/\lambda$. If $S_n = 2\sin\theta_n/\lambda$ is the magnitude of the reciprocal space vector and I_{tot} is the observed total intensity of coherent and incoherent scattering at the scattering angle $2\theta_n$, then the fraction of incoherent scattering at the same angle $2\theta_n$ is obtained from the obvious relationship:

$$I_{inc}(S_n) = I_{tot}(S_n)\ (1 - \langle f^2 \rangle / \langle I_{inc} \rangle) \tag{5.53}$$

where $\langle I_{inc} \rangle$ and $\langle f^2 \rangle$ are taken at the same value S_n.

As soon as I_{inc} is established for large reciprocal vectors S_n $(2\theta_n)$, I_{inc} can be extrapolated to lower 2θ values by replicating the variations $\langle I_{inc} \rangle$ with S taken from the International Tables. The curve I_{inc} (2θ) so obtained is subtracted from the observed total intensity I_{tot} (2θ) point by point.

The critical step is, of course, definition of minimal S_n (minimal $2\theta_n$) when equation (5.52) is valid. It seems that values of $S = 1$ Å^{-1} ($2\theta \approx 100°$ for CuK_α radiation) can be taken as a first approximation. At that angle, the intensity of incoherent scattering for light elements greatly exceeds the intensity of coherent scattering, making the separation according to equation (5.53) viable.

Vonk (1973) suggested a modified scaling procedure based on integrated intensities in reciprocal space, rather than on positional intensities, as was discussed earlier. This procedure has been applied with modified techniques for the determination of crystallinity, as discussed in Section 4.12.

It must be emphasized that equation (5.53) is valid only on the condition that no extraneous scattering, such as fluorescence or white radiation, is present. Monochromatization is assumed in the implementation of this technique.

5.7 Determination of sample absorption

In a number of QXRD techniques, such as those involving internal and external standards, dilution, or spiking, sample absorption is eliminated from the analytical formulation. However, in many other techniques (outlined in Chapter 4), sample absorption plays a crucial role in the analytical procedure and must be measured or calculated. Very often, knowledge of phase absorption is needed in order to diminish microabsorption effects by appropriate sample preparation or by choice of the proper internal standard.

5.7.1 Calculation of mass absorption

If the chemical composition of the material is known, the mass absorption coefficient is easily calculated by:

$$\mu^* = \sum_{i=1}^{n} \mu_i^* C_i \tag{5.54}$$

where μ_i^* is the mass absorption coefficient of component (element) i and C_i is the weight fraction of component i, and the sum is taken over all the sample components. For example, the mass absorption coefficient of SiO_2 on CuK_α is equal to:

$$\mu_{SiO_2}^* = \mu_{Si}^* C_{Si} + \mu_O^* C_O = 60.3 \times 0.467 + 12.7 \times 0.533 = 34.9 \text{ cm}^2/\text{g}.$$

μ_i^* data for elements are to be found in numerous literature sources and in Table 2.1C of the International Tables for X-Ray Crystallography, Vol. IV (1974). Unfortunately, these data are subject to appreciable uncertainties, which are estimated to be less than 2% for the elements C, Mg, Al, Ti, Fe, Ni, Cu, Zn, Zr, Mo, Pd, Ag, Cd, Sn, La, Gd, Ta, W, Pt, Au, Pb, Th, and U and between 2% and 5% for the other elements. The same level of deviations can be expected for calculated μ^* values.

5.7.2 Measurements of absorption coefficients by transmission methods

If a monochromatic primary beam of intensity I_o strikes the sample at a 90° angle (Fig. 5.12), then the intensity of the transmitted beam is equal to:

$$I = I_o \exp(-\mu l) = I_o \exp(-\mu^* \widetilde{m}) \tag{5.55}$$

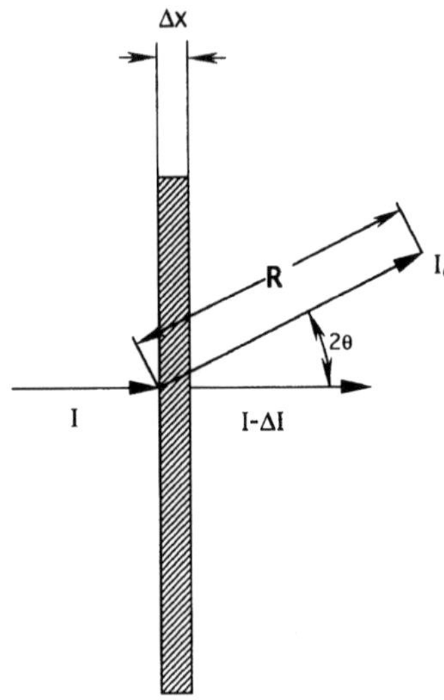

Figure 5.12. Scheme of the transmission technique.

where l is the sample thickness and $\tilde{m}$ is the mass of the sample per square centimeter of its area (loading), which is defined as:

$$\tilde{m} = G/S \tag{5.56}$$

where G is the sample mass and S is the sample surface. The mass absorption coefficient is equal to:

$$\mu^* = (1/\tilde{m}) \ln(I_0/I) = (S/G) \ln (I_0/I) \tag{5.57}$$

In order to compensate for the probable nonuniformity in $\tilde{m}$, the incident beam should illuminate a large sample surface and not only a narrow strip, which is the usual shape of the X-ray beam cross section in a powder diffractometer. The solution to this requirement can be found in sample rotation or translational movement. If we ignore the probable errors in the estimation of $\tilde{m}$, then the precision of μ^* is defined by the precision of the intensity measurements:

$$\frac{\sigma(\mu^*)}{\mu^*} = \left[(\ln (I_0/I))^{-1}(I_0\, t_0)^{-1} + (It)^{-1}\right]^{1/2} \tag{5.58}$$

where $\sigma(\mu^*)$ is the standard deviation, and t_0 and t are the counting times for incident and transmitted intensity, respectively, and it is assumed that counting statistics are the major source of errors in I_0 and I. Elementary analysis of this equation shows that precision is better if $I_0 >> I$. The same is true if thick absorbing layers are used. Measurement of I_o might be problematic if I_o exceeds the acceptable level for a detector with a finite dead time. Measurements for the incident beam can be avoided by using a standard sample with a known absorption coefficient μ_s^*. In this case:

$$\ln I_o = \tilde{m}_s \cdot \mu_s^* + \ln I_s \tag{5.59}$$

where the subscript s refers to the standard sample. Substitution of ln I_o from equation (5.59) into equation (5.57) gives:

$$\mu^* = (\tilde{m}_s/m)\, \mu_s^* + (\tilde{m})^{-1} \ln (I_s/I) \tag{5.60}$$

thus circumventing direct measurement of I_o. Usually, a powder sample is compacted under pressure in the circular recess in the metallic plate to a thickness of ~ 0.1-0.2 mm for medium-absorbing materials. Let us estimate the upper limit of μ^* that can be determined by this method. Taking a somewhat extreme value for $I_o/I = 10^3$, a layer thickness = 0.01 cm, and an apparent density of the material $\rho = 5$ g/cm^3, we come to a surface density of $\tilde{m} = 5.10^{-2}$ g/cm^2 and $\mu^* \approx 140$ cm^2/g. It seems that maximal μ^* closer to 100 cm^2/g is a realistic value.

Greater values of μ^* can be measured if a thin layer of powder is evenly distributed over a substrate transparent to X-rays, because the loading $\tilde{m}$ may be much lower than in self-supporting pressed tablets. Such samples are often found in environmental studies. The attenuation of X-rays is given by:

$$I/I_o = \exp(-\tilde{m}\mu^*)\ \exp(-\tilde{m}_{Sb}\, \mu_{Sb}^*) \tag{5.61}$$

where $\tilde{m}_{Sb}$ is the surface density of substrate and μ_{Sb}^* is the mass absorption coefficient of the substrate materials. The absorption coefficient μ^* is calculated as:

$$\mu^* = (\tilde{m}^{-1})\ [\ln (I_o/I) - \tilde{m}_{Sb}\mu_{Sb}^*)] \tag{5.62}$$

For most organic membrane filters, $\tilde{m}_{Sb}\ \mu_{Sb}^*$ is close to zero, and this quantity can be measured once for a certain type of filter and then used as a constant in equation (5.62). Fiberglass filters made of borosilicate glass exhibit appreciable absorption, making the second exponent in equation (5.61) comparable with the first one. Although this situation is less advantageous than that found for low-absorbing substrates, Davis (1988) reported excellent precision in absorption measurements for powders deposited on Whatman fiberglass filters.

The accuracy of the method can be affected by microabsorption effects. According to de Wolff (1947), the observed linear absorption coefficient is lower than the true value by:

$$\Delta\mu = (1/2) \sum_j V_j < d_j > (\mu_j - \mu)^2 \tag{5.63}$$

where V_j, $< d_j >$, and μ_j are the volume fraction, mean grain size and absorption coefficient for phase j, respectively, and μ is the absorption coefficient of the sample. For coarse powders $\Delta\mu$ defined here can exceed the precision of the measurements expressed in equation (5.58).

Some idea of the deviations obtained in practical measurements of the absorption coefficient may be gained from the data presented in Table 5.8. (Samples with different absorptions were prepared by pressing powders.)

In the practical application of this technique, we must be aware that the incident beam is monochromatic. In the simplest arrangement, monochromatization is achieved by placing a single crystal in the sample position and setting the goniometer at an angle of $2\theta_m$, where θ_m is the Bragg angle for the single crystal. The sample being analyzed is placed in the reflected beam. In many cases the single crystal can be replaced by a powdered material giving a strong reflection, such as the $10\bar{1}1$ reflection of quartz.

5.7.3 *Attenuation of reflection from a crystalline substrate*

This is a special case of the transmission technique. Suppose that a layer of powder sample is placed on a crystalline substrate, for example, to fill up a recess in a metallic plate. If I_o' is the intensity reflected by the free substrate (without the sample being placed on it) and I' is the intensity of the same reflection attenuated by the sample layer, then I' is related to I_o' by the obvious equation:

$$I'/I_o' = \exp(-2\mu t/\sin\theta') = \exp(-2\mu^*\tilde{m}/\sin\theta') \tag{5.64}$$

where θ' is the Bragg angle for substrate reflection and $2t/\sin\theta'$ is the absorption path of the primary and reflected beams inside the sample layer. The mass absorption coefficient is given by:

$$\mu^* = [(\sin\theta')/2\tilde{m}] \ln I_o'/I' \tag{5.65}$$

In this technique, the same sample mount is used for absorption and sample diffraction measurements, giving it a certain advantage over the common transmission technique, in which a special sample mount is needed for absorption measurements. Moreover, the sample can be thick enough to be considered infinitely thick for diffraction measurements but still transparent enough to allow absorption measurements according to equation (5.65). Indeed, if $\tilde{m} \geq \tilde{m}_{cr}$, where $\tilde{m}_{cr}$ is the critical value for an infinitely thick sample, defined in equation (4.208),

Table 5.8. Mass absorption coefficients determined by the transmission technique

	Mass absorption coefficient, cm^2/g for different samples							
	1	2	3	4	5	6	7	8
Observed	5.30	9.90	53.2	59.4	58.5	66.8	85.0	123.10
Calculated	5.10	10.0	55.0	61.0	62.0	64.5	84.5	115.0
Precision, $S(\mu^*)/\mu^*$, %	2	1	6	4	1	2	0.5	3

$S(\mu^*)$ is the standard deviation for ten replicate measurements of μ^*.

Table 5.9. Determination of mass absorption coefficient by attenuation of reflection from a crystalline substrate

	Mass absorption coefficient, cm^2/g				
	BeO	Al_2O_3	SiO_2	SiC	$CaCO_2$
Observed (mean value)	8.9	33.4	35.3	44.4	75.4
Calculated	8.6	32.2	34.9	43.9	75.5
Precision $S(\mu^*)/\mu^*$, %	7	4.5	6.5	5.5	3.0

then $\widetilde{m}$ from expression (5.65) can be replaced with $\widetilde{m}_{cr}$ to give the equation:

$$I_o'/I = 100 (\sin\theta/\sin\theta') \tag{5.66}$$

Usually, an angle θ' is chosen greater than the θ Bragg angle for sample diffraction. The ratio $I_o'/I' < 100$ is realistic and measurable, facilitating absorption measurements for a sample considered to be infinitely thick.

The range of μ^* determined by this technique may be extended to greater values if the absorption path for substrate diffraction is reduced by using the back-reflection range. Single-crystal (Brehler 1957), bulk metal (Williams 1959), and silver membrane (Altree-Williams et al. 1977) filters are used as reflecting substrates.

This technique was applied to powders with a wide spectrum of μ^*, from BeO to $CaCO_3$. The substrate was a single NaCl crystal; reflection (600); $2\theta = 110.0°$; CuK_α radiation. Measurements were repeated 10 times, with reloading of the sample holder; the results are given in Table 5.9. It seems that 5% is a fair estimate of the precision of the method discussed.

5.7.4 Determination of the absorption coefficient by diffraction techniques

The absorption coefficient can be determined if diffraction measurements are carried out for infinitely thick samples and samples of lesser mass. The diffracted intensity for the latter is expressed as:

$$I_i = I_\infty [1 - \exp(- 2\mu^* \tilde{m}_1/\sin \theta_i] \qquad (5.67)$$

where I_∞ is the intensity diffracted by the infinitely thick sample and $\tilde{m}_i$ is the sample loading (surface density). In fact, μ^* can be evaluated from equation (5.67). Combining two equations (5.67) for the samples of different $\tilde{m}_2$, we obtain a compact expression for μ^*:

$$\mu^* = \frac{\ln (L_1/L_2)}{2(\tilde{m}_2 - \tilde{m}_1)} \qquad (5.68)$$

where $L_l = I_\infty - I_1$ and $L_2 = I_\infty - I_2$.

Several diffraction peaks can be utilized to determine the mass absorption coefficient. The reported precision of the method is within 5%. A similar idea was used by Werner and Ekström (1980) in transmission mode analysis. Ergun and Tiensuu (1958) suggested that transmission mode and reflection mode diffraction be combined for evaluating μ^*. If the same sample of thickness l and loading $\tilde{m}$ is measured twice, once in a transmission arrangement and once in a reflection arrangement, then, assuming constant intensity and divergence of the primary beam, we arrive at the following ratio:

$$I_{tr}/I_{ref} = 2Z \exp(-Z)/[1 - \exp(- 2Z \cot \theta)] \qquad (5.69)$$

where $Z = \mu^*\tilde{m}/\cos \theta$.

For the case in which $2Z\cot \theta > 5$, equation (5.69) is simplified to:

$$I_{tr}/I_{ref} = 2Z \exp(-Z) \qquad (5.70)$$

which is solved for μ^*.

Pure diffraction techniques are rather tedious, demanding either multiple samples or multiple techniques. They do not offer any improvement of precision, although they can be useful in special cases when simpler transmission measurements are not viable.

5.7.5 Determination of the mass absorption coefficient by means of Compton scattering

Attenuation of incoherently scattered radiation by the sample material is subject to the same law as coherent scattering (Section 2.1).

For the symmetrical Bragg case, the intensity of Compton scattering is expressed as:

$$I_{inc} = \frac{I_o K (I_{inc})_o}{\mu_{in}^*} \tag{5.71}$$

where I_o is the intensity of the incident beam; K is a constant factor including some geometric and some numeric parameters, such as beam cross section and scattering angle; μ_{in}^* is the mass absorption coefficient for incoherently scattered radiation; and $(I_{inc})_o$ is the intensity of Compton scattering by a unit mass of the sample material.

If $(I_{inc})_o$ is independent of the chemical composition of the sample, then equation (5.71) can be applied for calculating μ^*, assuming that I_{inc} is measured. However, the assumption that $(I_{inc})_o$ is constant is correct only to a certain extent. $(I_{inc})_o$ in electronic units may be obtained by:

$$(I_{inc})_o = (I_{inc})_{at} / (M/A) \tag{5.72}$$

where M is the atomic weight, A is Avogadro's number, and $(I_{inc})_{at}$ is the incoherent scattering per atom, given in the International Tables for X-Ray Crystallography, Vol. IV (1974). $(I_{inc})_o$ varies less at large sin θ/λ values, but even in this case, light elements show significant increases of $(I_{inc})_o$. However, it can be expected that for mineral matter with major components, such as O, Na, Al, Si, Ca, and Fe, the variations of $(I_{inc})_o$ will be confined to within a few percent. Sahores (1972) estimated the maximal deviation as 3.8%.

If we measure the coherent scattering (for QXRD analysis) and the incoherent scattering of the same radiation, say CuK_α, then the difference in wavelength between incoherent and coherent scattering μ^*_{inc}/μ^*_{coh} can be substantial. However, a direct proportionality undoubtedly exists, and

$$\mu^*_{coh} = \alpha' \mu^*_{inc} \approx \alpha'' / I_{inc} \tag{5.73}$$

where α' and α'' are calibration constants that can be evaluated by measurements of incoherent scattering by materials with known mass absorption coefficients. This technique was proposed by Sahores (1972), who incorporated a special monochromator into a diffractometer to separate incoherent scattering. However, there are certain drawbacks to using relatively longwave radiation, like CuK_α, for measurements of Compton scattering, as a result of the sharp drop of intensity with wavelength (Blokhin 1957) and the greater sensitivity of $(I_{inc})_o$ to chemical compositions at low sin $(\theta)/\lambda$ values. [In this technique, shortwave radiations like MoK_α (Bar On et al. 1981) or AgK_α (Battaglia and Leoni 1977) are preferable.] In this case, however, there is a real gap between the wavelength of diffraction measurements λ_{coh} (usually CuK_α or longer) and the wavelength of incoherent scattering λ_{inc}. The relationship expressed in equation (5.73) is also valid in this case, if the wavelength dependence of μ^* for all major sample components can be described by the same function of λ in the spectral range λ_{coh}-λ_{inc}. This requirement is fulfilled if the sample components do not have an absorption edge in this spectral range, which is the case for most geological samples.

We used incoherent scattering of MoK_α radiation to determine mass absorption coefficients in a series of clays and sandstones (Bar On et al. 1981; Fang and Zevin 1985). The intensity of Compton scattering was measured with a Philips fluorescent spectrometer with a LiF 200 analyzer. A group of standards of known chemical composition was selected to cover a range of μ^* characteristics for geological samples, as was shown in Table 4.3. The linear regression constructed from these data gives:

$$\mu^* = 3.04 + 0.706T$$

where T is the time in seconds for collecting 10^5 counts in an incoherently scattered MoK_α peak and μ^* is the mass absorption coefficient for MoK_α radiation. The standard deviation from the regression line is 1.4 cm^2/g. The uncertainty in the slope and in the intercept point of the regression line causes the expected relative errors in estimated μ^* values of 4%-5% for μ^* = 30-50 cm^2/g and 2%-3% for μ^* = 70-100 cm^2/g. These errors are comparable with errors in other methods of determining μ^*. The technique described here is, of course, not universal. It cannot be applied to organic materials or to materials containing elements with an absorption edge between λ_{coh} and λ_{incoh}. It requires additional equipment to separate Compton scattering and usually greater sample mass in order to meet the "infinite" sample condition for more penetrating MoK_α or AgK_α radiation. Nevertheless, in many cases it can prove to be very convenient, accurate, and fast.

5.8 Pattern decomposition and simulation

5.8.1 *Introduction*

Pattern decomposition and simulation—two somewhat opposite procedures—are very important in QXRD, especially if the whole diffraction pattern rather than a few peaks is used for analytical purposes.

Decomposition of a whole diffraction pattern into separate peaks is needed when peak overlapping occurs and we want to separate peaks of an individual phase prior to analysis. In principle, quantitative analysis can be carried out with overlapping peaks, as was discussed in Section 4.8, but peak separation undertaken as an independent procedure will stabilize to the system being analyzed by diagonalizing the matrix of the type expressed in equation (4.63). Peak separation is a highly recommended step in standardless analysis (Section 4.10). In addition, in any QXRD technique employing the whole diffraction pattern, a weighting procedure can be initiated after peak separation, giving less weight to less reliable peaks susceptible to sample-related effects, such as preferred orientation. The analytical procedure of pattern decomposition is divided into two steps:

1. Peak separation (decomposition of clusters into individual peaks); and
2. Application of conventional QXRD methods referring to separate peaks.

The second procedure—pattern simulation—is needed when calculated data are used for calibration purposes; here again, preference is given to the whole diffraction pattern in the continuous-pattern mode.

The difference between the two methods is easily recognized: in pattern decomposition, a continuous pattern is decomposed into individual peaks and the integrated intensities of these peaks are measured. In pattern simulation, we begin with calculated integrated intensities and then simulate the continuous pattern. The two procedures have a common base, known as *profile fitting*, which is no more than an adjustment of any analytical function to the observed profile of a diffraction peak.

Suppose that the profile of any diffraction peak i of phase j is approximated by the function:

$$(I_{ij})_k = (I_m)_{ij} \ P\,(x_{ij}, \alpha_{ij}, \beta_{ij}, \gamma_{ij} \ldots) \tag{5.74}$$

where $(I_{ij})_k$ is the peak intensity of the scattering angle $2\theta_k$; $x_{ij} = 2\theta_k - (2\theta_m)_{ij}$, in which $2\theta_k$ is the running angle, $(2\theta_m)_{ij}$ is the angle corresponding to the maxima of the peak i, $(I_m)_{ij}$ is the peak intensity at $2\theta_k = (2\theta_m)_{ij}$; and α_{ij}, β_{ij}, γ_{ij} are parameters defining the shape of the function P. This function is usually defined analytically, and in this case the same class of functions is used to fit the profiles of different phases. Shape parameters α_{ij}, β_{ij}, and γ_{ij} depend on peak location, denoted by subscript i, which can be different for different phases due to probable phase-related peak broadening effects. In this definition, shape functions are normalized so that $P = 1$ at $x_{ij} = 0$.

Suppose now that $(I_{obs})_k$ is the intensity scattered by a multiphase specimen at a scattering angle $2\theta_k$. $(I_{obs})_k$ must be matched by the sum of the calculated intensities of the diffraction peaks and the background.

The former is the sum of $(I_{ij})_k$, as defined in equation (5.74), taken over all peaks i and phases j. The polynomial form is usually assigned to the background intensity $(I_b)_k$:

$$(I_b)_k = \sum_{s=o}^{n} b_s (2\theta_k)^s \tag{5.75}$$

where b_s are variable parameters and n is the order of the polynomial.

The best fit is achieved by minimizing the sum of the squared differences:

$$\sum_k W_k \left[(I_{obs})_k - \sum_{i,j} (I_m)_{ij} P(x_{ij}, \alpha_{ij}, \beta_{ij}, \gamma_{ij}) + \sum_{s=o}^{n} b_s (2\theta_k)^s \right] \tag{5.76}$$

where W_k is the weight assigned to the intensity measured at $2\theta_k$. The k summation is taken over the whole diffraction pattern or any part of it. Maximum intensities $(I_m)_{ij}$, peak positions $(2\theta_m)_{ij}$, shape parameters, and parameters of background polynomials are considered variables in minimizing the sum of the squares (5.76). As soon as optimal values are established for these parameters, clusters are separated into individual peaks, and integrated intensities are easily calculated.

The efficiency of fitting or, in other words, agreement between observed intensities $(I_{obs})_k$ and the calculated intensity $(I_{calc})_k$ is estimated by two closely related factors:

$$R = \frac{\sum_k \left[(I_{obs})_k - (I_{calc})_k \right]^2}{\sum_k (I_{obs})_k} \tag{5.77}$$

$$R_W = \left[\frac{\sum W_k \left[(I_{obs})_k - (I_{calc})_k \right]^2}{\sum W_k (I_{obs})_k^2} \right] \tag{5.78}$$

For good fitting techniques, the factors are within a few percent.

Note that in the minimization procedure expressed in equation (5.76) neither phase abundances nor structural data for the phases present are involved: clusters are separated into individual peaks solely on the basis of preliminary knowledge of approximate peak positions and the shape of the peak profile.

The profile shape function P from expression (5.74) is certainly a nonlinear function of shape parameters, and any nonlinear optimization algorithms must be applied in order to evaluate all variable parameters. A detailed description of the relevant mathematics and computer programming is not within the scope of this book. Two of the most popular algorithms are those of Nelder and Mead (1965) and Marquart (1963).

5.8.2 Profile-fitting functions

The observed diffraction profile $H(2\theta)$ results from the convolution of three functions:

$$H(2\theta) = F*G*S \qquad (5.79)$$

where S is the "pure" diffraction profile, obtained from the spectral distribution by applying Bragg's law; G is the convolution of all instrument- and sample-related aberrations; and F is the contribution of the sample's microstructure, such as short coherent length or any non-uniformity of the lattice.

Parrish et al. (1976) suggested that S be determined with a standard specimen that has a high angle reflection but is free of line broadening effects. For CuK_{α} radiation, they used the 444 reflection of silicon and the 400 reflection of tungsten. In this case, F and G can be considered as δ-functions, and the observed profile H is identical to S. Function S revealed a very weak K_{α_3} component in addition to the two obvious K_{α_1} and K_{α_2} peaks. Experimental points were fitted to the sum of seven Lorentzian functions—three each for the α_1 and α_2 peaks and one for the α_3 peak. If the α_2 and α_3 components are not removed by the monochromator or by computer-aided stripping, then three peaks α_1, α_2 and α_3 should contribute to any observed reflection. In profile-fitting techniques for QXRD, we do not need to fit the functions S, G, and F separately, but instead we fit their convolution H. In many cases, the weak α_3 component can be ignored, and the two peaks α_1 and α_2 must then be fitted to any observed reflection.

In the beginning of this discussion, we ignored sample-related broadening and supposed that observed diffraction peaks are symmetrical, leaving sample contribution and peak asymmetry to subsequent subsections. There is no simple analytical presentation of the diffraction peak profile, and it can only be approximated by a suitable analytical function. Among these functions, the two best known are the Gaussian and the Lorentzian functions.

Gaussian

$$H(x) = I_m \exp\left(-4\,(\ln 2)x^2/w_G^2\right) = I_m \exp\left(-\pi\, x^2/\beta_G^2\right) \qquad (5.80)$$

Lorentzian (Cauchy) function

$$H(x) = I_m\,(1 + 4x^2/w_L^2)^{-1} = I_m\,(1 + \pi^2\, x^2/\beta_L^2)^{-1} \qquad (5.81)$$

where $x = 2\theta - 2\theta_m$, $I_m = H(2\theta_m)$, $2\theta_m$ is the scattering angle, corresponding to the maximal intensity, w_G and w_L are full width and FWHM for Gaussian and Lorentzian functions, respectively, and β_G and β_L are integral peak breadths defined as $\left(\int_{-\infty}^{\infty} H(x)\,dx\right)/I_m$.

Each of these functions is even and is characterized by three parameters, which must be adjusted to the observed peak: the maximal intensity I_m, $2\theta_m$, and the peak width w or β. The principal difference between the Lorentzian and Gaussian functions lies in the rate of the decay in the tails. The Lorentzian function descends more slowly than the Gaussian function. A real diffraction peak, obtained with a conventional X-ray diffractometer and a monochromatic source, usually shows intensity decay intermediate between the two extremes—Gaussian and Lorentzian. Attempts to fit a diffraction profile by one of these functions revealed significant discrepancies between the observed and calculated profiles (Khattak and Cox 1977). The reason is that in the low-angle range, where instrumental aberrations dominate the peak profile, the peak shape is closer to Gaussian, whereas in the high-angle range, where the peak profile is dominated by chromatic dispersion, the peak shape approaches Lorentzian. Louër and Langford (1988) have clearly illustrated the monotonous increase of the Lorentzian component in the peak shape with increasing scattering angle.

In order to achieve better fitting, Taupin (1973) suggested that the diffraction peak be described by the algebraic sum of three or four Lorentzian functions, differing in terms of height, position, and width. The best fit is achieved by adjusting 9-12 variable parameters (three parameters for each Lorentzian function). The choice of Lorentzian functions confers a clear mathematical advantage, because these functions are easily convoluted.

These fitting algorithms were considerably improved and extensively used by Parrish and his co-workers (Parrish et al. 1976; Will et al. 1983). Each reflection, consisting of α_1, α_2, and α_3 components, was fitted by seven Lorentzian functions, three for the α_1 peak, three for the α_2 peak, and one for the α_3 peak. The total number of adjustable parameters was equal to 21.

Schreiner and Jenkins (1983) reduced the number of parameters to 12 in a similar model by taking into account the α_1–α_2 separation and approximating the α_1 and α_2 peaks as split Lorentzian functions with different half-width at half-maximum (HWHM) for low- and high-angle sides.

Despite the good fit usually achieved by a number of Lorentzian functions (R-factor as low as 1%-2%), a large number of refined plane shape parameters seem to be redundant. Taking into account the variations of peak shape with scattering angle, it is more natural to look for an analytical function with a shape that can vary between Lorentzian and Gaussian. Three such functions are extensively used in profile fitting, i.e., the Voigt function, the pseudo-Voigt function, and the Pearson VII function.

Voigt function—a convolution of Lorentzian and Gaussian functions

The Voigt function (Gupta and Anantharaman 1971; Langford 1978; Cox et al. 1988) is expressed as:

$$H(x) = \frac{I_m \beta}{\beta_G \beta_L} \int_{-\infty}^{\infty} L(u)\, G(x-u)\, du$$

$$= I_m (\beta/\beta_G) \text{ Real } \omega(\pi^{1/2} x/\beta_G + i\beta_L/\pi^{1/2}\beta_G) \qquad (5.82)$$

where G(x) and L(x) are the Gaussian and Lorentzian functions expressed in equations (5.80) and (5.81), respectively, supposing that in these equations $I_m = 1$, β is the integral breadth of the Voigt function, β_G and β_L are the integral breadths of the Gaussian and Lorentzian components of the Voigt function, and Real $\omega(Z)$ is a real part of the error function of the complex argument Z. The shape of the Voigt function is characterized by the Voigt parameter:

$$\phi = w/\beta$$

where w is the FWHM and β is the integral breadth of the Voigt function. There are two obvious limits for the Voigt parameter: $\phi_{min} = 2/\pi = 0.6366$ for a pure Lorentzian function and $\phi_{max} = 2(\ln 2/\pi)^{1/2} = 0.9394$ for a pure Gaussian function. Thus, in fitting a Voigt function to the observed profile, four parameters have to be adjusted, i.e., the peak height, the peak position, the peak width (w or β), and the Voigt parameter ϕ.

Pseudo-Voigt function

The pseudo-Voigt function (Hecq 1981) is the algebraic sum of the Lorentzian and Gaussian functions:

$$H(x) = I_m [\eta\, G(x) \cdot (1 - \eta)\, L(x)] \qquad (5.83)$$

where G(x) and L(x) are expressed in equations (5.80) and (5.81) supposing that $I_m = 1$ and η is the pseudo-Voigt parameter. It is usually assumed that the FWHM of the two components of the pseudo-Voigt function are equal and, naturally, equal to the FWHM of the pseudo-Voigt function itself. Thus, four parameters must be adjusted in fitting the pseudo-Voigt function, i.e., the peak height, the peak position, the FWHM, and the pseudo-Voigt parameter η, which defines the shape of this function.

Pearson VII function

The Pearson VII function (Hall et al. 1977; Brown and Edmonds 1980; Naidu and Houska 1982) is expressed as:

$$H(x) = I_m\, a \cdot b\left(1 + x^2 b^2\right)^{-m} \tag{5.84}$$

where

$$a = \pi^{-1/2} \frac{\Gamma(m)}{\Gamma(m - 1/2)}$$

$$b = \frac{2}{w}(2^{1/m} - 1)^{1/2}$$

in which w is the FWHM of H(x), and $\Gamma(m)$ and $\Gamma(m - 1/2)$ represent gamma functions of the corresponding arguments.

When m = 1, the Pearson VII function yields a pure Lorentzian function, and for $m \to \infty$ it approaches a Gaussian function. In fact, approximation to a Gaussian function is valid for m = 10-20. For m = 2, the Pearson VI function becomes a modified Lorentzian function, which is often used for profile fitting (Sonneveld and Visser 1975). There are four adjustable parameters in fitting the Pearson VII function, i.e., the peak height, the peak position, and two additional parameters—m, which can be considered a shape parameter, and a or b, which can be considered width parameters.

All three functions yield equal goodness of fit to symmetrical diffraction peaks, although the computational time is greater for the Pearson VII function (Young and Wiles 1982; Hill and Madson 1987). Voigt and pseudo-Voigt functions have the additional advantage of being easily decomposed into Gaussian and Lorentzian components. Each of them is supposed to bear information on specific microstructural effects, i.e., microstrains and coherent length, each depending in its own way on the scattering angle (de Keijser et al. 1982, 1983). In addition to the importance of studying microstructural effects, this separation is important in coding the angle dependence of the shape parameters in full-pattern fitting.

Among the other analytical functions suggested for profile fitting, let us examine the rational function of Pyrros and Hubbard (1983):

$$H(x) = I_m/(1 + A_1 x^2 + A_2 x^4) \tag{5.85}$$

where constants A_1 and A_2 describe the shape of the profile. This function approximates the Pearson VII function well, with the parameter m lying between 1 and 3.

Another class of functions used in profile fitting are not expressed analytically but are obtained as numerical profiles of free-standing (separate) peaks of analyzed phases. In order to perform the decomposition of the actual patterns, learned profiles must be extended to the sections of overlapping peaks. This can be done only by assuming a certain angular dependence of the peak width, which will be discussed later. Mortier and Costenoble (1973) and later Hepp and Baerlocher (1988) used learned peak shape functions for profile fitting in patterns of zeolites.

Finally, functions suitable for profile fitting can be synthesized by convoluting the spectral distribution S with the instrumental function G and the specimen-related function F [equation (5.79)]. Kogan and Kupriyanov (1992) suggested an effective technique for the relevant calculations, resulting in an even better fit than that achieved with the Pearson VII function.

5.8.3 Fitting of asymmetrical profiles

The observed profiles of diffraction peaks are very often asymmetrical. At high diffraction angles, where peak shape is defined by chromatic dispersion, peak asymmetry is related to the natural shape of the spectral line. However, the most obvious asymmetry is observed at low scattering angles, where it results from convolution with asymmetrical instrumental functions, the most prominent of which is axial divergence of the primary and diffracted beams. An asymmetrical profile is easily fitted by a sum of symmetrical functions, for example, Lorentzian functions displaced along the scattering angle axis. However, single symmetrical functions that have only a few adjustable parameters, such as the Voigt or Pearson VII functions, have a serious deficiency in fitting low-angle diffraction peaks. This problem is usually approached by splitting any of these functions onto two parts with different shapes and width parameters. Toraya (1986) obtained a very good fit with a split Pearson VII function, with different shape (decaying) parameters m and different widths of the half-maximum for the low-angle and high-angle parts of the profile. The asymmetry of the peak width was modeled by an asymmetry factor A, the ratio of the widths of the low-angle and high-angle parts of the function. Both the exponent m and the factor A varied considerably with the scattering angle.

Another way to account for profile asymmetry, which seems to be much more natural, although more complex in computation, is the convolution of any symmetrical fitting function with an asymmetrical instrumental function.

The two approaches to creating an asymmetrical profile fitting function—splitting and convolution—differ from one another in that the former affects the position of the centroid but not that of the peak maximum, whereas the latter affects both the peak maximum and the peak centroid. Diffraction peak asymmetry in the low-angle range is largely an outcome of vertical divergence. Unfortunately, in contemporary diffractometers with long tube focus, Soller slits, and monochromators, the relevant instrumental function cannot be expressed analytically. Only in one particular case, that in which the sample height is much greater than the height of the tube focus, the receiving slit instrumental function due to axial divergence may be expressed as (Kheiker and Zevin 1963):

$$G(x) = \begin{cases} a\,(|x| \cot\theta)^{-1/2} & x_m \leq x \leq 0 \\ 0 & x_m \geq x \geq 0 \end{cases} \tag{5.86}$$

where $x = 2\theta - 2\theta_0$ and $2\theta_0$ corresponds to the maximum of the peak; a = constant; and

$$x_m = (H_p^2/4R^2) \cot\theta \tag{5.87}$$

If the same function (5.87) is used to approximate the vertical divergence in a diffractometer with two sets of Soller slits, then x_m must be taken as:

$$x_m = \left[(R\delta)^2/(4R^2)\right]\left[\cot\theta + \cot 2\theta\right] \tag{5.88}$$

where δ is the divergence of the Soller slits and R is the radius of the goniometer.

The centroid of function (5.87) is located at $x_c = (1/3)\,x_m$. If x_m is expressed by equation (5.88), then the centroid position of the approximate function (5.87) is equal to the centroid position of the instrumental function when the specimen, receiving slit, and tube focus heights are equal to $R\delta$. This geometry can approximate a diffractometric arrangement with two sets of Soller slits (Zevin et al. 1963).

5.8.4 *Adjustable parameters and constraints*

If structural information on the phases is not available or has not been utilized in the profile-fitting routine expressed in equation (5.76), then the following parameters must be optimized for each phase: maximum intensities H_{im}, peak positions $2\theta_{im}$, peak shape parameters inherent in the chosen profile function, and background parameters. Reasonable constraints imposed on the adjustable parameters can improve the fit, reduce the number of parameters, and make the fitting more reliable. If both $\alpha_1 - \alpha_2$ peaks are present, constraints can be imposed on the position, peak intensity, and width of the α_2 component. The angular distance between the components is easily calculated from the well-documented wavelengths of the K_{α_1} and K_{α_2} spectral lines. The ratios of intensities and peak widths depend to some extent on the monochromator setting. They can be established by examining the well-resolved high-angle $\alpha_1 - \alpha_2$ reflection from a well-crystallized sample (Parrish et al. 1976).

In the absence of specimen-related peak broadening [function F in expression (5.79) is considered a σ-function], peak shape parameters are smooth and are usually a monotonous function of the scattering angle. The same is true if specimen-related broadening is isotropic, i.e., if it depends only on the value but not on the direction of the reciprocal vector. It is very attractive to model peak width and other shape-related parameters by a function of the scattering angle that has only a few constants. During the profile fitting for the whole pattern, we have refined only these constants instead of the shape parameters for many peaks.

Caglioti et al. (1958) suggested the following expression for the FWHM in neutron diffractometry:

$$\mathrm{FWHM} = (u \tan^2 \theta + v \tan \theta + w)^{1/2} \tag{5.89}$$

where u, v, and w are optimized constants. This model is often accepted for X-ray diffractometry. Moreover, a proper fitting function, in addition to FWHM, is characterized by at least two more parameters, such as for example, the exponent m and the asymmetry parameter for the Pearson VII function. These parameters must also be modeled. It seems that the only way to do so is to fit the chosen profile function to well-resolved peaks of any standard material that does not exhibit microstructural effects. The LaB_6 Standard Reference Material of the NBS seems to be a suitable candidate.

The presence of anisotropic microstructural effects considerably complicates the fitting procedure. A detailed strategy for dealing with this case is to be found in the paper of Delhez et al. (1993).

Available structural information brings additional constraints to optimized parameters. If lattice constants are known, then peak positions are obviously a constraint. In such a case, we have to optimize a few lattice constants instead of a much larger number of peak position parameters $2\theta_{ij}$. In this case, profile fitting also results in a refined unit cell. This type of procedure was developed by Pawley (1981) and improved by Jansen et al. (1992). The latter suggested a two-step fitting procedure with intensities and background parameters being optimized in the first stage and lattice constants and peak shape parameters in the second. This two-step procedure differs from the mathematical approach in that it is linear, whereas the mathematical approach is nonlinear. These steps are repeated until a statistically reasonable fit is obtained.

Peak intensities $(I_m)_{ij}$ are constrained if crystal structure data are involved in profile fitting. The corresponding optimization procedure leads to simulation of patterns of pure phases, which are used as analytical standards. There are two approaches to dealing with structural data.

1. Crystal structure data are used as such to calculate integrated intensities of the diffraction peak $(I_{int})_{ij}$. Maximal intensities from expression (5.74) are equal $(I_m)_{ij} = (I_{int})_{ij}/B_{ij}$, where the integrated breadth B_{ij} is one of the refinable peak shape parameters. Thus, intensities are excluded from the list of optimized parameters. However, scaling factors have to be added to each phase to account for its abundance, as was discussed in Chapter 4. As soon as peak shape parameters are established in an optimization routine, they can be used together with calculated integrated intensities to simulate the diffraction pattern of a pure phase, which corresponds to a chosen diffractometer setting.
2. Crystal structure data, such as atomic positions, occupational parameters, and lattice constants, are refined in the same optimization procedure as peak shape parameters. Intensities are excluded from the variable parameters, although new variables—structure parameters and phase scaling factors—are added. This is the Rietveld technique, as discussed in Section 4.9.3. Quantitative phase analysis using the Rietveld method was recently summarized by Hill (1993).

The final group of variable parameters in the optimization procedure are those of the background polynomial. In fact, if the background is well-defined on the observed pattern, it can be removed from $(I_{obs})_k$ data, and the background term can then be excluded from expression (5.74). However, the great virtue of the pattern decomposition technique is the determination of the true background in the case of severely overlapping peaks.

5.8.5 *Number of separated peaks and angular resolution*

Suppose that we intend to decompose a cluster of overlapping peaks. In order to initiate the optimization procedure (5.76), the peak position angle $2\theta_{i0}$ must be assigned to each peak of the cluster. Thus, the peak hunting routine must include cluster decomposition. Taupin (1973) suggested that the optimization procedure be attempted with 0, 1, 2,..., peaks in the cluster in turn until the fit could not be improved by adding more peaks. Each added peak is positioned at the angle at which the difference between the observed intensity and the background level is maximal. Better peak hunting techniques are based on numerical derivation of observed intensity data. In such techniques, the peak position is associated with the minimum of the second derivative. Success in locating closely overlapping peaks depends largely on the quality of the raw data, because derivative curves are strongly influenced by counting statistics. Naidu and Houska (1982) analyzed errors in the second derivative and came to the conclusion that 10^4-10^5 peak counts are needed to resolve overlapping peaks with a separation of 0.5 FWHM or even less.

In the situation usually encountered in QXRD, the approximate peak position in a cluster can be deduced from such available data as diffraction patterns of analyzed phases or lattice parameters of these phases. Nevertheless, an independent peak search is encouraged, because it can reveal peaks of an undetected phase.

For readers with a special interest in whole pattern simulation and refinement, additional information may be found in the book edited by Young (1993).

5.9 Methodology of corrections for preferred orientation

5.9.1 *Introduction*

Preferred orientation of particles is one of the most serious problems in QXRD. It may even give rise to doubts as to the validity of this technique as an analytical tool. However, in many cases the problem is not one of preferred orientation itself but rather of variations in the preferred orientation within replicates of the same sample and between samples. In the variation of the preferred orientation phenomenon, there is a component of a random nature, and the problem may be solved by increasing the number of replicates and averaging the measured peak intensities. However, the average result usually has a systematic character that depends on the morphology of the particles and the symmetry of the crystals in different phases.

In polycrystalline materials, the preferred orientation has its origin in certain material-processing technologies, such as the rolling of metals or the extrusion of polymers. In such cases, plastic flow proceeds in strongly preferred crystallographic directions, resulting in preferred orientation of the grains. Another example that is often encountered is directional crystal growth in thin films, in which case the final crystal orientation is dictated by the minimal free energy concept. Preferred orientation is also common in the preparation of powder samples if the particles are not spherical but are predominately platelike or needle-like. Typical examples are platelike clay particles and fibrous minerals. A more detailed account of texture-susceptible processes and materials is to be found elsewhere (Bunge 1982).

Preferred orientation of crystallites is easily detected by a systematic and usually significant deviation of peak intensities from the normal sequence obtained either by calculation or experimentally—by running a number of random samples.

The intensity of the hkl reflection for a polycrystalline sample is proportional to the pole density $P^{hkl}(\psi,\phi)$ in the direction (ψ,ϕ) of the diffraction vector $\overline{S}$ (see also Section 3.3, Fig. 3.9). Since the diffracted intensity is proportional to the scattering volume, it is reasonable to define the pole density as the volume fraction of the crystallites in the sample, which are in reflection position for the sample direction (ψ,ϕ):

$$P^{hkl} = \frac{dV}{V} \Big/ \frac{d\Omega}{4\pi} \tag{5.90}$$

where V is the total volume and dV is the volume of all the crystallites where the direction of the plane normal (hkl) has the orientation (ψ,ϕ) within the solid angle element $d\Omega$.

For a random sample, the pole density is constant, i.e., it does not depend on direction in a sample-bound coordinate system. From definition (5.90), it follows that:

$$P^{hkl}_{ran}(\psi,\phi) \equiv 1 \tag{5.91}$$

for samples of random orientation distribution. For a textured sample, the pole density is not constant, and the diffracted intensity is expressed as:

$$I^{hkl}_{tex} = I^{hkl}_{ran} P^{hkl}(\psi,\phi) \tag{5.92}$$

where the term I^{hkl}_{ran} represents the general expression for diffracted intensity by a polycrystalline randomly oriented material given in equation (2.19).

Under certain circumstances (high axial divergence and strong texture), some integration does take place (Zevin 1990). The intensity is not proportional to P^{hkl} itself but to some average value of P^{hkl} taken over a certain angular area around the diffraction vector. In the following discussion, we will set aside this special case and assume that equation (5.90) is valid, so that the intensity diffracted by random sample is given by:

$$I_{ran}^{hkl} = I_{tex}^{hkl} / P^{hkl}(\psi,\phi) \tag{5.93}$$

There are three methodological approaches to dealing with preferred orientation:

1. Avoiding preferred orientation of particles by selecting the optimal sample preparation technique.
2. Reducing preferred orientation effects by proper analytical procedures, i.e., selecting analytical diffraction peaks that do not vary strongly with preferred orientation, utilizing the pattern for analytical work, and so on.
3. Correction of measured intensities for preferred orientation.

The first approach is applicable to powder samples for which preferred orientation may be reduced or completely eliminated by the special handling techniques discussed in Section 5.3. For bulk samples exhibiting preferred orientation, intensity correction seems to be the only way to account for preferred orientation.

5.9.2 Presentation of the pole distribution

The pole distribution is usually presented as a pole figure, i.e., a stereographic projection of the pole distribution. Suppose that we examine a flat sample. The Z axis of the rectangular external (specimen-bound) coordinate system is chosen to be normal to the sample plane. The two other axes may be chosen arbitrarily if the pole distribution has axial symmetry. If there are special directions in the sample plane (rolling or transverse direction in a rolled plate), these directions together with the OZ axis constitute the coordinate system. In a system of spherical coordinates, the pole density P^{hkl} is a function of two variables, the polar angle ψ and the azimuthal angle ϕ (Fig. 5.13). The function $P^{hkl}(\psi,\phi)$ must follow the obvious normalization:

$$\frac{1}{4\pi}\int_0^{\pi}\int_0^{2\pi} P^{hkl}(\psi,\phi)\sin\psi \; d\psi \; d\phi = 1 \tag{5.94}$$

The (hkl) pole figure can be obtained by measuring the intensity of the reflection hkl with the diffraction vector $\bar{S}$ in different directions (ψ,ϕ) in the sample. The data are normalized according to equation (5.94), and the pole distribution $P^{hkl}(\psi,\phi)$ is obtained. Suppose that in Bragg-Brentano geometry, OZ ($\psi = 0$, $\phi = 0$) is the direction of the

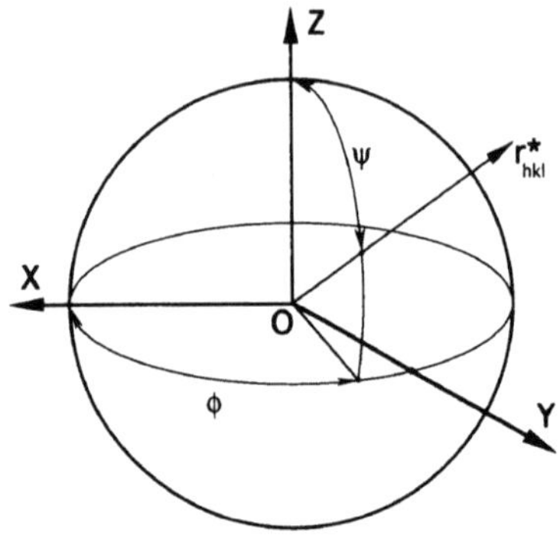

Figure 5.13. Sample-bound coordination system XYZ.

diffraction vector and XOZ is the equatorial plane. Then, in order to bring any point with spherical coordinates ψ and ϕ to the point $\psi = 0$, $\phi = 0$, two sample rotations are required. One rotation must be around the OZ axis because this is the simplest rotation to execute. The second rotation may be around the OX axis in the most commonly used Schultz method (Schultz 1949) or the OY axis (the less acceptable ω-scan method). A detailed discussion of these experimental methods may be found elsewhere (Bunge 1982; Klug and Alexander 1974).

Although the procedure looks rather simple, in practice it is a formidable time-consuming technique, usually requiring special equipment. Moreover, the outer part of the pole distribution ($70° \le \psi \le 90°$) cannot be obtained in a reflection mode diffractometer, and complementary measurements with a transmission mode scheme are needed to complete the distribution.

The experimental work may be simplified considerably by spinning the sample around the polar axis OZ and by measuring the time average of the diffracted intensity. The pole distribution calculated with these data depends only on the polar angle ψ. The normalization equation (5.94) now becomes:

$$\frac{1}{2}\int_{-\pi}^{\pi} P(\psi)\sin\psi \; d\psi = 1 \tag{5.95}$$

Nevertheless, even with this simplification, the texture correction must include the following steps:

1. Drawing the pole distribution curve by measuring the intensity of the chosen reflection at polar angles from $\psi = 0$ up to a maximal value ψ_{max};
2. Completing the distribution curve at $\psi_{max} < \psi < 90°$ by extrapolation or transmission mode measurement;

3. Performing normalization according to equation (5.95);
4. Evaluating $P^{hkl}(\psi,\phi)$ and the intensity correction.

If more than one analytical peak is used for quantification or if the whole-pattern approach has to be used, the appropriate correction must be applied to every peak of the phase exhibiting preferred orientation. However, we must remember that in QXRD we are not interested in the pole figure itself but in the diffraction peak intensity integrated over all the sample orientations, i.e.:

$$I_{in}^{hkl} = \int_0^{\pi/2} I^{hkl} \sin\psi \, d\psi \quad \text{(sample spinning activated)}$$

It would be very attractive to automate this integration, and in the 1960s several analog devices were proposed for this purpose. Ruland and Dewaelheyns (1967) proposed an analog integration circuit controlled by the current from a conventional rate meter. Desper and Stein (1967) suggested an analog servomechanism that continuously changes the velocity of the ψ-rotation in such a way that time integration would simulate the sin ψ term in equation (5.95). Currently available computer-controlled diffractometers with special geometry provide a solution to this problem. If the ψ motion is obtained in a step-wise mode, then the required sin ψ correction can be easily introduced into the computer data processing together with all the required corrections for variations in the diffraction geometry. Nevertheless, even automatic integration over each analytical peak seems to be too time and labor consuming. In attempting to reduce experimental work, we should remember that pole distributions are interdependent and thus only a few are needed to evaluate all the others.

The general problem has been discussed in detail by Bunge (1982). Roe and Krigbaum (1964) have dealt with axial texture and have shown how to derive the pole distribution of any pole after a sufficient number of distributions have been measured experimentally. Our problem is thus limited to the evaluation of the pole density $P^{hkl}(\psi,\phi)$ in the direction taken over a certain angular area around the diffraction vector. In the subsequent discussion in this section, we will consider methods of texture correction, which can reduce the volume of experimental work in the evaluation of the pole distribution.

5.9.3 *Orientation distribution function*

We will restrict our discussion to the axial (fiber) texture and follow the formalism given by Roe and Krigbaum (1964). Suppose that XYZ is the external sample-bound system of mutually orthogonal axes (Fig. 5.13), XOY is the sample plane, and OZ is the texture axis. Then,

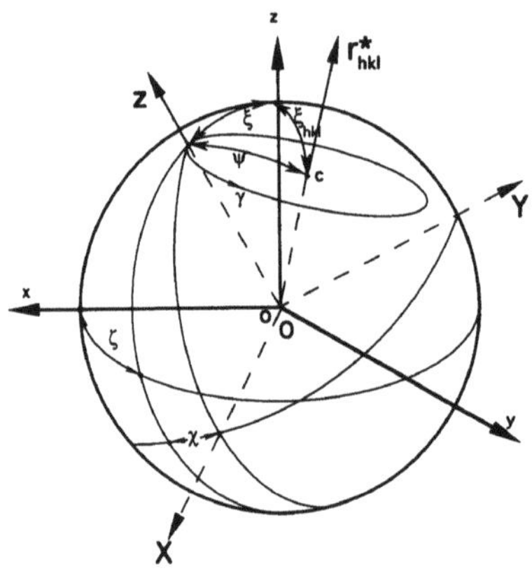

Figure 5.14. Crystal-bound coordination system xyz.

axial symmetry in the pole distribution is either present naturally or ensured by sample rotation around the OZ axis.

Suppose also that xyz is an internal or crystallite-bound system of orthogonal axes. Orientation of any crystallite will be defined if we first fix the axis OZ in the xyz system (O and o coincide) and then the rotation around OZ. The direction OZ in the internal coordinate system is defined by the polar angle ξ and the azimuthal angle ζ (Fig. 5.14), and the crystallite rotation is defined by the angle χ between plane zOZ and XOZ. Thus, in the general case, crystallite orientation is described by the function $W(\xi, \chi, \zeta)$ with three variable angles ξ, χ, and ζ. For an axially symmetrical texture with OZ as the texture axis, W is obviously independent of χ, and the number of crystallites having orientation within the angular ranges of $d\xi$, $d\zeta$ is proportional to $W(\xi, \zeta) \times \sin\xi \, d\xi \, d\zeta$.

The distribution function for any reciprocal vector, e.g., the vector r^*_{hkl} (Fig. 5.13), is also axially symmetrical and is a function of only one variable, the polar angle ψ. The corresponding pole density function is $P^{hkl}(\psi)$. The distribution functions $W(\xi, \zeta)$ and $P^{hkl}(\psi)$ are interdependent. If a number of pole distributions $P^{hkl}(\psi)$ are obtained experimentally, then the crystalline distribution function $W(\xi, \zeta)$ can be calculated.

Within the framework of QXRD, our objective is limited, i.e., evaluating $P^{hkl}(\psi)$ for various (hkl) in the directions of the diffraction vector, or especially at $\psi = 0$ for Bragg-Brentano geometry. The values of the pole orientation function $P^{hkl}(0)$ are needed for intensity correction to expression (5.93).

In the case of axial texture, the relationship between $W(\xi, \zeta)$ and $P^{hkl}(\psi)$ is easily established. As can be seen in Fig. 5.13, the pole density $P^{hkl}(\psi)$ receives a contribution from each crystal whose reciprocal vector $\bar{r}^*_{hkl}$ makes the angle ψ with the OZ axis. In Fig. 5.14, the pole of $\bar{r}^*_{hkl}$ is the point c and its coordinates are (ξ_{hkl}, ζ_{hkl}). All the OZ axes that make the angle ψ with $\bar{r}^*_{hkl}$ intersect the unit sphere on the circumference of the small circle c in Fig. 5.14. Thus, the pole density is proportional to the mean value of the orientation distribution function $W(\xi, \zeta)$ on the circle whose radius corresponds to the angle ψ and whose center lies at a point on Oc.

$$P^{hkl}(\psi) = \langle W(hkl, \psi) \rangle = \frac{1}{2\pi \sin\psi} \oint W(\xi, \zeta) d\gamma \tag{5.96}$$

where the integration is carried out over the small circle and γ is the running angle.

We suppose that $W(\xi, \zeta)$ is normalized and that:

$$\frac{1}{4\pi} \int_{4\pi} W(\xi, \zeta) \ d\Omega = 1 \tag{5.97}$$

For $\psi = 0$, equation (5.96) is simply:

$$P^{hkl}(0) = W(\xi_{hkl}, \zeta_{hkl}) \tag{5.98}$$

5.9.4 *Determination of the orientation distribution function $W(\xi, \zeta)$*

The orientation distribution function $W(\xi, \zeta)$ cannot be measured directly but is usually calculated from experimentally obtained pole density distributions, pole figures or inverse pole figures. The conventional texture analysis technique used for pole-figure measurements was discussed in Section 5.9.2. The Schultz method, in which the sample is tilted around the ψ axis (Fig. 5.1), is more versatile than the ω-scan, in which the sample is rotated around the goniometer axis. Both techniques are, however, geometrically hindered, resulting in limitation of the range. $P^{hkl}(\psi)$ can be evaluated for the angle of tilt $\psi < 60°$-$70°$ for the Schultz technique and $\omega < \theta_{Bragg}$ for the ω-scan. On the other hand, the area of $\psi > 60°$ is very important in the normalization procedure (5.95) due to the higher $\sin\psi$ term. It seems that in a number of cases $P^{hkl}(\psi)$ can be extrapolated to $\psi = 90°$ by one of the approximating functions discussed in the next section. The unknown parameters of the function are found by the best fit to the experimentally measured part of the polar axis distribution. Nevertheless, it must be remembered that both techniques require special geometry with low axial divergence and special equipment, i.e., a multiaxis goniometer for the Schultz method and separate θ,2θ rotation for the ω-scan.

In addition to the pole-figure data, inverse pole-figure data can be obtained as basic experimental information about texture. If Bragg-Brentano geometry is used in intensity measurements, then $\psi = 0$ and $\omega = 0$, and some discrete values for $W(\xi, \zeta)$ can be calculated from equation (5.98). According t equation (5.92), $P(0) = I_{tex}/I_{ran}$, so the value of $W(\xi, \zeta)$ at reflection {hkl} is:

$$W(\xi_{hkl}, \zeta_{hkl}) = KI_{tex}^{hkl} / I_{ran}^{hkl} \tag{5.99}$$

where I_{tex}^{hkl} is the intensity of reflection hkl in a textured sample, I_{ran}^{hkl} is the intensity of the same reflection in a random sample, and K is a scaling constant. It is assumed that a random sample is not available (otherwise we would not need the preferred orientation correction). I_{ran}^{hkl} may be acquired from calculation from known crystal structure. The coefficient K is evaluated through normalization (5.97). It must be remembered that only a limited number of reflections hkl are available and that the orientation distribution function $W(\xi, \zeta)$ is defined in only a limited number of directions (ξ_{hkl}, ζ_{hkl}). For more efficient use of this limited data, a further assumption is usually made that (ξ_{hkl}, ζ_{hkl}) is independent of the angle ζ. This condition seems to be fulfilled adequately in the case of uniaxial crystals. Thus, in carrying out normalization (5.97), we have to interpolate between adjacent angles. An example of distribution of the 001 poles for hexagonal barium iron oxide obtained by the described method and normalized according to equation (5.97) is shown in Fig. 5.15.

The distribution $W(\xi)$ is easily extended up to $\xi = 90°$ due to the presence of hk0 reflections from the planes normal to [001]. A typical drawback of this method is the black area around $\xi = 0$, which usually surrounds the major crystallographic directions. The behavior of $W(\xi)$ in the beginning, when it descends rapidly, becomes somewhat uncertain. Fortunately, the contribution of this part of $W(\xi)$ into the integral in equation (5.97) is not decisive, due to the sin ξ term. Nevertheless, if the initial part of $W(\xi)$ is needed, it may be obtained by any other method of texture analysis.

From the measured pole figures or inverse pole figure data, the orientation distribution function $W(\xi, \zeta)$ can be determined from equation (5.96). There are various methods to calculate the solutions (Bunge and Esling 1982), but usually some approximate analytical function is used:

$$W(\xi, \zeta) = F(\xi, \zeta, a, b, c, \ldots.) \tag{5.100}$$

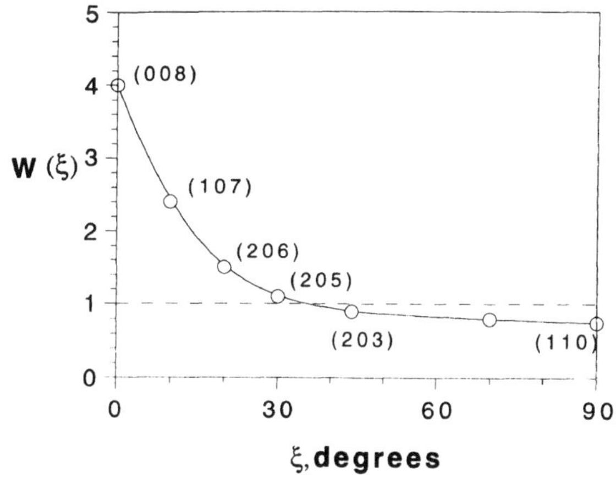

Figure 5.15. Inverse pole figure for $BaFe_{12}O_{19}$.

where a,b,c are unknown parameters of the approximation function, which must be adjusted in accordance with the experimental values $P^{hkl}(\psi_i)$. As usual, this is achieved by minimizing the sum of squares:

$$\sum_{hkl,i}[P^{hkl}(\psi_i)-\langle W(hkl,\psi_i)\rangle]^2 \tag{5.101}$$

where the summation is carried out over all reflections hkl and all sample directions ψ_i. The approximation function $F(\xi, \zeta, a, b, c....)$ obtained by the minimization procedure (5.101) is subjected to the obvious normalization (5.97), and then the properly scaled pole density $P^{hkl}(0)$ values are obtained by equation (5.98). The intensity diffracted by a random sample is given by equation (5.93).

A simplified semiempirical approach to scaling the intensity data collected for a textured sample may be found in works devoted to the analysis of retained austenite in steels (Arnell 1968; Dickson 1969) and in a more recent study of the composition of textured coatings (Gu and Marder 1991). In this approach normalization by equation (5.97) is replaced by summation over all available $P^{hkl}(0)$:

$$V_i = \frac{1}{n_i}\sum_{hkl}[P^{hkl}(0)]_j \tag{5.102}$$

$$P^{hkl}(0) = I^{hkl}_{tex} / I^{hkl}_{ran}$$

where V_j is the volume fraction of phase j, n_j are the number of reflections of phase j, and I^{hkl}_{ran} is the calculated intensity. Gulberg and

Lagneborg (1966) showed that the sum in expression (5.102) is more or less stabilized for $n > 5$-6, when the major reflections of austenite and martensite are included in the summation. Although their method lacks a sound theoretical basis, it is suitable for the analysis of austenite, but there is no proof of its virtue in the analysis of other textured materials.

5.9.5 *Crystallite orientation distribution—the general case of axial texture*

As a consequence of the structure or preparation of the sample, there can be approximate axial symmetry—known as *fiber symmetry*—but the rotation of the sample ensures the real axial symmetry as time average (as mentioned in Sections 5.9.2 and 5.9.3). In this case, the crystallite orientation distribution is a function of two angles, ξ and ζ (Fig. 5.14). $W(\xi, \zeta)$ may be calculated from expression (5.96). Järvinen (1993) suggested that $W(\xi, \zeta)$ can be expanded in terms of symmetrized harmonics Y_{mn}:

$$W(\xi,\zeta) = \sum_{n,m} C_{nm} Y_{nm}(\xi,\zeta) \tag{5.103}$$

When $n = 0$ and $m = 0$, then $C_{00}=1$ due to the normalization (5.97) and the other C_{nm} are adjustable parameters.

The symmetrized harmonics are orthogonal functions that obey point-symmetry operations in crystals. The basic functions are real spherical harmonics (used in quantum mechanics for the orbitals of hydrogen atoms):

$$Y_{nmp}(\xi,\zeta) = N_{nm} L_n^m(\cos\xi)_{\sin m\zeta}^{\cos m\zeta} \tag{5.104}$$

where N_{nm} is a constant for normalization, L_n^m are associated Legendre polynomials of the order n, and m runs for different terms in order $n \le m$. The ζ-variance depends on parity p, so that for even parity the function is cosine and for odd parity it is sine. The selection rules for spherical harmonics in different Laue classes were established by Kurki-Suonio (1977). The selection rules for the indices nmp in different Laue classes are given in the literature (Järvinen 1993).

There are two different families of symmetrized harmonics: The first type are functions corresponding to the noncubic site symmetries, which can be called axial harmonics (Järvinen 1993). Only harmonics with even order are used in expansion (5.103), and it is sufficient to examine merely the relevant Laue class of the crystal. The sym-

Table 5.10. First even-order zonal Legendre polynomials

Order	Legendre polynomial
0	1
2	$1/2\,(3\cos^2\psi - 1)$
4	$1/8\,(35\cos^4\psi - 30\cos^2\psi + 3)$
6	$1/16\,(231\cos^6\psi - 315\cos^4\psi + 105\cos\psi - 5)$

metrization annihilates those functions that do not suit the Laue class. The second type are functions for cubic site symmetries, which are known as cubic harmonics (Järvinen 1993). The symmetrization of cubic harmonics needs an extra threefold axis in the [111] direction.

According to Järvinen (1993), the integration of equation (5.96) gives:

$$P^{hkl}(\psi) = T^{hkl} \sum_{n,m} C_{nm} Y_{nm}(\xi_{hkl}, \zeta_{hkl}) L_n(\cos\psi) \qquad (5.105)$$

where $Y_{nm}(\xi_{hkl}, \zeta_{hkl})$ is the value of the harmonics in the direction of the plane normal (hkl), $L_n\cos\psi$ is an n-order Legendre polynomial (Table 5.10), and T^{hkl} is a scaling factor. The left-hand side of equation (5.105) is the experimentally obtained pole density distribution $P^{hkl}(\psi)$, and the right-hand side includes the unknown expansion coefficient C_{nm} and the scaling factor T^{hkl}, which has to be evaluated by the best fit of the right-hand side sum of the observed $P^{hkl}(\psi_i)$.

When the pole densities $P^{hkl}(\psi_i)$ of different reflections hkl with different orientations ψ_i are measured, a set of linear equations (5.105) can be obtained. If the number of these equations exceeds the number of parameters, they can be solved by the well-known least-squares method. The amount of experimental data needed in these calculations depends on the symmetry of the crystal, the degree of texture of the sample, and the desired accuracy.

The question in the application of this technique is the minimum number of harmonics required for adequate representation of $W(\xi, \zeta)$. There is no unambiguous answer, although it is clear that the stronger the texture, the greater the number of harmonics required. The conventional number is $n \leq 10$ and $m = 0,1$, resulting in four to six adjustable coefficients C_{nm}. In a study of preferred orientation of pressed Mg powders with maximum $P^{001}(0) = 1.759$, identical corrections were obtained with $n = 4$ and $m = 0$ (two C_{nm} parameters) and with a more extended series $n \leq 10$ and $m = 0,1$ (eight C_{nm} parameters) (Pesonen et al. 1973). The practical limitation of the series (5.103) by the maximum angle $\psi = 60°$ in the experimental curve $P^{hkl}(\psi)$ can be compensated for by increasing the number of reflections used to produce $P^{hkl}(\psi)$ curves and to evaluate expansion coefficients.

Examples of the application of the method with a detailed discussion of the form of the symmetrized harmonics, the number of needed reflections, and the precision are to be found in the works of Järvinen et al. (1970) for cubic crystals, Pesonen et al. (1973) for hexagonal crystals, Pesonen (1979) for tetragonal crystals, Paakkari et al. (1988) for monoclinic crystals, and Järvinen (1993) for trigonal crystals.

5.9.6 *Approximation of the orientation distribution function W(ξ,ζ)*

We will limit this discussion to a needlelike texture type for which the orientation distribution is a function of only one variable, ξ. The analytical approximation of W(ξ) is needed in a number of circumstances:

1. to extrapolate to $\xi = 90°$ the experimentally obtained curve that is limited in the reflection mode to $\xi \leq 60°$-$70°$;
2. to interpolate between discrete experimental points if the inverse pole figure approach is used to create W(ξ); and
3. to account for the preferred orientation in the whole pattern approach to QXRD, especially in combination with structure refinement.

The most general W(ξ) may be expanded by an appropriate function, the Legendre polynomial apparently being the most suitable. In fact, we have here a particular case of the general crystallite orientation function expanded in a series of spherical harmonics [equation (5.103)]. In this case:

$$W(\xi) = \sum_{n} C_n L_n(\cos\xi) \tag{5.106}$$

where $L_n(\cos\xi)$ is an n-order zonal Legendre polynomial (Table 5.10) and C_n represents an unknown coefficient. Since the Legendre polynomial coefficients are orthogonal, C_n can be calculated from the following equation:

$$C_n = \int_0^{\pi/2} W(\xi) L_n(\cos\xi) \sin\xi \; d\xi \tag{5.107}$$

For normalization, $C_0 = 1$. The other C_n can be chosen in terms of best fit of experimental data [(left-hand side of equation (5.106)]. Due to even parity of W(ξ), only even-rank polynomials are used in the expansion

(5.106). The number of terms in the expansion needed for adequate approximation of $W(\xi)$ apparently grows with the strength of the texture (only one $P_0 = 1$ is needed for a random sample). de Wolff and Sas (1969) supposed that no less than five terms are needed, even for a weak texture. Välvoda (1986) obtained a rather good approximation of the pole-distribution function of pressed cubic TiC with four terms n = 4, 6, 8, and 10.

In uniaxial crystals with an exclusive axis preferably oriented normal to the sample plane, $W(\xi)$ is maximal at $\xi = 0$ and descends monotonously to $\xi = 90°$. Many attempts have been made to produce an analytical approximation of such a distribution with a minimal number of adjustable parameters (Dollase 1986). Some of these functions are presented in Table 5.11.

In all analytical forms, G is an adjustable parameter that defines the central tendency of the corresponding distribution. The distribution of March (1932) has the advantage of being normalized according to expression (5.97). For other functions, the scaling parameters W(0) must be varied to fit the normalization condition after each adjustment of the dispersion (central tendency) parameter G. In addition, the March distribution has—in a way—a theoretical basis, since it was developed to describe the compaction of platelike grains under uniaxial stress. The coefficient G in the March distribution has a physical meaning—that of the compaction ratio. For a distribution with a maximal density at $\xi = 0$, the coefficient $G < 1$.

If the preferred orientation is very strong, so that $W(0) > 10$ and the distribution density at $\xi > 30°$ is negligible, a normal distribution fits the experimental data rather well. In the angular range, the distribution of Uda (1967) is very similar to that of Capkova and Valvoda (1974) (Table 5.11). In this case, the March model is less compatible with the experimental data.

All functions presented in Table 5.11 are equally good for a mild texture, in which $W(0) < 1.2$-1.3. In the intermediate case of moderate texture for which the pole density is still appreciable at $\xi = 90°$ and the normal distribution function thus descends too fast, two functions, the

Table 5.11. Analytical approximation of the crystallite distribution $W(\xi)$

Analytical form	Reference
$W(0) \exp(-G\,\xi^2)$	Uda (1967)
$W(0) \exp(-G \sin^2 \xi)$	Capkova and Valvoda (1974)
$W(0) \exp(G \cos^2 \xi)$	Sturm and Lodding (1968)
$(G^2 \cos^2 \xi + G^{-1} \sin^2 \xi)^{-3/2}$ $(G< 1)$	March (1932), Dollase (1986)

Table 5.12. Angles ξ_0 of unit density of the crystallite orientation function $W(\xi)$

	Angle ξ_0, degrees for various approximation functions		
$W(0)$	$\exp(-G\xi^2)$	$\exp(-G \sin^2 \xi)$	$(G^2\cos^2 \xi + G^{-1}\sin^2 \xi)^{-3/2}$
2	54	57	43
3	51	54	41
4	48	50	39
5	46	48	38

March distribution and the function suggested by Capkova and Valvoda (1974), are superior. Dollase (1986) tried various approximations in Rietveld structure refinement for minerals exhibiting preferred orientation and found the March distribution and the $\exp(-G \sin^2 \xi)$ function superior for mild [$W(0) \sim 1.4$], medium [$W(0) \sim 2$], and strong [$W(0) \sim 6$] textures.

5.9.7 W(ξ) of diffraction peaks not dependent on preferred orientation

For any normalized crystallite orientation, the distribution $W(\xi)$ value $W(0) > 1$ and the value $W(90°) < 1$. For a continuous function $W(\xi)$, there must by an angle ξ_0 such that $W(\xi_0) = 1$. If there is a "pole making" angle $\xi_{h_0k_0l_0} \approx \xi_0$ with the polar axis, then the reflection $h_0k_0l_0$ will be less dependent on the texture when reflections with angle ξ_{hkl} are closer to 0 or 90°. It is clear, however, that angle ξ_0 is not constant but will deviate with the degree of the preferred orientation. Angles ξ_0 are given in Table 5.12 for various degrees of preferred orientation, characterized by the density of orientation distribution function $W(\xi)$ at $\xi = 0$ and three different approximation functions from Table 5.11.

5.10 Estimation of analysis errors

5.10.1 Introduction

Any analytical technique is characterized by:

- the accuracy of the analysis
- the precision of the analysis.

The accuracy of an analysis may be checked by repeating the analysis on samples with a known composition, which has been established by any independent reliable technique, or on artificially prepared mixtures. The accuracy of the analysis is defined as the difference between the mean values of the measured analytical data and those of known phase abundances. Errors in analysis that influence the accuracy of phase determination are usually of a systematic character, particularly with regard to interlaboratory factors. The main sources of systematic errors in QXRD are the crystallochemical inadequacy of the analyzed phases and analytical standards, absorption effects that have not been taken into account, peak overlapping, and an erroneous definition of the background.

The precision of the analysis is determined by the scattering of the results around a mean value, i.e., by the reproducibility of measurements. Scattering of analysis data is caused by inevitable random errors encountered at various stages of analysis. The main factors affecting precision in QXRD techniques are counting statistics, crystallite statistics, sample nonhomogeneity, and instability of the X-ray source.

Mathematical statistics are used to estimate the precision and accuracy of analyses, to compare different analytical techniques, and to identify the major causes of errors. A full discussion of the application of mathematical statistics in analytical practice is to be found elsewhere (International Tables for X-Ray Crystallography, Vol. II 1959). Here, we will give only some elementary examples.

It is usually supposed that under multiple repetition of an analysis, the frequencies of single results (phase abundances c_{jk}, where k is the analysis number) are distributed according to a normal, or Gaussian, distribution:

$$\phi(c_j) = G(c_j)\,\Delta c_j = \{(1/\sqrt{2\pi}\,\sigma)\exp[-(c_j - \langle c_j \rangle)^2/2\sigma^2]\}\Delta c_j \qquad (5.108)$$

where $G(c_j)$ is a Gaussian function, $\phi(c_j)$ is the frequency for the phase abundance lying between c_j and $c_j + \Delta c_j$, $\langle c_j \rangle$ is the mean value of the phase abundance, σ is the standard deviation (dispersion), and σ^2 is the variance of distribution. The Gaussian distribution defined in equation (5.108) is normalized in the same way as any probability distribution $\int_{-\infty}^{\infty} G(c_j)\,dc_j = 1$. Strictly speaking, the normal distribution in our case is not necessarily correct, although it is probably valid when the observed results are affected by a large number of random factors, none of which is individually particularly important. Deviation from the Gaussian distribution is observed if the scattering of the results is defined by one of a number of factors. The hypothesis of a normal distribution may be verified.

The final outcome of any quantitative analysis technique is the determination of the abundances of the component phases c_j. The abundances of the phases are acquired through the measurement of certain experimental quantities, such diffraction peak intensities, background intensities, or mass absorption coefficients. All experimental values are distributed around a mean value with certain dispersion σ. According to the general rule of error propagation, the dispersion of c_j is calculated as:

$$\sigma(c_j) = \sqrt{\sum_{n=1}^{N} \left[(\partial c_j/\partial I_n)\,\sigma(I_n)\right]^2 + \left[\partial c_j/\partial \mu^* \sigma(\mu^*)\right]^2} \tag{5.109}$$

where I_n (n being 1,2, . . ., N) are the intensities of diffraction peaks or background, which are found in the formulations for phase abundance c_j derived in Chapter 4. We used this expression in Chapter 4 in order to compare various techniques and to find the optimal amount of doping, dilution, and internal standard. More sophisticated error propagation schemes are to be found in the literature (Szabó 1980; Kolotov and Chertock 1981; Davis 1986). However, we encounter certain difficulties in attempting to obtain numerical values for $\sigma(c_j)$ due to the uncertainty in the dispersion $\sigma(I_n)$. If we assume that counting statistics constitute the sole source of fluctuations in I_n, then $\sigma(I_n)$ is easily expressed (see Section 5.2), and we can proceed with the calculation of σ/c_j. Most works deal with precision in QXRD in this way. Unfortunately, there are other powerful, but less well-defined, contributions to the dispersion of intensity data. These factors include particle statistics, variations in preferred orientation from sample to sample, and variations in microabsorption from sample to sample. Estimation of $\sigma(I_n)$ by counting statistics may also often give unjustified low values. On the other hand, in any practical application of the QXRD technique, the precision of the analysis can be estimated experimentally by repeated measurement of one or several samples. We will now discuss elementary statistical techniques to estimate experimental data.

5.10.2 Calculation of reproducibility

The width of the Gaussian distribution defines the scattering of the results around a mean value. Reproducibility of the analysis is estimated in terms of the standard deviation σ which is defined by equation (5.110):

$$\sigma^2 = \int_{-\infty}^{\infty} G(c_j)(c_j - \langle c_j \rangle)^2 \, dc_j \tag{5.110}$$

The points $[(c - c_j) = \pm \sigma]$ on G are the points of inflection. The standard deviation is estimated by repeated measurements for one sample. The data obtained represent a random sampling from the general population, which is characterized by a Gaussian curve. The parameters $<c_j>$ and σ^2 of the latter are estimated by the values $\bar{c}_j$ and S^2 calculated for a random sampling as follows:

$$\bar{c}_j = \frac{1}{n}\sum_{k=1}^{n} c_{jk}\,; \quad S^2 = \frac{\sum_{k=1}^{n}(c_{jk} - \bar{c}_j)^2}{n-2} \tag{5.111}$$

where n is the number of measurements (replicate samples). We have to differentiate clearly between the mean and the variance of the general population (general mean and general variance), $<c_j>$ and σ^2, respectively, and the corresponding values for the sampling, $\bar{c}_j$ and S^2 (the sampling mean and the sampling variance, respectively). The former are constant values for a given general population, and the latter are random variables that vary from sampling to sampling.

An important measure of data scattering is the coefficient of variation:

$$\varepsilon = S\,\bar{c}_j \tag{5.112}$$

Suppose that we analyze m different samples with n replicates for each sample. If the abundances $(c_j)_l$ are similar in the different samples (l being the sample number), then the variances S_l^2 obtained for the different samples are in fact estimations of the same general variance s. Then, we obtain the sampling dispersion by averaging all S_l^2:

$$S^2 = \frac{1}{m}\sum_{l=1}^{m} S_l^2 = \frac{\sum_{k=1}^{n}\sum_{l=1}^{m}\left[(c_{jk})_l - (\bar{c}_j)_l\right]^2}{m(n-1)} \tag{5.113}$$

where $(\bar{c}_j)_l$ is the sampling mean for the m*th* sample and $(c_{jk})_l$ is the phase abundance obtained in the k*th* replicate for the *l*th sample.

Let us calculate the sampling dispersion and standard deviation of given intensity measurements of the 002 peak of chrysotile asbestos in asbestos-cement sheets. It was supposed that the variation in the preferred orientation of asbestos fibers during sample preparation might increase the standard deviation to a level that is unacceptable for analysis. Eight samples were studied with range of asbestos weight fractions of from 9% to 20%. There were four replicates for each sample (Table 5.13).

The sampling dispersion calculated from equation (5.113) is equal to $S^2 = 40$, and the sampling standard deviation $S = 6.3$ cps. The coefficient

Table 5.13. Intensity of the 002 diffraction peak of asbestos in asbestos-cement

Replicate number k	Intensity (counts/sec) in sample number: 1	2	3	4	5	6	7	8
1	220	210	135	140	120	105	115	85
2	220	190	145	150	120	120	95	90
3	230	200	145	145	115	110	100	95
4	220	200	130	155	110	105	105	90
$(\bar{I})_l$	220	200	145	150	120	110	100	90

$(\bar{I})_l$ is the mean value for sample l

of variation ε was 3%-6%, depending on $(I)_1$. In this case, ε was low enough to enable reasonable quantification of the asbestos.

5.10.3 *Scattering range of results*

Suppose that only a single determination of c_j is carried out for each sample. If the dispersion of the general population σ is known, then we can easily estimate the closeness of the obtained result to the general mean $\langle c_j \rangle$:

$$P(c_j) = \int_{\langle c_j \rangle - a}^{\langle c_j \rangle + a} \phi(c_j)\, dc_j \tag{5.114}$$

where $P(c_j)$ is the probability that c_j will be found between $\langle c_j \rangle - a$ and $\langle c_j \rangle + a$, and $\phi(c_j)$ is the Gaussian distribution defined in equation (5.108). If $a = \sigma$, then $P = 68.3\%$, and if $a = 2\sigma$, then $P = 95\%$. The meaning of the latter figure is that if single measurement gives the phase abundance c_j, then the general mean with a 95% confidence level is equal to:

$$\langle c_j \rangle = c_j \pm 2\sigma \tag{5.115}$$

For a 99% confidence level, the range "a" increases to $a = 2.58\sigma$.

In practical work, the dispersion of the general population $\sigma(c_j)$ is not known, but the dispersion of sampling $S(c_j)$ can be calculated from equations (5.111) and (5.113). In this case, the scattering range for a given confidence level P is increased by Student's coefficient $t(P, f)$, where P is the confidence level and f is the degree of freedom in calculations of the sampling dispersion S. [Values of $t(P, f)$ may be found in the International Tables for X-Ray Crystallography, Vol. II 1959.]

$$\langle c_j \rangle = c_j \pm t(P, f)\, S \qquad (5.116)$$

For a confidence level of P = 99% and f = 3, $\langle c_j \rangle = c_j \pm 6S$, i.e., the scattering range is equal to ±6S instead of 2.56σ, if σ is known.

Conventional analytical practice is to carry out a finite number of replicate measurements and to calculate the sampling means $\bar{c}_j$. Let us suppose that $\bar{c}_j$ are also distributed by a normal law but that this distribution is narrower than the distribution of single measurements c_j. If σ is the dispersion for the distribution c_j, then the dispersion of the distribution of mean values $\bar{c}_j$ is equal to $\sigma/\sqrt{n}$, where n is the number of replicates. Thus, in the case of repeated measurements, the general mean is given by:

$$\langle c_j \rangle = \bar{c}_j \pm t(P,f)\, S/\sqrt{n} \qquad (5.117)$$

The scattering range decreases twofold if we make four repetitions for each sample. Of course, in order to take all the probable errors into account each replicate should be carried out with reloading of the sample holder and not simply by repeated counting with a loaded sample holder.

5.10.4 Comparison of observed and known phase abundances

We meet this problem in the estimation of the accuracy of an analysis when we have to compare the observed and known abundances in the analysis of a standard sample. In this procedure, the difference between the mean observed value $\bar{c}_j$ and the known value $\langle c_j \rangle$ is compared with the sampling dispersion $S/\sqrt{n}$ by means of Student's coefficient *t*:

$$t = \left| \bar{c}_j - \langle c_j \rangle \right| / \left(S \sqrt{n} \right) \qquad (5.118)$$

where S is calculated by equation (5.111) or (5.113) and n is the number of repeated measurements.

The two results $\bar{c}_j$ and $\langle c_j \rangle$ are considered identical if *t* calculated by equation (5.118) is lower than Student's criterion for the chosen confidence level *P* and the given number of degrees of freedom f (International Tables for X-Ray Crystallography, Vol. II 1959).

To estimate the merits of various experimental techniques, we compare the reproducibility (dispersion) of these techniques and the mean values of the abundances. In practice, in the comparison of two different techniques, we have to estimate the statistical significance of the differences between two dispersions S and two mean values. If S_1 and S_2 are the sampling dispersions for the two techniques, and f_1 and f_2 are the degrees of freedom in the determination of S_1 and S_2, then we have to determine the ratio of the general dispersions σ_1 and σ_2 for the two techniques. The analysis is carried out on the basis of the Fisher criterion. The equality $\sigma_1 = \sigma_2$ is rejected with confidence level $P = 1 - p$, if:

$$\frac{S_1^2}{S_2^2} > F_{P/2}(f_1, f_2) \tag{5.119}$$

where F is the Fisher criterion for the degrees of freedom f_1 and f_2 at $P = 1 - p$ confidence level. In equation (5.119), S_1 is chosen to be greater than S_2. Table 5.14 presents the results of repeated measurements of the quartz abundance by two QXRD techniques—the dilution method and a method with experimental determination of the mass absorption coefficient μ^*.

The Fisher criterion is equal to 3.45 at the 90% confidence level and 6.0 at the 98% confidence level, i.e., less than the ratio $(S_1/S_2)^2$. Thus, the null hypothesis $\sigma_1 = \sigma_2$ is rejected, and the precision of the second technique is higher.

The next step in the comparative study of the two techniques is the comparison of the mean values $\bar{c}_j'$ and $\bar{c}_j''$. The question is whether the difference $(\bar{c}_j' - \bar{c}_j'')$ is statistically significant, or, in other words, whether any systematic deviation is manifested in either of the techniques. The comparison of two mean values is carried out in terms of Student's coefficient. If S_1 and S_2 are the sampling dispersions for the two techniques and n_1 and n_2 are the corresponding number of replicates, then the weighted mean of the two dispersions is given by:

Table 5.14. Quartz weight fractions in sandstone, %

	No. of replicates								
Method	1	2	3	4	5	6	7	8	9
Dilution	61.5	57.5	55.4	59.4	56.5	60.2	61.5	62.4	60.4
μ^* determined experimentally	60.6	60.8	59.3	60.8	59.5	58.8	59.3	58.8	60.3

Degrees of freedom f = 8 in both cases; $S_1^2 = 6.0$; $S_2^2 = 0.69$; $S_1^2/S_2^2 = 8.7$.

$$\overline{S}^2 = \frac{(n_2 - 1)\, S_1^2 + (n_1 - 1)\, S_2^2}{n_1 + n_2 - 2} \tag{5.120}$$

and the value of the t coefficient is given by:

$$t = \frac{\overline{c}_j' - \overline{c}_j''}{\sqrt{\overline{S}^2}} \sqrt{\frac{n_1 \cdot n_2}{n_1 + n_2}} \tag{5.121}$$

where degrees of freedom $f = n_1 + n_2 - 2$.

The difference $|\overline{c}_j' - \overline{c}_j''|$ is not statistically significant if t calculated by equation (5.121) is less than t (P,f) obtained for a given number of degrees of freedom f at a chosen confidence level P, usually 95%. As an example, let us compare the results of quartz determination by two techniques—experimental determination of the mass absorption coefficient μ^* and the doping method. The results of repeated determinations for the two methods are:

exp. μ^*= 6.9%, 6.5%, 7.2%, 7.0%, 7.1%;
$\overline{c}_j' = 6.9\%$; $S_1 = 0.27\%$; $n_1 = 5$
doping μ^*= 7.5%, 7.3%, 7.6%, 7.3%, 7.3%;
$\overline{c}_j'' = 7.3\%$; $S_2 = 0.14\%$; $n_2 = 5$.

The value of $\overline{S}^2 = 0.04$, and $t = 3.96$.

Student's coefficient at the 95% confidence level for f = 8 is equal to 2.31. The calculated value $t > 2.31$ and the difference $\overline{c'_j} - \overline{c''_j}$ is not random but statistically significant. Both techniques should be examined thoroughly to determine the probable source of systematic deviation.

5.10.5 *Inter- and intralaboratory precision*

Suppose that a test material is distributed to l = 1,2,...,m laboratories and $\tilde{c}_{jl}$ is the mean value of the phase abundance obtained in laboratory l. Interlaboratory precision is characterized by standard deviation $\tilde{S}$, which is calculated by the same rules as the intralaboratory value:

$$\tilde{S}^2 = \frac{\sum_{l=1}^{m} (\overline{c}_{jl} - \tilde{c}_j)^2}{m - 1} \tag{5.122}$$

where $\tilde{c}_j = \frac{1}{m} \sum_{l=1}^{m} \tilde{c}_{jl}$ is the mean interlaboratory value and m is the number of participants. The dispersion $\tilde{S}$ may carry very important information on systematic errors in analytical procedures employed in different laboratories. The intralaboratory means $\tilde{c}_{jl}$ are distributed with a dispersion estimated by $S_l/\sqrt{n}$, where n is the number of replicate samples and S_l is calculated by equation (5.111). The mean interlaboratory dispersion is given by:

$$\bar{S} = \frac{\sum_{l=1}^{m} (S_l/\sqrt{n})}{m} \tag{5.123}$$

The interlaboratory $\tilde{S}$ and intralaboratory $\bar{S}$ are compared by means of the Fisher criterion, as was shown in Section 5.10.4. If the interlaboratory dispersion $\tilde{S}$ is significantly greater than the mean intralaboratory dispersion $\bar{S}$ ($\tilde{S} > \bar{S}$), then at a certain confidence level, systematic errors are present in the techniques used in the participating laboratories. Unfortunately, interlaboratory cooperation is rare in QXRD practice, certainly in comparison with the cooperation for general chemical or spectroscopic techniques. Perhaps the lack can be explained by the great variety of potential QXRD subjects, amounting to thousands of different compounds versus the 109 elements that comprise the subjects of analytical chemistry.

One of the very few materials that has been subjected to interlaboratory study is Portland cement (Moore 1964; Zevin et al. 1977; Aldrige 1982a). The results of three interlaboratory studies are presented in Table 5.15. It should be noted that analytical standards were not exchanged in the three projects. As was to be expected, the interlaboratory standard deviation significantly exceeds the intralaboratory values, emphasizing the presence of systematic errors in laboratory practice. The most probable reason for the large dispersion of interlaboratory data is the lack of analytical standardization. Cement components—alite, belite, and aluminate phases—occur in many polymorphic modifications, which are stabilized by minor impurities. Although the basic structural features of these polymorphs are similar, and large variations in integrated intensities are unlikely, the symmetry does vary, causing peak splitting, overlapping, and so on. It can be assumed that each laboratory uses analytical standards suitable for specific types of cement minerals and not specified for any arbitrary substance. The results of the three projects once more emphasize the necessity for proper analytical standardization. Further development of analytical techniques undoubtedly demands wider interlaboratory cooperation and round-robin tests. It is preferable that in the early stages of such collaboration the test material should have a well-established phase composition for well-specified component phases.

Table 5.15. Interlaboratory study of Portland cement samples

	Ordinary Portland cement			Sulfate-resistant Portland cement			White Portland cement		
Mineral	c_j,%	$\tilde{S}$,%	$\bar{S}$,%	c_j,%	$\tilde{S}$,%	$\bar{S}$,%	c_j,%	$\tilde{S}$,%	$\bar{S}$,%
Moore (1964), 9 participants									
Alite		3.1			5.8				
Aluminate phase		2.8			1.8				
Zevin et al. (1972), 8 participants									
Alite	59.6	4.6	2.3	45	3.9	2.2	50.1	6.6	2.5
Belite	21.5	3.5	2.5	34	4.6	2.3	36.4	5.7	3.0
Aluminate phase	4.9	0.7	0.3	2.6	1.1	1.0	9.7	1.6	1.2
Ferrite phase	12.5	1.5	1.1	14.5	2.2		-	-	
Aldrige (1982a)									
Alite*	62.0	7.2	3.1						
Belite*	20.7	5.4	3.7						
Aluminate phase*	7.2	2.1	1.3						
Ferrite phase*	6.6	1.5	1.1						

* Mean values for six cement samples.

$\bar{S}$ is the mean intralaboratory dispersion.

$\tilde{S}$ is the mean interlaboratory dispersion.

5.10.6 Parameters of the calibration graph—estimation of standard deviation

In many QXRD techniques (e.g., internal standard or doping), the linear regression is of the type:

$$I = a + bc \tag{5.124}$$

where a and b are regression constants and c is the weight fraction of the analyte. The parameters a and b are evaluated by running standard mixtures with known weight fractions of the analyte. To evaluate the precision of the analysis, we need to know the standard deviations of the regression constants a and b, which are given by:

$$S_a = S_o \sqrt{\frac{<c^2>}{\sum_{k=1}^{n} [(c)_k - <c>]^2}} \tag{5.125}$$

$$S_b = S_o \sqrt{\frac{1}{\sum_{k=1}^{n} [(c)_k - <c>]^2}} \qquad (5.126)$$

where k is the running number of the point on the regression line, n is the total number of experimental points, $<c> = (1/n) \sum_{k=1}^{n} (c)_k$ and $<c^2> = (1/n) \sum_{k=1}^{n} (c^2)_k$ are mean values, and $S_o = \sum_{k=1}^{n} [(I_{obs})_k - I_k)]^2/(n-2)$ is the standard deviation of the observed data points $(I_k)_{obs}$ from the corresponding point on the regression line I_k.

5.11 Detection limit

The term detection limit may be used in two senses. In one sense, the detection limit is the minimal concentration of analyte that can be detected with acceptable reliability in a small mass of material in which it is supposed to be present in abundance. In the second sense, the same term defines the minimal mass of analyte that can be reliably detected when the analyte is the major or even the only phase in the sample. (A related problem of analysis of low-mass samples is discussed in Chapter 4.) Thus, we have two different cases, requiring special measures, that require detection of a weak diffraction peak:

1. Low mass of sample material with high analyte abundance in the sample.
2. Adequate mass of sample material with low analyte abundance in the sample.

5.11.1 *Basic formulation*

The intensity of a diffraction peak I_p (integrated or at maximum point) is defined as:

$$I_p = I - I_b \qquad (5.127)$$

where I is the total intensity of the peak plus the background and I_b is the background intensity.

In a blank sample not containing any analyte the mean value of the diffracted intensity $<(I_p)_{bl}> = 0$. If $(I_p)_{bl}$ is distributed around its mean value with standard deviation σ_0, then the abundance of analyte, at least in the low-concentration range that is of interest here, is given by:

$$c = \frac{I - I_b}{m} = \frac{I_p}{m} \qquad (5.128)$$

where m is the slope of the linear graph, which is expressed as counts/sec per 1% of analyte weight fraction. However, the very fact of obtaining $I_p > 0$ cannot prove the presence of the analyte, if $I_p \sim \sigma_0$. It is thus possible to make an incorrect decision by:

a. Taking a false signal as a true one, or

b. Taking a true signal as a false one.

If $P(I_p)_{bl}$ is the distribution of I_{pb} in the blank sample, and measurements of the peak intensity in the analyzed sample gave the value $I_p - I$, then the probability that the error will be of the first type will be:

$$\alpha = \int_{I_{p1}}^{\infty} P(I_p)_{bl}\, dI \tag{5.129}$$

In this expression, we suppose that I_{pl} and $(I_p)_{bl}$ have similar distributions with the same variance, σ_0^2, which is true, because I_{pl} and I_{pb} must be very close if we consider the detection limit. If $I_{pl} - 2\sigma_0$, as is generally accepted for the detection limit, and I_{pl} has a normal distribution, then the probability of error *a* will be less than 5%. However, the abundance $c_l = 2\,\sigma_0/m$ cannot be considered as the lower detection limit since the reliability of detection of the corresponding signal $I_{pl} = 2\,\sigma_0$ is low. Indeed, if the mean value of the signal is $\langle I_{pl}\rangle = 2\,\sigma_0$, then in ~50% of events the detected intensity will be less than $2\sigma_0$ and the hypothesis of analyte detection will be rejected. In other words, in this case the probability of error *b* is too high. If we want to keep this error at the same low level as an error of type *a*, then the mean intensity value corresponding to the detection limit must be:

$$(I_p)_{min} = I_{pl} + 2\sigma_0 = 4\sigma_0 \tag{5.130}$$

and the detection limit will be given by:

$$c_{min} = 4\sigma_0/m \tag{5.131}$$

If the concentration range is close to the detection limit, usually $I_p << I_b$, and the total intensity I is close to the background intensity I_b, then allowance must be made for the fact that the two measurements I and I_b are always required to evaluate I_p. Thus, the variance $\sigma_0^2 \approx 2\sigma_b^2$ and $\sigma_0 \approx \sqrt{2}\sigma_b$, where σ_b^2 is the variance of the distribution of I_b.

Taking into account equation (5.131), we will have the following expression for the detection limit:

$$c_{min} = 4\sqrt{2}\,\sigma_b/m \tag{5.132}$$

Let us suppose that in the case discussed here the variations of background intensity I_b and total intensity I are due to two major factors: counting statistics and instability of the equipment, mainly instability of the high-voltage generator. If I_b is the background counting rate and T_b is the time spent on counting the background, then the standard deviation σ_b' due to counting statistics is equal to $\sqrt{I_b/T_b}$. Instability of the equipment is characterized by its relative value Δ, which is obtained by repeated measurements of diffracted intensity at constant experimental conditions. The value of Δ depends on time averaging at each measurement. If a certain value of Δ corresponds to time T_b, then the standard deviation σ_b'' corresponding to the instability of the equipment is equal to $\Delta\cdot(I_b)$. The total standard deviation is given by:

$$\sigma_b = \sqrt{I_b/T_b + (\Delta I_b)^2} \tag{5.133}$$

The standard deviation σ_b is reduced and the detection limit is improved by increasing the counting time T_b. However, increasing the counting time is feasible only up to a certain limit, at which the intensity fluctuations due to counting statistics will be equal to the intensity fluctuations due to instability of the X-ray source. The maximal time in which it is practical to reach the lower detection limit is approximated as:

$$T_{max} \approx 1/(\Delta^2 I_b) \tag{5.134}$$

and the accumulated number of counts as:

$$N_{max} = I_b T_{max} = \Delta^{-2} \tag{5.135}$$

In poorly stabilized diffractometers, instability is rather high and may be close to 1%. In this case, $N_{max} = 10^4$ counts and for any mean background counting rate of 300 counts/sec, $T_{max} \approx 30$ sec. Thus, in a realistic case in which the time spent for counting exceeds 30 sec, the detection limit for a poorly stabilized diffractometer is defined by the instability of the X-ray source:

$$c_{min} = \frac{4\sqrt{2}\,\Delta\,(I_b)}{m} \tag{5.136}$$

The stability of contemporary X-ray generators is much better. Short-time instability within a few tenths of a percent has been reported (Brown et al. 1987). In order to match the 0.1% error level with counting statistics, we have to accumulate $N_m = 10^6$ counts. For some average background counting rate of 300 counts/sec, the corresponding counting time $T_{max} = 3 \cdot 10^3$ sec. This time is long enough for substantial variations in ambient conditions, such as temperature and air pressure. Brown et al. (1987) showed that the diffracted intensity increases at a rate of 0.3% per 1°C of ambient temperature and at an even greater rate of 0.6% per 1°C with temperature of the tube tower of a Philips diffractometer. Variations of pressure affect diffracted intensity at rates of ~0.03% per torr of air pressure for CuK_α radiation and of 0.05% per torr for FeK_α, assuming that the path length of X-rays in air is 300 mm.

This short survey shows that attaining a low detection limit by increasing the counting time demands careful control of ambient conditions, including stabilization of the room temperature. Intensity measurements may be started only after the diffractometer temperature has reached equilibrium. Brown et al. (1987) found that the intensity is stabilized in ~2 h after switching on the generator.

If the counting time is lower than T_{max} defined by equation (5.134) and short enough not to be affected by significant changes in ambient conditions, then the standard deviation σ_b and the detection limit will be defined by counting statistics. According to equation (5.136), the detection limit is defined by:

$$c_{min} = \frac{4\sqrt{2I_b/T_b}}{m} \tag{5.137}$$

5.11.2 *Reduction of the detection limit*

The detection limit is inversely proportional to the square root of the analysis time. The detection limit drops sharply with counting time up to 100 sec and then at a much more gradual rate. Indeed, increase of analysis time from 1 to 100 sec in order to reduce the detection limit 10 times can be considered quite reasonable, whereas the next decrease of the detection limit by 10 times demands an increase in time up to 10,000 sec, which seems to be too long for a single measurement.

The next factor enabling a reduction in c_{min}, which is hidden in equation (5.137), is the overall increase of the scattered intensity, which may be achieved, for example, by increasing the intensity emitted by the X-ray source I_0 or by increasing the mass of the irradiated sample. Although both peak intensity I_p and background intensity I_b are proportional to I_0, and the peak to background ratio is constant, the slope coefficient m is proportional to I_0, and according to equation (5.137) the detection limit decreases as $(I_0)^{-1/2}$. The trend in the slope m (or peak intensity I_p) as well as background intensity must be taken into account in any attempt to decrease c_{min}. For example, monochromatization that leaves I_p intact, such as that obtained with a diffracted-beam graphite monochromator, is undoubtedly an excellent way to improve the detection limit by decreasing the background intensity. However, the merits of a monochromator that decreases I_p must be checked according to the trend in ratio $I_p/\sqrt{I_b}$ (or $m/\sqrt{I_b}$). Thus, the ratio $I_p/\sqrt{I_b}$ can be considered a measure of the merit of any diffractometric arrangement with regard to its sensitivity in the detection of weak diffraction peaks.

Although in the discussion leading to equation (5.137) we considered $I_p = I - I_b$ as the intensity at the peak maximum, the time dependence of the detection limit will be equally valid if I_p is taken as the integrated intensity. In this case, the net background count is equal to $I_b T_b$, and the standard deviation $\sigma_b = \sqrt{I_b T_b}$ increases with T_b. However, the slope coefficient, which is expressed as the net peak count per 1% of analyte, is proportional to the counting time, leaving the detection limit with the same dependence on $(T_b)^{-1/2}$.

After the theoretical discussions in this section, it is interesting to become acquainted with real figures for the detection limit. For many years, 1% was regarded as the rule-of-thumb value (Klug and Alexander 1974). However, in highly stabilized and computerized diffractometers, time has become a powerful tool in improving the detection limit. Thus, more realistically, a detection limit of 0.1% or a few tenths of a percent reflects the contemporary situation. The actual detection limit depends on the X-ray beam source and on the analyzed phase and matrix. The detection limit will be lower for high-symmetry phases, giving a few diffraction peaks of high intensity and a larger slope coefficient m. A low detection limit is also possible in the analysis of a heavy phase in a light matrix, which is also associated with high slope coefficient m. There are several examples in the literature of excellent detection limits, exceeding the 0.1% limit mentioned above.

In our practice, a satisfactory detection limit of ~0.1% was obtained in the analysis of the minor phases in glass ceramics and cements. The calibration graph for the analysis of rutile in glass-ceramics showed a detection limit of ~0.1%. In the analysis of gypsum in Portland cement, the slope m was 120 counts/sec per 1% and $I_b \approx 200$ counts/sec (Kheiker and Zevin 1969); the short-time instability was $\Delta = 0.7\%$, and the detection limit was estimated as ~0.07%. In order to achieve this sensitivity, the counting time T_b must be greater than 100 sec. The analysis of MgO in Portland cement was performed with a detection limit of ~0.1%.

Robie and Scalzo (1986) compared the ability of two detection systems—semiconductor detector and scintillation detector combined with a reflected beam monochromator—in the detection of trace phases and found the semiconductor detector to be superior. They found that with the conventional combination of a scintillation counter and monochromator, the detection limit for Cr_2O_3 in an alumina matrix was close to 0.1% for a counting time of 100 sec.

5.11.3 Detection limit in the sample mass approach

Suppose that a single-phase material is available in minute quantities, and we want to define the minimal detectable sample mass. In a low-mass sample, absorption is negligible, and the diffracted intensity may be expressed by equation (4.216). Combining equations (4.216) and (4.218), we came to an expression for the intensity diffracted by a low-mass sample, I_p (Zevin and Zevin 1987):

$$I_p/I_\infty = G/(B/2\mu^*) \qquad (5.138)$$

where G is the mass of the small sample, I_∞ is the intensity diffracted by an infinitely thick sample of the same material, and B is the cross section of the primary beam at the sample location. For fixed divergence in the Bragg-Brentano case, the sample material is spread over an area $S = B/\sin\theta$. The sample mass is expressed as:

$$G = \frac{I - I_b}{r} \qquad (5.139)$$

where the coefficient $r = I_\infty/(B/2\mu^*)$.

Equation (5.139) is analogous to equation (5.128), and by repeating all of the intermediate operations leading to the definition of the detection limit, we will come to a similar expression for the minimal detectable sample mass:

$$G_{min} = 4\sqrt{2}\,\sigma_b/r \qquad (5.140)$$

where σ_b is the standard deviation of the background measurements. If background fluctuations are only due to counting statistics, then:

$$\sigma_b = \sqrt{I_b/T_b}$$

and

$$G_{min} = \frac{4\sqrt{2}\,\sqrt{I'_b}}{I_\infty\sqrt{T_b}}(B/2\mu^*) = \frac{4\sqrt{2}\,\sqrt{N_b}}{N_\infty}(B/2\mu^*) \qquad (5.141)$$

where the coefficient r is replaced by the expression for r from equation (5.140), N_b is the net number of counts accumulated in the background counting, and N_∞ is the net number of counts accumulated for an infinitely thick sample at time T_b.

Considering the probable ways of reducing G_{min}, we naturally come to the time dependence of G_{min}. By increasing the counting time for the total (peak and background) and background intensities, we will decrease G_{min} as $(T_b)^{-1/2}$.

The best way of reducing G_{min}—the decrease of background intensity—demands particular attention. If the sample material is available in quantities suitable for preparing an infinitely thick sample, then the background level is largely an intrinsic property of the sample material itself. In contrast, for low-mass and low-absorbing samples, the overwhelming contribution to the background comes from the sample support and not from the sample itself. Three factors enable an adequate decrease of background: choice of a low-background sample support, optimal distribution of the sample over the support, and proper collimation of the primary beam. To meet the first condition, a single-crystal support is recommended, and today such supports are available commercially. The single crystal is cut so that the angle between the surface of the cut and any crystallographic plane is no less than several degrees. The second requirement—the optimal spread of the powder sample over the support—must be met to satisfy the two somewhat contradictory demands of high peak intensity and low background. The intensity of diffraction is increased if the sample is spread over the greater surface, thus reducing the absorption correction, but the background intensity increases almost linearly with irradiated surface due to an increased volume of the irradiated sample surface. Zevin and Zevin (1987) have defined the optimal spreading area S as:

$$S \approx 10\,G\,\mu^*/\sin\theta \qquad (5.142)$$

where G is the sample mass.

The primary beam should be collimated so that it illuminates only the area covered by the sample. An automatic divergence slit (Jenkins and Paolini 1974) is very useful. The effect of proper collimation is shown in Fig. 5.16. In the upper part of Fig. 5.16 conventional collimation was used (1° constant divergence slit and 4.5° divergence Soller slit), and in the lower part, only the area occupied by the sample was irradiated.

In the example given in Fig. 5.16, a microgram sample of calcite produced a diffraction peak comparable with the background fluctuations when recorded by a chart recorder with a time constant of 4 sec. The peak-to-background fluctuations ratio improved dramatically with increasing counting time (Fig. 5.17). It seems that the detectable mass of calcite may be as low as ~0.1 μg if adequate time is allowed for counting. Acquisition of a full diffraction pattern for microgram range samples can take many hours with a conventional "point" detector. This time may be reduced by one order of magnitude with a position-sensitive detector (PSD).

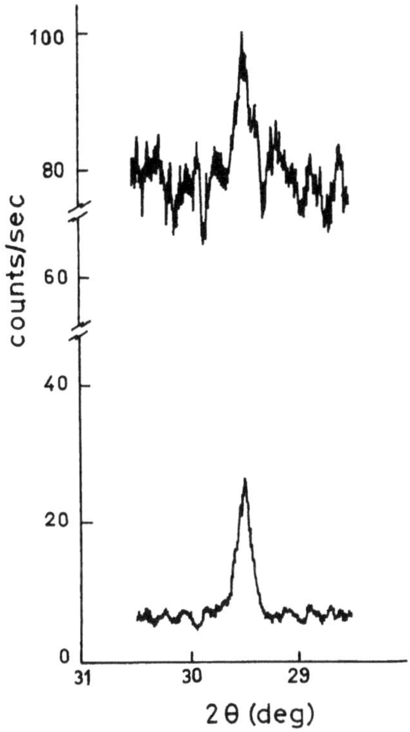

Figure 5.16. Effect of collimation on the measurement of a small amount of crystalline powder at 1° divergence (top) and optimal 2° divergence (bottom).

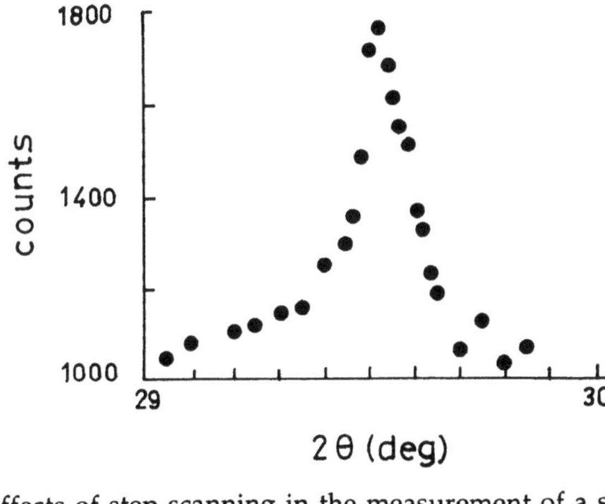

Figure 5.17. Effects of step scanning in the measurement of a small amount of crystalline powder (100 sec/step).

The acquisition of powder patterns in the 1-10 μg range has been reported in a number of works (Baldock and Parker 1973; Zevin and Zevin 1987). The use of PSDs for such samples offers a great advantage with these samples since they facilitate a significant reduction in time. A low mass sample is usually spread over a relatively small area, thus reducing the serious defocusing effects inherent in the implementation of PSD in focusing diffractometric schemes.

5.12 Crystallite statistics

5.12.1 Intensity error

Suppose that $<n_0>$ is the mean number of properly oriented crystallites of any phase j reflecting X-rays. The real number of reflecting crystallites in replicate samples fluctuates around $<n_0>$, according to the Poisson distribution. If $<n_0> \gg 1$, this distribution is close to Gaussian, with a standard deviation $\sigma = \sqrt{\langle n_0 \rangle}$ and a coefficient of variation $\varepsilon = \sigma/<n_0> = <n_0>^{-1/2}$. It is reasonable to suppose that diffracted intensity will follow these fluctuations and coefficients of variation, as follows:

$$\varepsilon_I = (<n_0>)^{-1/2} \tag{5.143}$$

In equation (2.69), we defined the probability f for a crystallite to reflect X-rays. If N_j is the number of crystallites in phase j in the irradiated volume of our sample, then:

$$<n_0> = f\,N_j \tag{5.144}$$

For a nonabsorbing sample N_j is defined simply as $VV_j/<\upsilon>$, where V is the sample volume, V_j is the volume fraction of phase j, and $<\upsilon>$ is the mean volume of crystallites. If we assume Bragg-Brentano focusing geometry, then V = St, where $S = B/\sin\theta$ is the irradiated area of the sample, B is the cross section of the primary beam, and t is the sample thickness. For an absorbing sample, the effective thickness is given by $t_{eff} \cong \sin\theta/\mu$, and the effective sample volume by (Kheiker and Zevin 1963):

$$V_{eff} = B/\mu \tag{5.145}$$

Combining equations (5.142) and (5.143), we arrive at:

$$\varepsilon_I = (fVV_j/<\upsilon>)^{-1/2} = (fBV_j/\mu<\sigma>)^{-1/2} \tag{5.146}$$

Integration of equation (5.146) will give for f:

$$f = (m\,\alpha\,\varphi_m)/2\pi \tag{5.147}$$

where m is a multiplicity factor for the corresponding reflection and α and φ_m are the maximal deviations of the diffraction vector from the sample normal in the focusing and axial planes, respectively. From the geometry described in Section 3.2.1, f can be expressed as:

$$f = \frac{(w_f + w_c)}{8\pi R \cdot \sin\theta} \tag{5.148}$$

where w_f is the apparent width of the focal spot, w_c is the receiving slit width, d is the divergence of the Soller slits, and R is the diffractometer radius. Finally, the intensity error is given by:

$$\varepsilon_I = \sqrt{\frac{8\pi \sin\theta\,\mu^*\,\rho_j\,R\,c_j}{U_0 \cdot m(w_f + w_c)\delta}\langle v\rangle} \tag{5.149}$$

If the phase abundance is expressed as the weight fraction $c_j = V_j\rho_j/\rho$, then:

$$\varepsilon_I = \sqrt{\frac{8\pi \sin\theta\,\mu^*\,\rho_j\,R\,c_j}{U_0 \cdot m(w_f + w_c)\delta}\langle n\rangle} \tag{5.150}$$

where ρ_j is the density of phase j.

Assuming that $2\theta = 30°$; R = 17.3 cm; $c_j = 1$; $U_0 \cong 0.45\ cm^2$, which corresponds to a 1° divergence slit and 1.5 cm height of the irradiated area; m = 12; $w_f = w_c = 0.2$ mm; and $\delta = 2.3°$, the calculated error is substantial, even for a relatively low-absorbing phase like quartz and a grain size as low as 10 µm. This result was illustrated in Fig. 5.6, in which intensity fluctuations are shown for various grain-size fractions of quartz (Zevin and Kheiker 1962). In this experiment, the sample was rotated slowly in its plane (around the sample normal) at a speed of 1 revolution in 10 min. The detector remained stationary at an angle of 2θ with respect to the chosen reflection. Rotation of the sample by a few degrees brings a new group of crystallites into the reflection position. The number of reflecting crystallites fluctuates according to the rules outlined above, resulting in observed intensity fluctuations that are registered by a chart recorder.

The same outcome will, of course, be obtained by multiple reloading of the sample holder and taking intensity measurements for a stationary sample. However, intensity fluctuations that more or less correspond to equation (5.150) are usually obtained only for well-crystallized materials, such as quartz or silicon. In many cases, especially for natural materials that are not so well crystallized, intensity fluctuations are lower than those estimated by equation (5.150). The reason is that the length measure used in equation (5.150) to estimate grain volume is actually the coherent length and $<\upsilon>$ is the mean volume of the coherent domain, which can be much lower than the volume of the grain. In this case, intensity fluctuations will be subdued. Experimental determination of ε as a function of grinding time is recommended in order to define the required degree of grinding (see Section 5.3).

The problem of reducing the intensity error due to crystallite statistics is acute for phases of low abundance. The error in ε_I increases in inverse proportion to $(c_j)^{1/2}$, and for an abundance of 1% it is increased 10 times in comparison to a sample of a pure phase.

There are, in fact, two reasons for low precision in the analysis of minor phases: poor crystallite statistics and poor counting statistics. If the background level is greater than the peak intensity (the usual situation in the analysis of trace phases), then the intensity error due to counting statistics increases in inverse proportion to the square root of the peak intensity or, as a first approximation, in inverse proportion to the square root of the phase abundance. The same is true for crystallite statistics. As a result, the observed errors in intensity measurements for minor phases often exceed values expected solely from counting statistics. Figure 5.18 shows how increasing of the total counting of

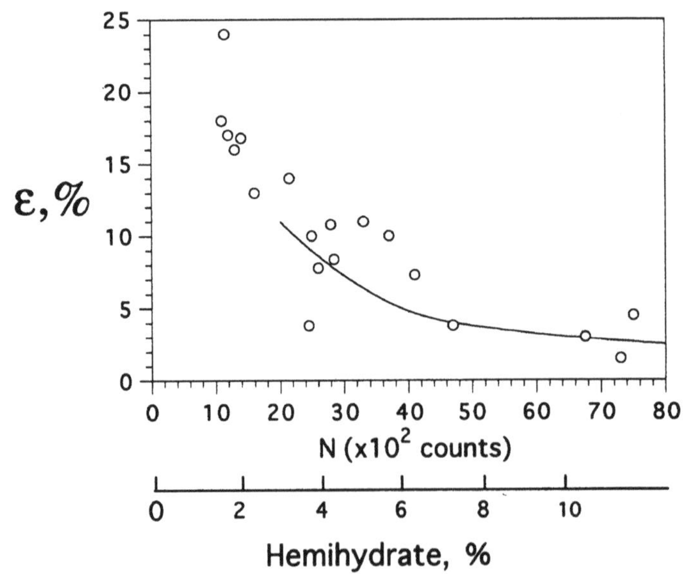

Figure 5.18. Reproducibility of intensity measurements for hemihydrate.

hemihydrate in Portland cement due to higher phase abundance is accompanied by a reduction of measurement error (expected error is given as a solid line).

5.12.2 *Methods of decreasing intensity fluctuations other than sample diminution*

In addition to mean grain volume $<\upsilon>$ [equation (5.149)], there are two more terms affecting the intensity error ε_I, i.e., the irradiated sample volume (V) and the probability for a crystallite to be in the appropriate orientation f. Increasing both terms leads to a decrease of ε_I. There are several ways of increasing V: increasing beam divergence in the focusing plane, increasing X-ray penetration, or giving the sample translational movement. These techniques are, however, not very effective, except for some cases of very high absorption coefficients. A two- to threefold increase of V will reduce the intensity error about 1.5 times. The increase of the irradiated volume is limited by sample availability, the size of the sample holder, and the inevitable increase of geometric aberrations with an increase of beam divergence.

The increase of the probability f is a more versatile and promising approach. Maximal deviations of the diffraction vector from the sample normal increase with increased axial beam divergence (divergence of the Soller slits δ) and increased widths of the tube focus and the receiving slit ($w_f + w_c$). These measures lead to an increase of the diffracted intensity and an unwanted increase of instrumental aberrations (Zevin and Kheiker 1963a). Usually, consideration of the necessary angular resolution is the main criterion for the implementation of wider slits and tube focus.

There is, however, an elegant technique for increasing the probability of a crystallite being in the reflecting position—that of sample rotation (de Wolff 1958). Taking into account the time that the crystallite remains in the reflection position, we come to the following expression for ρ (de Wolff 1958):

$$\rho = \sqrt{\frac{\pi}{4} \frac{C}{0.75 \pm 0.5 \ln C}} \tag{5.151}$$

where $C = \varphi_m/2\alpha$. For any conventional experimental arrangement with Soller slit divergence $\delta = 2.3$, apparent tube focus and receiving slit width $w_f = w_c = 0.2$ mm, $R = 175$ mm, and $2\theta = 45°$, the ratio $C \cong 8.5$, and sample rotation reduces the intensity error by a factor of 4.8, which is equivalent to an approximately threefold reduction of grain size.

The sample rotation must be fast enough so that at least one revolution is completed during the counting time that can be equal to the time per step in the step scanning mode or the time constant of the integrating circuit if a chart recorder is used. Naturally, sample rotation does not increase diffracted intensity, because increasing the number of reflecting crystallites in this case means that each crystallite spends proportionally less time in the reflecting position. This technique has the unequivocal virtue of not affecting instrumental aberration. This advantage, together with a substantial decrease of intensity fluctuations, makes sample rotation the most attractive technique for improving crystallite statistics. We should bear in mind, however, that sample rotation is effective only on the basis of substantial axial divergence of the primary and reflected beams, ensuring the condition $\varphi_m/2\alpha >> 1$.

A further increase of the number of reflecting crystallites by a factor of 2-3 can be achieved by rotating the sample around axes tilted in the axial plane by a small angle $\rho \cong \varphi_m/2$ (Zevin and Kheiker 1963a). However, sample tilt causes symmetrical broadening of the diffraction peak, which can be quite substantial at low scattering angles.

Another technique, based on the same formal idea of sweeping a larger area on a stereographic projection, is sample oscillation around the goniometer axis. In this case, the strip $2\alpha\,\varphi_m$ covers the area $(2\beta + 2\alpha)$, where β is the amplitude of oscillation. The number of reflecting crystallites in comparison with a stationary sample is increased $(2\beta/2\alpha + 1)$ times, and the intensity error is decreased by a factor given by:

$$\rho = \sqrt{1 + \beta/\alpha} \tag{5.152}$$

If $2\beta \cong 4°$, then for $1\alpha \cong 0.1°$ $\rho \cong 6$, i.e., it is substantial. However, sample oscillation, although as effective as sample rotation, is not aberration free. During oscillation the sample plane cuts the focusing circle and causes peak distortion. As a first approximation, the peak position does not change, but the peak profile is broadened. The magnitude of broadening is estimated by the standard deviation of the corresponding instrumental functions (Zevin and Kheiker 1963a):

$$\sigma_g = \frac{\beta\, \gamma \cot\theta}{3} \tag{5.153}$$

where γ is the beam divergence in the focusing plane.

As in the case of sample rotation, the period of oscillation must be smaller than the counting time, for example, the time per step in the step-scanning mode.

Effective techniques to reduce errors caused by crystallite statistics, such as sample rotation or oscillation, are especially important for minor phases.

6
Industrial applications

6.1 Ceramics and glass ceramics

Ceramics are particularly suitable materials for phase quantification. From a technical point of view, phase quantification of ceramic materials is not difficult to perform, since crystalline phases in ceramic materials usually have well-established structures.

Ceramic materials can be synthesized from pure chemical precursors, and structural variations in the phases are thus less marked than in natural minerals. In many cases, the pattern of solid solution is well established, and the relevant interpretation can be made on the basis of diffraction data like peak position. Thus, calculated patterns may be used for calibration in the quantification of ceramic phases, for example, for partially stabilized zirconia and for silicon nitride.

Toraya et al. (1984) calculated the ratio of the intensities of the cubic and monoclinic phases in partially stabilized ZrO_2, taking into account the concentrations and atomic positions of the substituents (MgO, CaO, $YO_{1.5}$), the variation of lattice constants with a corresponding change of Lorentz polarization factor, and variations in the cation temperature factor. The error in the phase abundances due to uncertainty in the calibration constant was estimated to be less than 0.7% in the worst case. Hill et al. (1988) analyzed magnesia-stabilized ZrO_2 subjected to different heat treatments. Rietveld analysis was used to quantify three ZrO_2 polymorphs (cubic, tetragonal, and monoclinic) and to determine strain parameters and coherent domain size. Bulk properties were characterized by neutron diffraction and surface properties by X-ray diffraction. The analysis was characterized by a very acceptable R_B factor, which was usually within a few percent for the major phases and increased to 5%-6% for the minor (monoclinic) phase. The results of the analysis are physically meaningful and constitute a very good example of phase quantification.

Another very important system in engineering ceramics that has been subjected to an intensive phase-relations study is silicon nitride, Si_3N_4 (Gazzara and Messier 1977; Mencik et al. 1980; Devlin and Amin 1990).

Silicon nitride exists in two polymeric forms, α and β. The α-phase transforms into the β-form during sintering and densification. The ratio of the two polymorphs is a very important characteristic of both the precursor powder and the sintered material. If sintering aids are used to facilitate sintering, then additional minor phases may appear in the sintered product. For example, the addition of yttria results in the formation of yttrium silicates and yttrium silicate-nitride (Devlin and Amin 1990). The calibration constant β_{jr} [j is the α-phase; r is the β-phase; equation (4.62)] can either be obtained experimentally (Mencik et al. 1980) or calculated by crystal structure data (Gazzara 1977). It should also be noted that for this system samples with various proportions of the two polymorphs are usually available, which indicates the successful application of the standardless technique (see Section 4.10). Some elements of this technique were used by Devlin and Amin (1990) for correcting the calculated calibration constant.

QXRD can, of course, also be applied to the quantification of traditional ceramics (e.g., porcelain and earthenware); this involves the determination of the following major phases: mullite, corundum, quartz, crystalloballite, and tridimite.

Glass ceramics are also substances that can be analyzed by QXRD. They are polycrystalline materials produced by high-temperature nucleation and crystallization of glass. Although the degree of crystallization of the final product is usually high, the material retains appreciable amounts of glass. Due to the catalytic action of special additives and the hindrance of diffusion in the solid state, the crystalline phase is characterized by a very fine structure, which confers excellent mechanical properties on the materials. The mechanical and physical properties of glass ceramics depend on the overall content of the crystalline phases or the degree of crystallinity.

QXRD techniques that are used for polymers can also be applied to glass ceramics. Ohlberg and Strickler (1962) determined the degree of crystallinity as:

$$C_c = 100 \frac{I_g - I_x}{I_g - I_B}$$

where I_g, I_x and I_B are the noncrystalline intensities at a particular scattering angle 2θ within the principal amorphous halo measured for the parent glass, the glass ceramic being analyzed, and a mechanical mixture of crystalline phases having the same chemical composition as the parent glass, respectively. Measurement of I_B accounts for incoherent scattering, scattering by air, and any other parasitic scattering. The angle of measurements $2\theta_B$ must, of course, fall between the diffraction peaks in the mixture of crystalline peaks. It is assumed that scattering by the parent glass and scattering by the residual glass in glass ceramics differ only in intensity. Strictly speaking, this assumption is not valid because parent and residual glasses usually have different chemical compositions, which can influence the shape and intensity of the amorphous halo. Although Ohlberg and Strickler did not detect such compositional variations in their particular case, such variations can affect the analytical results.

The same criticism can be addressed to the standardless method of Hermans and Weidinger (1961a), in which the set of samples being analyzed must include samples with significantly different degrees of crystallinity, i.e., samples with residual glass of various compositions.

Theoretically sound methods based on the law of conservation of intensity may also be applied to glass ceramics (Kagan and Snovidov 1965; Vonk and Fagherazzi 1983). Vonk and Fagherazzi (1983) modified the method of Ruland (1961), which was discussed in Section 4.12, for application to glass ceramics. The modification mostly concerned accounting for different chemical compositions of the residual glass and the crystalline phase, which is not the case in polymers. A comparison of the crystallinity determined by Ruland (1961) and by Hermans and Weidinger (1961a,b) and techniques for the direct quantification of crystalline phases revealed that the former two techniques underestimated the amounts of crystalline phases. A probable explanation is the decay of intensity of diffraction peaks due to crystal imperfection, which is taken into account only by the Ruland method. Kagan and Snovidov (1965) attributed these imperfections to thermal atomic motion. In any further application of the Ruland method, it is worthwhile to compare the obtained disorder parameter K with the Debye-Waller factor, which can be approximately assigned for atoms in the crystalline phase. Moreover, a combined (static and dynamic) disorder parameter can be estimated studying the decay of diffracted intensities as a function of $(\sin \theta/\lambda)^2$ (Baram and Zevin 1990). Benedetti et al. (1985) applied a modified Ruland technique to two glass ceramics systems and found that comparable results (viable for kinetic study) were obtained by simple determination of the amounts of crystalline phases.

We believe that quantification of the crystalline phases in glass ceramics is the simplest and most straightforward way of determining the degree of crystallization. This technique has been used extensively to determine the degree of crystallinity in glass-ceramics systems with a major crystalline phase of barium feldspar (Bogdanova et al. 1966), cordierite (Bogdanova et al. 1971), or β-spodumene (Bogdanova et al. 1973, 1975). The chemical composition of the analyzed sample, which has the same composition as the parent glass, is usually known. Thus, the absorption coefficient can be easily calculated.

Pure crystalline phases may be synthesized and mixed in various proportions with the parent glass to prepare calibration mixtures. Calibration graphs are then prepared by regressing integrated intensities of several diffraction peaks multiplied by the sample mass absorption coefficient against the phase abundance in the calibration mixture. This technique constitutes an excellent tool for studying the kinetics of crystallization and the dependence of the physical properties of glass ceramics on the abundance of crystalline phases. It should also be noted that a study of the kinetics is greatly facilitated by *in situ* measurements with a position-sensitive detector (Zorn et al. 1988).

6.2 Naturally occurring (geologic) samples

QXRD is a suitable technique for the analysis of natural materials, particularly in the modal analysis of geologic samples; in the analysis of ore deposits having technological significance; and sometimes in the analysis of separate phases, such as quartz or chrysotile-asbestos Substances present in nature for prolonged periods are affected by special chemical, hydrothermal, and mechanical environments, resulting in structural and chemical variations. Thus, geologic samples are multiphase materials with different degrees of chemical substitution and crystal perfection. As a rule, these samples contain appreciable amounts of amorphous material. Moreover, most naturally occurring phases belong to medium- or low-symmetry groups, producing complicated X-ray patterns with inevitable peak overlapping, if a number of such phases are present in the sample. To these complications, we have to add preferred orientation problems for phases with platy or fibrous particlelike phyllosilicates, calcite, asbestos minerals, and so on.

If a single diffraction peak or a number of peaks can be separated for each phase, then the most popular QXRD technique in modal analysis is phase analysis (external standard technique described in Section 4.6). If the presence of an amorphous phase can be assumed, then the mass-balance equation cannot be used, and equation (4.58) will give only provisional phase abundances. To convert them into real phase concentrations, each sample must be spiked with an internal standard in order to obtain the conversion factor (Section 4.6).

In the past few years, we have witnessed the growing popularity of the full-pattern approach to modal analysis of geologic samples. Two techniques, one based on experimental patterns of analytes and the other on calculated patterns, are equally promising (see below).

We now come to the crucial aspect of any analytical technique—its accuracy—which is of great importance for geologic samples. Comparison of QXRD with other methods of phase analysis seems to be appropriate; among these, bulk chemistry and optical point counting for thin rock sections are often used for comparison with QXRD.

6.2.1 *Modal analysis of geologic samples*

The standard procedure for the modal analysis of rocks is point counting of minerals in thin sections. However, this technique has some obvious drawbacks, such as long analysis time, nonapplicability to fine-grained rocks, and probable errors due to incorrect identification of minerals. In light of these disadvantages, it is not surprising that QXRD was tried for modal analysis in geology.

The first works on modal analysis of rocks appeared in the 1960s. Tatlock (1966), in an analysis of felsic rocks, noted that variations of sample absorption were related to variations of iron content in the sample. The background, if diffraction is carried out with CuK_{α}, is also sensitive to the iron content. Thus, the background intensity served for absorption correction in a series of calibration graphs. The condition μ^* = constant has been used in the analysis of metamorphosed volcanic rocks and in the analysis of seven-phase geologic samples. A similar technique was used by Otálora and Hess (1969) for the modal analysis of igneous rocks. Numerous calibration graphs for two-phase mixtures of minerals detected in analyzed samples were plotted in somewhat unusual coordinates $I_1/(I_1 + I_2)$ versus the concentration of the first phase C_1 (I_1 and I_2 being the intensities of diffraction peaks for phases 1 and 2). In order to choose the proper analytical standards, Otálora and Hess first determined the forsterite:fayalite ratio in olivines and the albite:anorthite ratio in plagioclase using X-ray diffraction data. The ratios of the phase abundances in the analyzed samples were determined by calibration graphs. Thereafter, phase concentrations were brought to 100% in the absence of a glassy phase or to 100% abundance of glass if the latter was present. Otálora and Hess (1969) made an attempt to determine the glass phase from the broad halo produced by the glass.

In the 1970s, we witnessed the rapid development of QXRD techniques for complete analysis of multiphase samples, which are suitable for the modal analysis of rocks. The work of Karlak and Burnett (1966) on treating patterns with overlapping peaks, the rebirth of the internal standard technique in the works of Moore (1968) and Chung (1974a), the implementation of the full-pattern approach in QXRD by Smith et al. (1987), and the numerous works of Davis and his co-workers (1977, 1980) on the implementation of RIR in QXRD are relevant in this context.

The external standard technique in its "pure" form was used by Pawloski (1985) to analyze volcanic tuffs and alluvium from the Nevada Test Site. Quartz, which was always present in the samples examined, was chosen as the reference material for the determination of β_{jr} constants (Section 4.6). Thus, constants were obtained as slopes of the linear regression of the ratio of the integrated intensities versus the ratio of the concentrations in two-phase mixtures, in which one phase was quartz. A list of calibration constants is provided in the paper. Thirteen minerals were commonly found; in the analysis, the X-ray scan covered a range of 2°-45° 2θ with a 0.04° step and 4 sec/step. Thus, the duration of scanning was appreciable, being 73 min. The accuracy of the analysis was checked by running mixtures of minerals. The maximum error was 7 wt%, and the standard deviation about 3 wt%. The detection limit was about 5% for clay minerals and 1% or less for the others. However, glass could be detected only if its abundance exceeded 40%.

Martinez and Planа (1987) used the external standard technique to analyze carbonate sediments, including low-Mg calcite, high-Mg calcite, dolomite, aragonite, quartz, and the amorphous phase. The analytical problem in this case was the overlapping of the diffraction peaks of the first three carbonate phases. Integrated intensities of the overlapping peaks were determined by fitting them to Lorentzian profiles. A reference material (fluorite) was added to each analyzed sample in order to keep the 100% constraint for detected crystalline phases and to estimate the amount of the amorphous phase. Good accuracy (error within 4 wt%) was demonstrated on artificial mineral mixtures, and good correspondence was found between total carbonate content obtained by QXRD and quantitative chemical analysis.

Maniar and Cooke (1987) and Davis and Walawender (1982) compared two quantification techniques—QXRD and point counting. Maniar and Cooke (1987) analyzed three major components of granitoids—quartz; plagioclase, including exsolved albite in perthite; and orthoclase, including microcline of perthite. Quartz constituted the reference material, and β_{jr} constants (quartz) were obtained from two-phase mixtures of the minerals with quartz. The sum of the phase abundances was brought to 100% [the technique is similar to that proposed by Drits (1958), described in Section 4.6]. Three to 16 replicate analyses were carried out for each of six granitoids. The correspondence between mean QXRD and point counting analysis can be considered very good: only on rare occasions did the difference exceed 2 wt%. The precision of the analysis (standard deviation between replicates) was better for point counting. Maniar and Cooke (1987) noted the obvious advantages of QXRD over point counting: QXRD is relatively inexpensive and fast; QXRD can be used for very fine-grained rocks; the powder sample used in QXRD is more representative than thin sections; and errors due to mineral misidentification are practically excluded. Davis and Walawender (1982) performed a comprehensive study of the mineralogical composition of five granitoid rocks. QXRD was carried out by a variation of the external standard technique, in which calibration constants β_{rj} were expressed though RIR values (Section 4.7). The list of phases detected by the technique included 14 minerals. For diopside, enstatite, augite, and olivine, RIR values were obtained from calculated intensities, and RIRs for the remaining minerals were determined experimentally. The mineralogical composition was also obtained by an optical point counting method. The study of Davis and Walawender thus provides an excellent and relatively rare opportunity to compare these two phase-related techniques. Remarkable correspondence between QXRD and optical point counting was demonstrated for most of the quantitative data, except perhaps in some cases, those in which RIRs were calculated. Good agreement was also observed between chemical composition obtained by X-ray fluorescent analysis and that deduced from QXRD phase quantification.

Further development of QXRD techniques tackled the problem of overlapping peaks. The group of Smith used the method of Karlak and Burnett (1966) to analyze Colorado oil shales. The method was checked on multiphase mixtures, revealing good accuracy. It is interesting to note that implementation of chemical constraints in this case did not bring any appreciable improvement of analytical results. The subsequent publications of Smith and his co-workers (Smith et al. 1988, 1989) presented excellent analytical results for the full-pattern approach for mineral mixtures of up to ten minerals.

During the past few years, we have witnessed growing interest in QXRD with structure refinement in the quantification of rocks. Good examples are given by Taylor and Pecover (1988) for the analysis of zeolite-bearing rocks.

A standardless technique was used by Moore (1968) for the quantification of geologic samples. A more recent application of a technique that is similar in principle is to be found in the work of Fang and Zevin (1985) on the quantification of carbonate rocks. A set of 22 limestones with various proportions of the five phases, quartz, calcite, dolomite, anhydrite, and pyrite, were analyzed by the standardless method outlined in section 4.10. Indirect estimation of the precision is given by the standard deviation of the sum of the concentrations from 100%, which was equal to 5 wt%.

6.2.2 Analysis of bauxites

Bauxite is the raw material used primarily for the manufacture of aluminum, and its quantification is therefore of great practical importance. In addition to the principal alumina-bearing minerals, boehmite AlOOH, gibbsite $Al(OH)_3$, diaspore AlOOH, and corundum Al_2O_3, bauxites also contain a number of accessory minerals in different proportions (Table 6.1). This list includes minerals with alternative chemical compositions and minerals that, being found in different deposits, show variations in relative peak intensities. The quantification of this complicated system is thus a real challenge. In an analysis of six bauxite minerals—gibbsite, hematite, goethite,

Table 6.1. Mineral composition of bauxites

Mineral	Chemical formula
Boehmite	$AlOOH$
Diaspore	$AlO(OH)$
Gibbsite	$Al(OH)_3$
Corundum	Al_2O_3
Goethite	$FeOOH$
Hematite	Fe_2O_3
Pyrite	FeS_2
Siderite	$FeCO_3$
Anatase	TiO_2
Rutile	TiO_2
Calcite	$CaCO_3$
Dolomite	$CaCO_3.MgCO_3$
Crandallite	$CaAl_3(PO_4)_2(OH)_5.H_2O$
Lithiophorite	$(Al,Li)MnO_2(OH)_2$
Kaolinite	$Al_2(Si_2O_5)(OH)_4$
Chamosite[1]	
Quartz	SiO_2

[1]A variety of chemical formulas appear in the literature.

boehmite, anatase, and kaolinite—by the external standard method (see Section 4.6), MgO was used as the reference material, and the calibration constants β_{jr} (r = MgO) were determined from 1:1 mixtures of the analytes with MgO. The accuracy of the analysis was estimated by a comparison of the phase abundances obtained by QXRD, by the chemical technique outlined in Section 4.11, and by thermal analysis. Good correspondence of the results was reported for metamorphic bauxites due to the high degree of crystallization. In contrast, QXRD of poorly crystallized bauxites of relatively younger deposits from Middle Asia gave abnormally low concentrations of Al-bearing minerals. It is supposed that some of these minerals are present in the amorphous state. An internal standard (fluorite) technique has been used to analyze bauxites from various deposits in Russia. Calibration constants β_{js} were determined from binary mixtures with an internal standard. An analysis of four-mineral mixtures (boehmite, gibbsite, hematite, and kaolinite) revealed that the difference between the observed and actual concentrations usually fell within a few percent.

A significant contribution to QXRD of bauxite was made by Bárdossy and his co-workers (1980). The main QXRD technique used by this group was the external standard technique. It was found that β_{jr} values [equation (4.58)] were practically constant for minerals from various deposits located in Hungary, except for clay minerals. For the latter the variations of β_{rj} were within 5%-10%. However, minerals taken from a wider range of bauxite deposits (former USSR, Greece, former Yugoslavia) showed much greater variations of β_{rj}—up to 20%. The most reliable results with a smaller probability of systematic errors were obtained by QXRD with chemical constraints (Bárdossy et al. 1980); the chemical composition of the phases was corrected by studying the diffraction peak shifts. Phase abundances obtained by QXRD were normalized by means of the chemical element-balance equation (4.57). Phase analysis was carried out in a fully automated mode. Bárdossy et al. (1980) achieved a remarkable detection limit for the major minerals—0.5% for the major Al-bearing minerals, 0.2% for quartz and 0.1% for rutile and anatase. QXRD has also proved to be useful in the analysis of bauxites to be processed by the Bayer process.

6.2.3 *Mineralogical analysis of coal and coal ash*

The abundance of mineral matter in coal ranges from a few percent up to a very substantial amount of 20%-25%. A knowledge of the phase composition of the mineral matter is undoubtedly important in studying the formation of coal, but it is essential in the technology of storage and burning.

The fly ash generated by the burning of coal causes slagging, fouling of equipment, and pollution. Although the chemistry of the mineral matter is of prime interest, a knowledge of the phase composition can be very helpful in the estimation of coal quality. For example, it is important to know the distribution of alkalis between phases, whether Ca carbonates are present, and whether Ca sulfates are likely to be formed during firing.

Using the formulas derived in Section 4.10, we can easily obtain the ratio of the diffracted intensity for any phase j in the mineral matter that has been separated from coal I_j and that in the whole coal:

$$\frac{I_j'}{I_j} = \frac{w\mu_m^*}{w\mu^*_m + w\mu_c^* (1 - w)}$$

where w is the weight fraction of the mineral matter in the coal, and μ_m^* and μ_c^* are the mass absorption coefficients of the mineral matter and carbonaceous matter, respectively.

Davis et al. (1986) estimated μ_c^* as 6 cm^2/g. For some average values of $w \approx 15\%$ and $\mu_m^* \approx 50$ cm^2/g, the ratio I_j'/I_j is equal to ~ 0.6, i.e., dilution of the mineral matter by the carbonaceous component leads to only a moderate reduction of diffracted intensities. However, poorly ordered carbon significantly increases the background and reduces the detection limit of crystalline phases. In order to improve the quality of the diffraction pattern, the carbonaceous material in the coal is oxidized in plasma at a low temperature of ~150°C, a temperature that is assumed not to cause any thermal destruction to the mineral matter.

The mineral matter obtained by this treatment is known as low-temperature ash (LTA). Thus, the major crystalline phases can be easily analyzed in the whole unseparated coal, but full analysis of the mineral matter should be carried out in LTA samples.

Two phases are of special interest in coal—quartz, which can be a hazardous component of the dust from coal heaps, and pyrite, which is the source of SO_2 in power station smoke. Both phases are analyzed in whole coal. Because the mineral matter (heavy phase) is distributed in a very low-absorbing matrix, any variation in the amount of mineral matter will produce significant variations in sample absorption. In this case, the internal standard technique seems to be appropriate.

Davis et al. (1986) devised an elaborate scheme to analyze mineral matter in coal, which included low-temperature ashing and measurements of sample mass-absorption coefficients. Both whole coal and its LTA were analyzed. More than 20 phases were detected, with quartz, kaolinite, calcite, gypsum, bassanite, and pyrite being the major components. The QXRD technique employed was the external standard method, in which all β_{rj} calibration constants were expressed through the RIR for corresponding minerals. The coefficient of variation for phase abundances, propagated from errors in RIR, absorption measurements, and intensity measurements, was seldom less than 10% and was sometimes as high as 20%-30%. Nevertheless, taking into account the large number of phases present, the complexity of phases, and the unfavorable "diffraction environment," the results are remarkable. Moreover, good correlation was generally found between the experimentally obtained chemical composition of the mineral matter and the chemical composition deduced from the QXRD mineral content.

Belloto and Cristiani (1991) analyzed the LTA from ten coals of various origins. The major phases were found to be quartz, kaolinite, illite, and vermiculite, with pyrite, carbonates, and sulfates constituting the minor phases. The full-pattern approach was used in this work, with lattice parameters, peak width, and preferred orientation parameters (for clay minerals) being refined by the Rietveld procedure. It was found that refinement of the structure and the composition of the phases was hampered by correlation between the variables. Each sample was spiked with an internal standard (α-Al_2O_3), and RIR values for the phases were calculated assuming some idealized crystal structure. As usual, implementation of the internal standard method enables us to overrule the mass-balance equation and to determine the overall amount of detected crystalline phases, which varies from 50% to 80% for analyzed coals. The detection limit was estimated as ~1%, and standard deviation of the phase concentrations varied between 3% and 15%.

Another coal-related inorganic product that is produced in enormous quantities in coal-burning power plants is fly ash. Even in such a small country as Israel, which has two coal-burning plants, fly ash amounts to many hundred thousands of tons annually. In the United States, this amount may reach 50 million tons. Some 20% of the fly ash generated is used in the cement industry as a substitute for Portland cement, and the remaining 80% of fly ash is stored to await some technological use. Knowledge of phase composition of fly ash would therefore be useful. McCarthy and Johansen (1988) found the following phases in fly ash: anhydrite, tricalcium aluminate, hematite, lime (crystalline CaO), melilite, mullite, periclase, quartz, a sodalite-type phase, ferrite spinel, dicalcium silicate, and glass. We have analyzed quartz (hazardous component) and CaO (useful in the cement hydration process) by the internal standard technique. The same technique can be used for the analysis of other constituents. The phases in fly ash show smaller structural and compositional variability than the mineral matter of coal and are thus easier to quantify.

A number of papers have appeared on the QXRD of fly ash. Schlorholtz and Demirel (1987) discussed the role of sample preparation, and Taylor and Clapp (1992) published a full-line profile-fitting computation for the QXRD of fly ash. Thedchanamoorthy and McCarthy (1989) applied the RIR technique to the QXRD of fly ash.

6.3 Analysis of Portland cement

Portland cement clinker, which is obtained by firing a raw mixture of fixed and usually constant chemical composition, is composed of four major phases:

- alite or tricalcium silicate $3CaO.SiO_2$
- belite or dicalcium silicate $2CaO.SiO_2$
- brownmillerite or aluminoferrite $\sim 4CaO.Al_2O_3.Fe_2O_3$
- tricalcium aluminate $3CaO.Al_2O_3$

and a few minor phases, of which free lime CaO is an indispensable component.

Portland cement is obtained from the clinker by milling and spiking the product with additional materials, such as gypsum, slags, fly ash, and pozzolanic materials. Stoichiometric tricalcium silicate is triclinic, but small (~1%) amounts of alumina and/or magnesia stabilize the monoclinic form (alite) usually found in clinkers. Tricalcium aluminate is cubic, but in the presence of alkalis, orthorhombic alkali aluminate may also be found in clinkers.

Severe peak overlapping and calibration (by preparation of proper analytical standards) are the two main problems in the quantification of Portland cement clinker. The absorption correction is not a special problem. Moreover, for regular clinker, containing 3%-5% of Fe_2O_3, the mass absorption coefficient (CuK_α) remains almost constant (~100 cm^2/g) within reasonable variations of clinker composition.

The importance of the quantification of Portland cement clinker may be viewed in light of several major considerations:

1. The phase composition is a very good indication of the performance of the complex technological route, which includes grinding, proportioning, mixing, and firing of raw materials.
2. The mechanical properties of cement mortar and concrete are positively correlated to the weight fractions of alite and calcium aluminate in Portland cement (Alexander 1972; Aldrige 1982b).
3. The stability of concrete to sulfate corrosion is inversely proportional to the weight fraction of calcium aluminate.

One of the first, and possibly the best, works on cement quantification was published at the end of the 1950s by Copeland et al. (1959). Clinker minerals were synthesized and used for the preparation of calibration mixtures. In order to approach the composition and crystal structure of alite and belite in real clinkers, 1% of Al_2O_3 and MgO were added to the raw mixtures. The set of synthesized alumino-ferrites with molar ratio n = Al_2O_3/Fe_2O_5 were used to determine the ratio Al_2O_3/Fe_2O_3 in real aluminoferrites. Copeland et al. used the internal standard method with Si (10%) as the intensity and peak position standard. The only available analytical peak for aluminoferrite is 141 ($2\theta \approx 33.9°$) and that for calcium aluminate 440 ($2\theta \approx 33.2°$), but even these peaks suffer from partial overlapping. A somewhat naive graphic technique was used to separate them. There are almost no non-overlapping diffraction peaks for silicate phases. The two peaks chosen in the work cited were the peak at ~51.8°, d ≈ 1.77 Å with an overwhelming contribution of alite and the peak at ~41.2°, d ≈ 2.20 Å with approximately equal contributions of alite and belite. From QXRD results, a ratio of phase abundances of alite and belite was obtained. The final scaling was achieved by using chemical data—known proportions of CaO to silicate phases. The latter was obtained by subtracting the CaO present in aluminate and aluminoferrite from the total weight fraction of CaO in the clinker analyzed. Analysis of 20 different cements revealed acceptable correspondence between QXRD and chemical analysis (Section 4.11) for all phases, except calcium aluminate. The latter was systematically overestimated by the chemical technique, due to the assignment of all the available Al_2O_3 to the calcium aluminate and aluminoferrite phases. In fact, an appreciable amount of Al_2O_3 is dissolved in the silicate phases.

Further development of the technique was related to technical improvements of diffractometric measurements, such as more effective detectors, monochromators, and the widespread use of the step-scanning technique. It was possible, for example, to analyze the weak, but overlapping, free 120 ($2\theta \approx 31.1°$) peak of belite by the internal standard technique (Zevin et al. 1977). A large number of works were published in the 1960s and 1970s, which differed mainly in the material of the internal standard—CaF_2 by Volkov and Kheiker (1963), MgO by Péter (1971), Al_2O_3 by Smolczyk (1961), and CeO_2 by Howard (1984). Mixtures of pure clinker minerals were used for preparation of the calibration graphs.

This traditional approach to the quantification of Portland cements was broken by Knudsen (1976), Bezjak and Jelenic (1971a), and especially Berger et al. (1966). Knudsen used optical point counting to establish the phase composition of 27 clinkers, which were later used for calibration in QXRD. Integrated intensities were measured in eight different scattering angle intervals. Increasing the number of analytical peaks over the minimal number of four and calibration of real clinkers undoubtedly contributed to the reliability of analysis. The work of Berger et al. (1966) was the first example of the full-pattern approach in QXRD. Phase abundances were obtained as the scaling factors in the best fit of the scaled sum of digitized patterns of the component phases to the digitized pattern of the clinker analyzed. The first step in this approach was, of course, the evaluation of the spectra of the component phases. They had been extracted from patterns of 20 clinkers, for which phase compositions had been established by chemical balance calculations. The technical procedure is expressed in equation (4.182), in which the columns in the matrix represent the digitized pattern of the *jth* component phase.

We should remember that in the techniques proposed by Knudsen (1976) and Berger et al. (1966) the evaluated calibration constants (or the patterns of the individual phases) were averages for the group of clinkers used for calibration. Thus, we can hardly expect them to be unequivocally suitable for the analysis of any particular clinker with its specific pattern of polymorphic phases. There are several ways to approach this problem: choose the pattern of the proper polymorph from an existing bank of digitized patterns (Bezjak and Jelenic 1971a; Mander et al. 1974); separate the component phases from the clinker (Tabikh and Weht 1971); or combine phase quantification with structure refinement.

The necessity to account for the crystal chemistry of specific clinker phases was clearly demonstrated by Aldrige (1982a) in his report on a round-robin examination of six Portland cements. The results revealed large interlaboratory standard deviation, which was exacerbated by within-laboratory errors.

In addition to the four major phases, Portland cement contains several minor phases, which may have a considerable influence on the properties of the cement. The most important of these minor phases are free lime and calcium sulfates. The amount of free lime in the clinker indicates the degree of digestion of CaO in the synthesis of alite in the clinker. This phase is particularly undesirable due to its volume expansion during hydration inside a set concrete block. In light of the exceptional importance of this phase, a special chemical technique for its determination has been developed and authorized (ASTMC114). However, this technique does not distinguish between CaO and $Ca(OH)_2$. If the anhydrous form is essential and the presence of $Ca(OH)_2$ is suspected, QXRD is an excellent tool for quantification. The strong 200 peak of CaO ($2\theta \approx 37.5°$) is relatively free of overlapping. We have analyzed CaO in clinkers by the spiking method. It is essential that the "spike" be "dead burned" CaO, which is less susceptible to hydration by atmospheric moisture. It is not difficult to achieve a detection limit of ~0.1%-0.2% and a precision of ~0.1% for a CaO abundance of less than 2%.

Gypsum is added to Portland cement to regulate the setting time. During milling with hot clinker, gypsum might dehydrate to the hemihydrate $CaSO_4.1/2H_2O$ or even to anhydrous $CaSO_4$. The presence of the latter two phases is undesirable, since they may cause false setting of cement mortar or concrete. We have carried out the analysis of calcium sulfates in Portland cement by QXRD, using the spiking technique. X-ray quantification is also used in the analysis of high-alumina cements (Midgley 1976; Redler 1991) and in the analysis of cement hydration (Zevin and Kheiker 1965; Butt et al. 1973; Lubimova et al. 1974; Crammond 1985).

6.4 Metallurgy

Metals and metal alloys may be easily prepared for light (opaque) microscopy analysis. Their high electrical conductivity also facilitates SEM-EDS characterization. However, qualitative and quantitative phase analysis must be done by X-ray diffraction (XRD).

The use of XRD to characterize metals and alloys began very soon after the counter spectrometer was introduced in 1945 (Klug and Alexander 1954). Samples from materials of this type are usually thick and flat, which fits Bragg-Brentano geometry; this geometry is therefore commonly used in metallurgical laboratories. The integration of XRD into metallurgy and related subjects is given in detail by Taylor (1952). Nowadays, many of the suggestions that appeared in the book have become routine procedures in materials research and characterization. In addition to qualitative phase analysis, XRD may be used to determine composition, texture, microstrain, size of coherent cells, and residual stresses.

The extensive use of light and electron microscopy in metallurgy has resulted in a situation in which the majority of quantitative analyses are performed by metallography rather than by XRD. The obstacles of applying QXRD in metallurgy are well known (Fiala 1991) and originate from the nature of metallic samples. The samples are made of coherent blocks (use of the internal standard technique is impossible); the grain size and shape are not uniform; the samples are ductile, and surface grinding or mechanical polishing causes severe distortion of XRD results; it is almost impossible to avoid texture; many metallic samples are solid solutions, with intensities that deviate from literature references to the nearest structure; alloys can be fully ordered, partially ordered, or completely disordered, and the degree of ordering affects the diffraction pattern; the existence of a high-volume fraction of stacking faults in some metallic systems may affect the relationships between the intensities and the phase abundance; and most metals have a close packed structure, which is related to high densities and attenuation of X-rays, resulting in sensitivity to microabsorption effects and limiting the XRD information to a thin layer near the surface (giving rise to the question about the relationship between the XRD data and the bulk properties).

On the other hand, the existence of polymorphism in many metals and alloys requires XRD methods for accomplishing complete phase analysis. The list of systems with similar compositions but different crystal structures in metallurgy is not a short one: for example, α and β in copper-zinc (brass), cobalt, titanium, zirconium, and uranium alloys; α and γ in iron and uranium alloys; and γ and γ' in nickel-based alloys. Many of these systems form metastable structures with relative amounts of the phases, depending on the history of heat and mechanical treatment and even on aging at room temperature. QXRD is considered one of the best methods of characterizing these structures. Some of the errors caused by surface effects can be minimized by proper surface preparation (electro-polishing) and avoidance of X-ray wavelengths close to the absorption edge. The texture problem can be reduced by rocking and spinning of the samples.

A typical case that is well covered in the literature is the relationship between the γ and α phases in iron alloys. The metallurgy of iron-based alloys can be divided into three systems: cast iron, carbon steels, and other alloyed steels. The phases that may be found in iron and steel systems are α-ferrite (cI2), γ-austenite (cF4), distorted α-martensite (tI2), and carbides. The QXRD of carbon steels—determination of the amount of retained austenite—was described by Gardner et al. (1943) soon after the Geiger counter was invented. The use of integrated intensities was introduced by Averbach and Cohen (1948), who compared the QXRD with quantitative light metallography. Later works include those of Averbach et al. (1950), Ogilvie (1959), Erard (1963), Miller (1964), and Lindgren (1965). A discussion of the factors affecting the accuracy of QXRD results in plain carbon steel (Giamei and Freise 1967) two decades after the report of Averbach and Cohen demonstrates the importance of QXRD in metallurgy. In industry, it is also necessary to reduce the measurement time, which is facilitated by using an energy dispersive technique (Voskamp 1974).

In cast iron the quantitative determination of retained austenite is more difficult because of the high volume of carbides, which makes the diffractogram more complicated. In order to increase the beam penetration and simultaneously to increase the angular dispersion, it is possible to use a chromium anode and a helium chamber (McCune 1962; Kelly and Short 1972). The texture problem has also been addressed (Kim 1979; Maratray and Poulalion 1982;Whelan and Moroz 1985).

The relative abundance of α (hP2) and β (cI2) cobalt has received similar QXRD treatment (Houske et al. 1960).

In contrast to the quantitative analysis of polymorphs, for phases with different absorption coefficients it is necessary to prepare calibration curves using samples with similar microstructures. The best candidates for QXRD analysis of dissimilar metal mixtures are composite materials like heavy metals (machinable tungsten alloys).

In addition to conventional metallurgical systems. there is a wide field of intersection between metallurgy and ceramics; for example, high-speed steels contain a high volume of carbides, such as M_6C and MC. In these systems the amount of carbides can be measured by QXRD. Another example is cemented carbides, which consist of WC sintered in cobalt. The volume ratio between the cobalt and WC can be measured by QXRD.

6.5 Thin films and coatings

Since the penetration of X-rays into solids is limited to less than 0.1 mm in most solids, the X-ray diffraction spectra obtained from solids come from a thin layer close to the surface. Therefore, even if the sample is a thin film, it still yields high counting rates, and very good QXRD results are to be expected.

In contrast to bulk samples, for which QXRD gives relative amounts of phases, in polycrystalline thin films other properties can also be quantified, such as thickness, density, and mass absorption (Battagalia et al. 1992).

Nevertheless, in practice there are many obstacles. The main problem is one of texture. Whereas in bulk samples texture is largely a function of crystallite orientation and shape, in thin films having a grain size of the same order of magnitude as the film thickness, preferred orientation is also a function of crystallite size due to anisotropic growth rate. In some cases, crystal orientation gives rise to differences in line width, and it is thus recommended that integrated intensities be used. If the grain size is smaller than the film thickness, the texture problem can be solved by using more than one external standard (Gu and Marder 1991).

A number of other problems are associated with QXRD of thin films: 1) some thin films are layered structures for which special care is required in using absorption correction terms (Hejdov and Cermak 1972); 2) most thin films are prepared by a manufacturer and not by an XRD laboratory; and 3) microabsorption effects, like surface roughness, may be different for the external standard and the thin layer. In the latter case, a multiple-line method without an external standard is suggested (Brandt and van der Vliet 1985; Kimmel and Kaplan 1990). Multiple-line methods have been successfully used in the characterization of laser-treated anodized aluminum samples (Kimmel et al. 1989), in kinetics studies of growth of WC thin layers (Kimmel and Kaplan 1989), and in an in-depth profile of hydrogen in steel surfaces (Zevin et al. 1984).

6.6 Air pollution (aerosols and airborne dusts)

The studies of aerosols are naturally based on methods for the characterization of powders, including X-ray powder diffraction. The first analyses of airborne particles in the 1970s dealt with the reactions of salts in a humid atmosphere (Tang et al. 1978) and with the identification of particles of specific origin in air pollution; most of these analyses were qualitative, and some of them were not even based on XRD (Biggins and Harrison 1979; Alpert and Hopke 1981; D'Almedia and Schultz 1983; Tani et al. 1983; Harrison and Sturges 1984; Mamane and Noll 1985).

The quantitative analysis of airborne particles collected on filters was developed by Davis. His technique, which used standardless and reference intensity methods, including RIR, is described in a number of articles, for example, Davis and Cho (1977) and Davis (1978, 1980, 1981, 1984), and in special review publications (Davis 1988, 1992). In addition, the quantitative analysis of calcium compounds in heat-treated airborne dusts was performed by Nakamura (1988) with two internal standards. The internal standard method (semiquantitative) was also used by Sturges et al. (1989).

6.7 Pharmaceuticals

Since many drugs are crystalline powders, it seems that QXRD can serve as an important tool both in research and development and in analysis at different stages of production. The method was first used by Christ et al. (1948) to quantify penicillin by QXRD, but it has not become a standard analytical tool in the pharmaceutical industry (Shell 1963). In many pharmaceuticals, the existence of polymorphism—as a result of more than one crystal structure for the same molecule or the presence of an amorphous structure, or of the formation of different hydrates (Laine et al. 1984)—makes structural analysis complicated. The crystal structures of many pharmaceuticals are not known, and if their patterns do appear in databases like ICDD-PDF, the diffraction lines are not indexed. Despite these difficulties, X-ray diffraction is one of the best methods to characterize polymorphs and distinguish between them. QXRD has been found to be a fast and accurate method for direct analysis (Kuroda et al. 1968, 1969, 1972). QXRD can also be used to characterize the constituents of pharmaceutical products in their final formulations, including the active ingredients and the excipients (Papariello et al. 1964; Durbetaki and Quail 1982). Finally, it should be noted that it is possible to use nonindexed powder diffraction patterns in QXRD analysis. Expansion of the list of pharmaceuticals with indexed and nonindexed powder diffraction patterns may promote the use of QXRD in the pharmaceutical industry.

References

Aldrige, L. P. (1982a). Cem. Conc. Res. **12**: 381-398.
Aldrige, L. P. (1982b). Cem. Conc. Res. **12**: 437-446.
Alexander, K. M. (1972). Cem. Conc. Res. **2**: 633-680.
Alexander, L. E., Klug, H. P. (1948). Anal. Chem. **20**: 886-889.
Alexander, L. E., Klug, H. P., Kummer, E. (1948). J. Appl. Phys. **19**: 742-754.
Alpert, D. J., Hopke, P. K. (1981). Atmos. Environ. **15**: 675-681.
Altree-Williams, S. (1978). Anal. Chem. **50**: 1272-1275.
Altree-Williams, S., Byrnes, J. G., Jordan, B. (1981). Analyst (London) **106**: 69-75.
Altree-Williams, S., Jordan, B. (1980). Anal. Chem. **52**: 1296-1300.
Altree-Williams, S., Lee, J., Mezin, N. V. (1977). Ann. Occup. Hyg. **20**: 109-126.
Arnell, R. D. (1968). J. Iron Steel Inst. **206**: 1035-1036.
Averbach, B. L., Castleman, L. S., Cohen, M. (1950). Trans. AIME **42**: 112.
Averbach, B. L., Cohen, M. (1948). Trans. AIME **176**: 401-415.
Azároff, L. V. (1955). Acta Crystallogr. **8**: 701-704.
Azároff, L. V. (1963). Rev. Mod. Phys. **35**: 1012-1022.
Azároff, L. V. (1968). *Elements of X-ray Crystallography*. New York, McGraw Hill.
Baker, W. E. (1968). Proc. Australas. Inst. Mining Met. **228**:. 23-29.
Baldock, P. J., Parker, A. (1973). J. Appl. Crystallogr. **6 (Pt. 2)**: 153-157.
Ballard, J. W., Oshry, H. I., Schrenk, H. H. (1940). U.S. Bur. Mines Reports Invest., Report No. 3520.
Bar-On, P., Zevin, L., Lach, S. (1981). X-ray Spectrom. **10**: 57-60.
Baram, J., Zevin, L. S. (1990). Scr. Metall. **24**: 1605-1609.
Bárdossy, G., Bottyán, L., Gadó, P., Griger, A., Sasvári, J. (1980). Am. Mineral. **65**: 135-141.
Battaglia, S., Franzini, M., Leoni, L. (1990). Powder Diffr. **5**: 101-103.
Battaglia, S., Franzini, M., Leoni, L. (1992). Powder Diffr. **7**: 194-196.
Battaglia, S., Leoni, L. (1977). Anal. Chem. **49**: 1168-1171.
Bein, E. C. (1924). Trans. AIME **70**: 25.
Bellotto, M., Cristiani, C. (1991). Mater. Sci. Forum **79-82**:745-750.
Benedetti, A., Bottarelli, M., Fagherazzi, G. (1985). J. Non-Crystal. Solids **74**: 245-257.
Berger, R. L., Frohnsdorff, G. J. C., Harris, P. H., Johnson, P. D. (1966). Application of X-ray diffraction to routine mineralogical analysis of Portland cement. Special Report 90. p. 234-253. Washington DC, Highway Research Board.
Bezjak, A. (1961). Croat. Chim. Acta **33**: 197-200.
Bezjak, A. (1975). Anal. Chem. **47**: 790-793.
Bezjak, A., Jelenic, I. (1971a). Croat. Chem. Acta **43**: 193-198.
Bezjak, A., Jelenic, I. (1971b). Cem. Conc. Res. **1**: 475-492.
Biederman, R. R., Bourgault, R. F., Smith, R. W. (1974). Adv. X-ray Anal. **17**: 139-149.
Biggins, P. D. E., Harrison, R. M. (1979). Atmos. Environ. **13**: 1213-1216.
Bish, D. L., Howard, S. A. (1988). J. Appl. Crystallogr. **21**: 86-91.
Bish, D. L., Reynolds, R.C. (1989). *In*: Bish, D. L., Post, J. E., eds. *Review in Mineralogy, Modern Powder Diffraction*. Washington DC, Mineralogy Society of America.
Blokhin, M. A. (1957). *The Physics of X-rays*. 2nd ed. Moscow, State Publishing House of Technical-Theoretical Literature. (English translation AEC-tr-4502).

Bogdanova, G. S., Orlova, E. M., Bessmertnaya, Z. G., Zevin, L. S. (1971). Bull. Acad. Sci. USSR, Inorganic Materials **7**: 893-895.
Bogdanova, G. S., Orlova, E. M., Zevin, L. S. (1966). Bull. Acad. Sci. USSR, Inorg. Mater. **2**: 326-328.
Bogdanova, G. S., Zevin, L. S., Orlova, E. M., Bessmertnaya, Z. G. (1973). Bull. Acad. Sci. USSR, Inorg. Mater. **9**: 1813-1815.
Bogdanova, G. S., Zevin, L. S., Orlova, E. M., Bessmertnaya, Z. G. (1975). Bull. Acad. Sci. USSR, Inorg. Mater. **11**: 279-281.
Bogue, R. H. (1955). *The Chemistry of Portland Cement.* New York, Reinhold Publishing Corp.
Boldyrev, A. K., Mikheiev, V. I., Kovalov, G. A., Dubinina, V. N. (1939). Ann. Inst. Mines Leningrad **13**: 1.
Borg, I. Y., Smith, D. K. (1969). Calculated X-ray Powder Patterns for Silicate Minerals. Memoir 122. Boulder, CO, The Geological Society of America.
Bowden, M. E., Ryan, M. J. (1991). Powder Diffr. **6**: 78-81.
Bragg, W. H. (1921). Proc. Phys. Soc. (London.) **33**: 222.
Bragg, W. H., Bragg, W. L. (1913). Proc. R. Soc. London.**A88**: 428.
Brandt, C. G., van der Vliet, G. H. (1985). Adv. X-ray Anal. **29**: 203-229.
Brehler, B. (1957). Z. Kristallog. **109**: 68-72.
Brentano, J. C. M. (1925). Proc. Phys. Soc. London **37**: 184.
Brentano, J. C. M. (1935). Proc. Phys. Soc. London **47**: 932.
Brentano, J. C. M. (1937). Proc. Phys. Soc. London **49**: 61.
Brindley, G. W. (1945). Philos. Mag. **36**: 347-369.
Brown, A., Edmonds, J. W. (1980). Adv. X-ray Anal. **23**: 361-374.
Brown, G., ed. (1961). *The X-ray Identification and Crystal Structure of Clay Minerals.* London, Mineralogical Society.
Brown, G., Wood, I. G., Nicholls, L. (1987). Powder Diffr. **2**: 7-21.
Buerger, M. J. (1942). *X-ray Crystallography. An Introduction to the Investigation of Crystals by their Diffraction of Monochromatic X-radiation.* London, John Wiley & Sons.
Bunge, H. J. (1982). *Texture Analysis in Materials Science, Mathematical Methods.* London, Butterworths.
Bunge, H. J., Esling, C. (1982). *Quantitative Texture Analysis.* Oberursel, Deutsche Gesellschatt für Metallkunde, Informationsgesellschaft.
Butt, Y. M., Timashev, V. V., Nestic, N. V., Zevin, L. S., Margolina, G. S. (1973). Industrial Lab. **30**: 1229-1301.
Byram, S. K., Bui, H., Rothbart, G. B., Samdahl, R. N., Sparks, R. A. (1977). Adv. X-ray Anal. **20**: 529-545.
Byström-Asklund, A. M. (1966). Am. Mineral. **51**: 1233.
Caglioti, G., Paoletti, A., Ricci, F. P. (1958). Nucl. Instrum. **3**: 223-228.
Calvert, L. D., Sirianni, A. F. (1980). J. Appl. Crystallogr. **13**: 462.
Calvert, L. D., Sirianni, A. F., Gainsford, G. F., Hubbard, C. R. (1983). Adv. X-ray Anal. **26**: 105-110.
Capkova, P., Valvoda, V. (1974). Czech. J. Phys. **24**: 891-900.
Cernik, R. J., Murray, P. K., Pattison, P., Fitch, A. N. (1990). J. Appl. Crystallogr. **23**: 292-296.
Christ, C. L., Barnes, R. B., Williams, E. F. (1948). Anal. Chem. **20**: 789-795.
Chung, F. H. (1974a). J. Appl. Crystallogr. **7**: 519-525.
Chung, F. H. (1974b). J. Appl. Crystallogr. **7**: 526-531.
Clark, G. L., Reynolds, D. H. (1936). Ind. Eng. Chem. Anal. Ed. **8**: 36-40.
Clark, N. H., Preston, R. J. (1974). X-Ray Spectrom. **3**: 21-25.
Cline, J. P., Snyder, R. L. (1983). Adv. X-ray Anal. **26**: 111-117.
Cline, J. P., Snyder, R. L. (1987). Adv. X-ray Anal. **30**: 447-456.
Cohen, J. B. (1966). *Diffraction Methods in Materials Science.* New York, The Macmillan Company.
Cooper, M. J. (1965). Philos. Mag. **11**: 969-975.
Copeland, L. E., Bragg, R. H. (1958). Anal. Chem. **30**: 196-208.
Copeland, L. E., Brunaer, S., Kantro, D. L., Schulz, E. G., Weise, C. H. (1959). Anal. Chem. **31**: 1521-1530.
Cox, D. E., Hastings, J. B., Thomlinson, W., Previtt, C. T. (1983). Nuclear Instr. Methods. **208**: 573-578.

Cox, D. E., Toby, B. H., Eddy, M. M. (1988). Aust. J. Phys. **41**: 117-131.
Crammond, N. J. (1985). Cem. Conc. Res. **15**: 431-441.
D'Almedia, G. A., Schultz, L. (1983). J. Climate Appl. Meteorol. **22**: 233-243.
Darwin, C. G. (1914). Philos. Mag. **27**: 315, 675.
Davis, B. L. (1978). Atmos. Environ. **12**: 2403-2406.
Davis, B. L. (1980). Atmos. Environ. **14**: 217-220.
Davis, B. L. (1981). Atmos. Environ. **15**: 291-296, 613-618.
Davis, B. L. (1984). Adv. X Ray Anal. **27**: 339-348.
Davis, B. L. (1986). Powder Diffr. **1**: 240-243.
Davis, B. L. (1988). *Reference Intensity Method of Quantitative X-ray Diffraction Analysis.* Rapid City, South Dakota School of Mines and Technology.
Davis, B. L. (1992). Quantitative phase analysis with reference intensity ratios, *In*: E. Prince, J. K. Stalick, eds. *Proceedings of the International Conference on Accuracy in Powder Diffraction II*, Gaithersburg, MD, National Institute of Standards and Technology, Special Publication 846. p. 7-16.
Davis, B. L., Cho, N.-K. (1977). Atmos. Environ. **11**: 73-85.
Davis, B. L., Johnson, L. R. (1982). Adv. X Ray Anal. **25**: 295-300.
Davis, B. L., Johnson, R., Metrahtu, T. (1986). Powder Diffr. **1**: 244-254.
Davis, B. L., Kath, R., Spilde, M. (1990). Powder Diffr. **5**: 76-78.
Davis, B. L., Smith, D. K. (1988). Powder Diffr. **3**: 205-208.
Davis, B. L., Smith, D. K., Holomany, M. A. (1989). Powder Diffr. **4**: 201-205.
Davis, B. L., Spilde, M. N. (1990). J. Appl. Crystallogr. **23**: 315-320.
Davis, B. L., Walawender, M. J. (1982). Am. Mineral. **67**: 1135-1143.
de Keijser, T. H., Langford, J. I., Mittemeijer, E. J., Vogels, A. B. P. (1982). J. Appl. Crystallogr. **15**: 308-314.
de Keijser, T. H., Mittemeijer, E. J., Rosendaal, H. C. F. (1983). J. Appl. Crystallogr. **16**: 309.
de Wolff, P. M. (1947). Physica **13**: 62-78.
de Wolff, P. M. (1956). Acta Crystallogr. **9**: 682-683.
de Wolff, P. M. (1958). Appl. Sci. Res. **7**: 102.
de Wolff, P. M. (1968). Norelco Rep. **15**: 44-49.
de Wolff, P. M., Sas, W. H. (1969). Acta Crystallogr. **A25**: 206-209.
de Wolff, P. M., Visser, J. W. (1964). Delft, Holland, Technisch Pysische Dienst. Report No. 641.109.
de Wolff, P. M., Visser, J. W. (1988). Powder Diffr. **3**: 202-204.
Debye, P., Sherrer, P. (1916). Z. Phys. **17**: 277.
Debye, P., Sherrer, P. (1917). Z. Phys. **18**: 219.
Delhez, R., de Keijser, T. H., Langford, J. I., Louër, D., Mittemeijer, E. J., Sonneveld, E. J. (1993). Crystal imperfection broadening and peak shape in the Rietveld method *In*: Young, R. A. ed. *The Rietveld Method*. Chap. 5. Oxford, Oxford University Press.
Delhez, R., Mittemeijer, E. J. (1975). J. Appl. Crystallogr. **8**: 609-611.
Desper, C. R., Stein, R. S. (1967). J. Polym. Sci., Part B **5**: 893-900.
Devlin, D. J., Amin, K. E. (1990). Powder Diffr. **5**: 121-124.
Dickson, M. J. (1969). J. Appl. Crystallogr. **2**: 176-180.
Dollase, W. A. (1986). J. Appl. Crystallogr. **19**: 267.
Dowling, P. H., Hendee, C. F., Kohler, T. R., Parrish, W. (1956). Phillips Tech. Rev. **18**: 262-275.
Drits, V. A. (1958). Ind. Lab. (USSR)**27**: 565-569.
Durbetaki, A. J., Quail, T. (1982). Adv. X-ray Anal. **25**: 383-388.
Dyakonov, J., Mishchenk, K., Hering, A., Unger, G., Korecky, J., Melka, K., Zouhkova, J., Raynov, N., Thekhlanova, N., Rischak, G., Sidorenko, G., Volkov, M. (1992). Powder Diffr. **7**: 137-141.
Engler, P., Iyengar, S. S. (1987). Am. Mineral. **72**: 832-838.
Enzo, S., Parrish, W. (1985). Adv. X-ray Anal. **28**: 37-44.
Erard, H. R. (1963). Adv. X-ray Anal. **7**: 256-264.
Ergun, S., Tiensuu, V. H. (1958). J. Appl. Phys. **29**: 946-949.
Ermrich, M., Hermann, H. (1990). Phys. Status Solidi **118**: K1-K4.
Evgrafov, A. A., Zevin, L. S., Komyak, N. I., Rayzin, V. P., Kheiker, D. M. (1972). Appar. Metody Rentgenovskogo Anal. **4**: 242-245.
Fang, J. H., Zevin, L. S. (1985). J. Sediment. Petrol. **55**: 267-271.

Feder, R., Berry, B. S. (1970). J. Appl. Crystallogr. **3**: 372-379.
Fiala, J. (1976). J. Appl. Crystallogr. **9**: 429-432.
Fiala, J. (1980). Anal. Chem. **52**: 1300-1304.
Fiala, J. (1991). Mater. Sci. Forum **79-82**: 27-34.
Flinter, B. H. (1973). Neues Jahrb. Mineral Monatsh. **H.5**: 216-221.
Flinter, B. H. (1975). Neues Jahrb. Mineral Abh. **125**: 243-277.
Flörke, O. W., Saalfeld, H. (1955). Z. Kristallog. **106**: 460-466.
Frevel, L. K. (1978). J. Appl. Crystallogr. **11**: 184-189.
Frevel, L. K., Roth, W. C. (1982). Anal. Chem. **54**: 677-682.
Friedman, H. (1945). Electronics **18**: 132-137.
Frohnsdorff, G. J. C., Harris, P. H. (1964). Dev. Appl. Spectrosc **3**: 58-68.
Fukasawa, T., Iwatsuki, M., Kawakubo, S., Miyazaki, K. (1980). Anal. Chem. **52**: 1784-1787.
Garbauskas, M. F., Goehner, R. P. (1981). Adv. X-ray Anal. **25**: 283-288.
Gardner, F. S., Cohen, M., Antia, D. P. (1943). Trans. AIME **154**: 306.
Gazzara, C. P., Messier, D.R. (1977). Am. Ceram. Soc. Bull. **56**: 777-780.
Giamei, A. F., Freise, E. J. (1967). Trans. Metall. Soc. AIME **239**: 1677-1685.
Gibbs, R. J. (1967). Clay Miner. **7**: 79-90.
Göbel, H. E. (1981). Adv. X-ray Anal. **24**: 123-138.
Goehner, R. P. (1978). Anal. Chem. **50**: 1223-1224.
Gonzalez, C. R. (1987). Acta Crystallogr. **A43**: 769-774.
Gu, M., Marder, A. R. (1991). Powder Diffr. **6**: 89-94.
Guinier, A. (1963). *X-ray Diffraction in Crystals, Imperfect Crystals and Amorphous Bodies*. San Francisco, WH Freeman.
Gulberg, R., Lagneborg, R. (1966). Trans. AIME **236**: 1482-1485.
Gupta, R. K., Anantharaman, T. R. (1971). Z. Metallkde. **62**: 732-735.
Hall, M. M., Veeraraghaven, V. G., Rubon, H., Winchell, P. G. (1977). J. Appl. Crystallogr. **10**: 66-68.
Hanawalt, J. D., Rinn, H. W., Frevel, L. K. (1938). Ind. Eng. Chem. Anal. **10**: 457-512.
Harrison, M., Sturges, W. T. (1984). Atmos. Environ. **18**: 1829-1833.
Harrison, R. J., Paskin, A. (1964). Acta Crystallogr. **17**: 325-331.
Hastings, J. B., Thomlinson, W., Cox, D. E. (1984). J. Appl. Crystallogr. **17**: 85-95.
Hecq, M. (1981). J. Appl. Crystallogr. **14**: 60-61.
Hejdov, H., Cermak, M. (1972). Phys. Stat. Solid **72**: K95-K98.
Hepp, A., Baerlocher, C. (1988). Aust. J. Phys. **41**: 229-236.
Hermann, H., Ermrich, M. (1987). Acta Crystallogr. **A43**: 401-405.
Hermann, H., Ermrich, M. (1989). Powder Diffr. **4**: 189-195.
Hermans, P. H., Weidinger, A. (1948). J. Appl. Phys. **19**: 491-506.
Hermans, P. H., Weidinger, A. (1961a). Tex. Res. J. **31**: 558-571.
Hermans, P. H., Weidinger, A. (1961b). Makromol. Chem. **44**: 24-36.
Hicho, G. E., Eaton, E. E. (1983). Adv. X-ray Anal. **26**: 137-140.
Hidalgo, R. V., Renton, J. J. (1970). The Use of Pelletized Samples for X-ray Diffraction Analysis of Clay Minerals in Oil Shale. Circular 12. Morgantown, West Virgina, Geological and Economic Survey.
Hill, R. J. (1991). Powder Diffr. **6**: 74-77.
Hill, R. J. (1993). Data collection strategies: fitting the experiment to the need. *In*:Young, R.A. ed. *The Rietveld Method*. Oxford, Oxford University Press. (Chap. 5A).
Hill, R. J., Howard, C. J. (1987). J. Appl. Crystallogr. **20**: 467-474.
Hill, R. J., Howard, C. J., Reichert, B. E. (1988). Mater. Sci. Forum **34-36**: 159-163.
Hill, R. J., Madson, I. C. (1987). Powder Diffr. **2**: 146-162.
Hodgson, M., Dudeney, A. W. L. (1984). Clays Clay Miner. **32**: 19-28.
Hooton, D. H., Giorgetta, N. E. (1977). X-ray Spectrom. **6**: 2-5.
Houske, C. R., Averbach, B. L., Cohen, M. (1960). Acta Metall. **8**: 81.
Howard, B. H. (1984). Cem. Conc. Res. **14**: 729-740.
Howard, C. J., Hill, R. J., Sufi, M. A. M. (1988). Chemistry in Australia Oct. 367-369.
Hubbard, C. R. (1980). Standard reference materials for quantitative analysis and d-spacing measurements *Proceedings of Symposium on Accuracy in Powder*

Diffraction, pp. 489-502.National Bureau of Standards. Special publication no. 567.
Hubbard, C. R., Evans, E. H., Smith, D. K. (1976). J. Appl. Crystallogr. **9**: 169-174.
Hubbard, C. R., Smith, D. K. (1977). Adv. X-ray Anal. **20**: 27-39.
Hubbard, C. R., Snyder, R. L. (1988). Powder Diffr. **3**: 74-77.
Hughes, R. E., Bohor, B. F. (1970). Am. Mineral. **55**: 1780-1786.
Hull, A. W. (1917). Phys. Rev. **10**: 661.
Hull, A. W. (1919). J. Am. Chem. Soc. **41**: 1168-1175.
International Tables for X-ray Crystallography. (1959). volume II: Lonsdale, K. general ed., Birmingham, Kynoch Press.
International Tables for X-ray Crystallography. (1968). volume III: Macgillary, C. H., Rieck, G. D. eds., Birmingham, Kynoch Press.
International Tables for X-ray Crystallography. (1974). volume IV: Ibers, J. A., Hamilton, W. C., eds., Birmingham, Kynoch Press.
Ivoilov, A. S., Denisova, L. L. (1962). Ind. Lab. **28**: 700-702.
Jackson, M. L. (1956). Soil Chemical Analysis—Advanced Course. Madison, WI, Department of Soil Science, University of Wisconsin.
Jahanbagloo, I. C., Zoltai, T. (1968). Anal. Chem. **40**: 1739-1741.
James, R. D. (1962). *The Optical Principle of the Diffraction of X-rays*. 3rd ed. London, G. Bell & Sons.
Jansen, J., Pechar, R., Schenk, H. (1992). J. Appl. Crystallogr. **25**: 237-245.
Järvinen, M. (1993). J. Appl. Crystallogr. **26**:525-531.
Järvinen, M., Merisalo, M., Pesonen, A., Inkinen, O. (1970). J. Appl. Crystallogr. **3**: 313-318.
Jenkins, R. (1983). Adv. X-ray Anal. **26**: 25-33.
Jenkins, R., Paolini, F. R. (1974). Norelco Reporter **21**: 9-14.
Jenkins, R., Schreiner, W. N. (1989). Powder Diffr. **4**: 74-100.
Jennings, L. D. (1984). Acta Crystallogr. **A40**: 12-16.
Kagan, A. S., Snovidov, V. M. (1965). Industrial Lab. **31**: 707-711.
Karlak, R. F., Burnett, D. S. (1966). Anal. Chem. **38**: 1741-1745.
Kelly, C. J., Short, M. A. (1972). Adv. X-ray Anal. **15**: 102-113.
Khattak, C. P., Cox, D. E. (1977). J. Appl. Crystallogr. **10**: 405-411.
Kheiker, D., Zevin, L. (1963). *X-ray Diffractometry*. 2nd ed. Moscow, Fizmatgiz.
Kheiker, D. M., Zevin, L. S. (1969). Apar. Metody Rentgenovskogo Anal., No. 17, pp. 60-63.
Kheiker, D. M., Zevin, L. S., Konstantinov, I. E., Alekseev, B. A. (1962). Bull. Acad. Sci. USSR Phy. Ser. **26**: 392-398.
Kim, C. (1979). J. Heat Treat. **1**: 43.
Kimmel, G. (1987). Powder Diffr. **2**: 22-27.
Kimmel, G., Kaplan, W. (1989). Tungsten-carbide coating of heavy metals. *In*: 12th Plansee Seminar, Reutte, Austria. May 1989. pp. 301-313.
Kimmel, G., Kaplan, W. (1990). Adv. X-ray Anal. **33**: 121-127.
Kimmel, G., Schreiner, W. N. (1989). Aust. J. Phys. **41**: 719-728.
Kimmel, G., Shafirstein, G., Bamberger, M. (1989). Adv. X-ray Anal. **32**: 295-301.
Klug, H., Alexander, L. E. (1974). *X-ray Diffraction Procedures*. 2nd ed. New York, John Wiley and Sons.
Klug, H. P., Alexander, L. E. (1954). *X-ray Diffraction Procedures for Polycrystalline and Amorphous Materials*. New York, John Wiley.
Klug, H. P., Alexander, L. E., Kummer, E. (1948a). Anal. Chem. **20**: 607-609.
Klug, H. P., Alexander, L. E., Kummer, E. (1948b). J. Ind. Hyg. Toxicol. **30**: 166-171.
Knudsen, T. (1976). Am. Ceram. Soc. Bull. **55**: 1052-1054.
Kogan, V. A., Kupriyanov, M. F. (1992). J. Appl. Crystallogr. **25**: 16-25.
Kolotov, A. Z., Chertock, I. A. (1981). Zavod. Lab. **47**: 61-63.
Kurjumov, G., Sachs, G. (1930). Z. Phys. **64**: 325.
Kurki-Suonio, K. (1977). Isr. J. Chem. **16**: 115-123.
Kuroda, K., Hashizume, G. (1972). Chem. Pharm. Bull. Tokyo **20**: 2059-2062.
Kuroda, K., Hashizume, G., Fukuda, K. (1968). J. Pharm. Sci. **57**: 250-254.
Kuroda, K., Hashizume, G., Kume, K. (1969). Chem. Pharm. Bull. **17**: 818-821.
Ladell, J., Zagofsky, A., Pearlman, S. (1975). J. Appl. Crystallogr. **8**: 499-506.

Laine, E., Tuominen, V., Jalonen, H., Haapaniemi, J., Ilvessalo, P. (1984). On the polymorphism of drugs. Thirteenth International Congress of Crystallography, 9-18 August 1984, Hamburg. p. C-73.

Lang, A. R. (1956). Rev. Sci. Instr. **27**: 17-25.

Langford, J. I. (1978). J. Appl. Crystallogr. **11**: 10-14.

LeGalley, D. P. (1935). Rev. Sci. Inst. **6**: 279-283. (reprinted in Powder Diffr. **6**:126-129, 1991).

Leroux, J., Lennox, D. H., Kay, K. (1953). Anal. Chem. **25**: 740-743.

Leroux, J., Mahmud, M. (1960). Appl. Spectrosc. **14**: 131-134.

Lindemann, R., Trost, A. (1940). Z. Phys. **115**: 456-468.

Lindgren, R. (1965). Metal Prog. **87**: 102-104.

Lonsdale, K. (1949). Mineral. Mag. **28**: 14.

Louër, D. (1988). Chem. Scr. **28**: 89-95.

Louër, D., Langford, J. I. (1988). J. Appl. Crystallogr. **21**: 430-437.

Lubimova, T. Y., Zevin, L. S., Gracheva, O. I. (1974). Bull. Acad. Sci. USSR, Inorganic Materials **10**: 942-946.

Mack, M., Parrish, W. (1967). Acta Crystallogr. **23**: 693-700.

Mack, M., Spielberg, N. (1958). Spectrochim. Acta **12**: 169-178.

Madsen, I. C., Hill, R. J. (1990). Powder Diffr. **5**: 195-199.

Majumdar, A. J., Vallance, L. S., Born, C. G. G. (1972). J. Appl. Crystallogr. **5**: 343-347.

Malinowski, E. R., Howery, P. G. (1980). *Factor Analysis in Chemistry*. New York, John Wiley and Sons.

Mamane, Y., Noll, E. K. (1985). Atmos. Environ. **19**: 611-622.

Mander, J. E., Adams, L. D., Larkin, E. E. (1974). Cem. Conrete Res. **4**: 533-544.

Maniar, P., Cooke, G. A. (1987). Am. Mineral. **72**: 433-437.

Maratray, F., Poulalion, A. (1982). AFS Trans. **82**: 795.

March, A. (1932). Z. Kristallogr. **81**: 285-297.

Marquart, D. (1963). J. Soc. Ind. Appl. Math. **2**: 431-441.

Martinez, B., Plana, F. (1987). Sedimentology **34**: 169-174.

McCarthy, G. J., Gehringer, R. C., Smith, D. K., Injaian, V. M., Pfoertsch, D. E., Kabel, R. L. (1981). Adv. X-ray Anal. **24**: 253-264.

McCarthy, G. J., Johansen, D. M. (1988). Powder Diffr. **3**: 156-161.

McCune, R. R. (1962). Adv. X-ray Anal. **6**: 85-89.

Mencik, Z., Short, M. A., Peters, C. R. (1980). Adv. X-ray Anal. **23**: 375.

Midgley, H. C. (1976). Cem. Conc. Res. **6**: 217-224.

Miller, R. L. (1964). Trans. Am. Soc. Metals **57**: 892-899

Moore, A. E. (1964). Analysis of calcareous materials. SCI monograph. no. 18: p. 372.

Moore, C. A. (1965). Southeast. Geol. **6**: 139-158.

Moore, C. A. (1968). Clays Clay Miner. **16**: 325-336.

Mortier, W. J., Costenoble, M. L. (1973). J. Appl. Crystallogr. **6**: 488-490.

Naidu, S. V. N., Houska, C. R. (1982). J. Appl. Crystallogr. **15**: 190-198.

Nakamura, T. (1988). Powder Diffr. **3**: 86-90.

Náray-Szabó, I., Péter, É. (1964). Chimia **18**: 275-277.

Náray-Szabó, I., Péter, É. (1967). Acta Geolog. Acad. Sci. Hung. **11**: 347-356.

National Bureau of Standards. (1969–1978). *Standard X-ray Diffraction Patterns*. Monograph 25. Washington DC., National Bureau of Standards.

Navias, A. L. (1925). J. Am. Ceram. Soc. **8**: 296-302.

Nelder, J. A., Mead, R. (1965). Computer J. **7**: 308.

Nelson, J. B. (1960). The Debye-Sherrer powder camera, *In*: Peiser, H. S., Rooksby, H. P., Wilson, A. J. C. eds. *X-ray Diffraction by Polycrystalline Materials*. pp. 77-121. Reinhold Publishing Co.

Newnham, R. E., de Haan, Y. M. (1962). Z. Krist. **117**: 235-237.

Nuffield, E. Y. (1966). *X-ray Diffraction Methods*. New York, John Wiley & Sons.

O'Connor, B. H., Chang, W. J. (1986). X-ray Spectrom. **15**: 267-270.

O'Connor, B. H., Raven, M. D. (1988). Powder Diffr. **3**: 2-6.

Ogilvie, R. E. (1959). Norelco Rep. **6**: 60-61.

Ohlberg, S. M., Strickler, D. W. (1962). J. Am. Ceram. Soc. **45**: 170-173.

Otálora, G., Hess, H. H. (1969). Am. J. Sci. **267**: 822-824.

Paakkari, T., Blomberg, M., Serimaa, R., Järvinen, M. (1988). J. Appl. Crystallogr. **21**: 393-397.
Paakkari, T., Suortti, P. (1968). Acta Crystallogr. **A24**: 701-702.
Papariello, G. I., Letterman, H., Heutteman, R. E. (1964). J. Pharm. Sci **53**: 663.
Parrish, W. (1965). *X-ray Analysis Papers*. Eindhoven, Centrex.
Parrish, W. (1974). Adv. X-ray Anal. **17**: 97-105.
Parrish, W. (1988). Aust. J. Phys. **41**: 101-112.
Parrish, W., Hart, M. (1985). Trans. Am. Crystallogr. Assoc. **21**: 51-55.
Parrish, W., Hart, M. (1987). Z. Kristallogr. **179**: 161-173.
Parrish, W., Hart, M., Huang, T. C. (1986a). J. Appl. Crystallogr. **19**: 92-100.
Parrish, W., Hart, M., Huang, T. C., Belloto, M. (1986b). Adv. X-ray Anal. **30**: 373-382.
Parrish, W., Huang, T. C., Ayers, G. L. (1976). Trans. Am. Crystallogr. Assoc. **12**: 55-73.
Parrish, W., Lowitzsch, K. (1962). Geometry, alignment and angular calibration of X-ray diffractometers. *In*: Parrish, W., ed. *Advances in XRD and XRS*. Eindhoven, Centrex.
Parrish, W., Mack, M. (1967). Acta Crystallogr. **23**: 687-692.
Parrish, W., Mack, M., Taylor, J. (1966). J. Sci. Instrum. **43**: 623.
Parrish, W., Roberts, B. W. (1962). Filter and crystal monchromator techniques. *In*: *International Tables for X-ray Crystallography*. volume III: Birmingham, Kynoch Press.
Parrish, W., Wilson, A. J. C. (1962). Precision measurement of lattice parameters of polycrystalline specimens. *In*: Parrish, W., ed. *Advances in XRD and XRS*. Eindhoven, Centrex.
Pawley, G. S. (1981). J. Appl. Crystallogr. **14**: 357-361.
Pawloski, G. A. (1985). Am. Miner. **70**: 663-667.
Pearson, M. J. (1978). Clays Clay Miner. **26**: 423-433.
Pearson, W. B. (1958). *Handbook of Lattice Spacings and Structures of Metals*, volume 1. New York, Pergamon.
Pearson, W. B. (1967). *Handbook of Lattice Spacings and Structures of Metals*, volume 2. New York, Pergamon.
Pesonen, A. (1979). J. Appl. Crystallogr. **12**: 460-463.
Pesonen, A., Järvinen, M., Kurki-Suonio, K. (1973). Phys. Fennica **8**: 81-91.
Péter, É. (1971). Cem. Conc. Res. **1**: 105-111.
Péter, É., Kálmán, A. (1964). Acta Chim. acad. sci. Hung. **41**: 413-422.
Polizzi, S., Fagherazzi, G., Benedetti, A., Battagliarin, M., Asano, T. (1990). J. Appl. Crystallogr. **23**: 359-365.
Popovic, S., Grzeta-Plenkovíc, B. (1979). J. Appl. Crystallogr. **12**: 205-208.
Pyrros, N. P., Hubbard, C. R. (1983). J. Appl. Crystallogr. **16**: 289-294.
Rachinger, W. A. (1948). J. Sci. Instrum. **25**: 254-255.
Redler, L. (1991). Cem. Conc. Res. **21**: 873-884.
Renault, J. (1987). Powder Diffr. **2**: 96-98.
Reynolds, R. C., Jr. (1985). *NEWMOD a Computer Program for the Calculation of One-dimensional Diffraction Patterns of Mixed-layered Clays*. Hanover, NH. R. C. Reynold
Richtmyer, F. K., Kennard, E. H., Lauritsen, T. (1955). *Introduction to Modern Physics*. 5th ed. McGraw Hill.
Rietveld, H. M. (1969). J. Appl. Crystallogr. **2**: 65-71.
Ritter, G. L., Lowry, S. R., Isenhour, T. L., Wilkins, C. L. (1976). Anal. Chem. **48**: 591-595.
Robie, S. B., Scalzo, T. R. (1986). Adv. X-ray Anal. **26**: 361-365.
Roe, R. J., Krigbaum, W. R. (1964). J. Chem. Phys. **40**: 2608-2615.
Rouse, K. D., Cooper, M. J., York, E. J., Chakera, A. (1970). Acta Crystallogr. **A26**: 682-691.
Ruland, W. (1961). Acta Crystallogr. **14**: 1180.
Ruland, W. (1964). Br. J. Appl. Phys. **15**: 1301-1307.
Ruland, W., Dewaelheyns, A. (1967). J. Sci. Instrum. **44**: 236.
Sabine, T. M. (1988). Acta Crystallogr. **A44**: 374-379.
Sahores, J. J. (1972). Adv. X-ray Anal. **16**: 186-197.
Schlorhotz, S., Demirel, T. (1987). Adv. X ray Anal. **30**: 251-256.

Schreiner, W. N. (1986). Powder Diffr. **1**: 26-33.
Schreiner, W. N., Jenkins, R. (1983). Adv. X-ray Anal. **26**: 141-147.
Schreiner, W. N., Kimmel, G. (1987). Adv. X-ray Anal. **30**: 352-356.
Schulz, L. G. (1949). J. Appl. Phys. **20**: 1030.
Scott, R. K. (1968). Fifty Years of Progress in Metallographic Techniques. ASTM STP. **430**: 201-203.
Segmüller, A. (1957). Z. Metallke. **48**: 448-454.
Shäffer, K. (1933). Z. Phys. **86**: 738.
Shell, J. (1963). J. Pharm. Sci. **52**: 24-29.
Shimazu, M. (1967). Mineral. J. **5**: 180-198.
Shimazu, M. (1968). Mineral. J. **5**: 239-248.
Smith, D., Johnson, G. G., Jr., Scheible, A., Wims, A. M., Johnson, J. L., Ullmann, G. (1987). Powder Diffr. **2**: 73-77.
Smith, D. K. (1991). J. Appl. Crystallogr. **24**: 369-402.
Smith, D. K. (1993). Adv. X-ray Anal. **35**: 1-15.
Smith, D. K., Barrett, C. S. (1979). Adv. X-ray Anal. **22**: 1-12.
Smith, D. K., Johnson, G. G., Jr., Kelton, M. J., Anderson, C. A. (1989). Adv. X-ray Anal. **32**: 489-496.
Smith, D. K., Johnson, G. G., Jr., Wims, A. M. (1988). Aust. J. Phys. **41**: 311-321.
Smith, D. K., Nichols, M. C., Zolensky, M. E. (1983). *POWD10. A Fortran IV Program for Calculating X-ray Powder Diffraction Patterns. Version 10.* Uninersity Park, College of Earth and Mineral Sciences, The Pennsylvania State University.
Smith, J. V. (1974). *Feldspar Minerals.* New York, Springer Verlag.
Smith, S. T., Snyder, R. L., Brownell, W. E. (1979a). Adv. X-ray Anal. **22**: 77-87.
Smith, S. T., Snyder, R. L., Brownell, W. E. (1979b). Adv. X-ray Anal. **22**: 181-191.
Smolczyk, H. G. (1961). Zement-Kalk-Gips no. 9, p. 391-399.
Snyder, R. L. (1992). Powder Diffr. **7**: 186-193.
Sonnefeld, E. J., Visser, J. W. (1975). J. Appl. Crystallogr. **8**: 1.
Starks, T. H., Fang, J. H., Zevin, L. S. (1984). Math. Geol. **16**: 351-367.
STOE. (1988) *Modern Fast Automatic Powder Diffractometry*, Darmstadt, STOE.
Sturges, W. T., Harrison, R. M., Barrie, L. A. (1989). Atmos. Environ. **23**: 1083-1098.
Sturm, E., Lodding, W. (1968). Acta Crystallogr. **A24**: 650-653.
Suortti, P. (1972). J. Appl. Crystallogr. **5**: 325-331.
Suortti, P., Jennings, L. D. (1971). J. Appl. Crystallogr. **4**: 37-43.
Swanson, H. E., McMurdie, H. F., Morris, M. C., Evans, E. H. (1966). *Standard X-ray Diffraction Powder Pattern.* NBS monograph 25, section 4. pp. 3-4. Washigton, DC, National Bureau of Standards.
Swanson, H. E., Morris, M. C., Stinchfield, R. P., Evans, E. H. (1962). *Standard X-ray Diffraction Powder Pattern.* Washington, DC, National Bureau of Standards.
Szabó, P. (1980). J. Appl. Crystallogr. **13**: 479-485.
Tabikh, A. A., Weht, R. J. (1971). Cem. Conc. Res. **1**: 317-326.
Tamaru, K., Sakito, K. (1931). Sci. Rpts. Tohoku Imp. Univ. **20**: 1.
Tang, I. N., Munkelwitz, H. H., Cantrell, J. G. (1978). J. Aerosol Sci. **9**: 505-511.
Tani, B., Siegal, S., Johnson, S. A., Kumar, R. (1983). Atmos. Environ. **17**: 2277-2283.
Tanner, B. K. (1990). Adv. X-ray Anal. **33**: 1-11.
Tatlock, D. B. (1966). Rapid Modal Analysis of Some Felsic Rocks from Calibrated X-ray Diffraction Patterns. United States Geological Survey Bulletin 1209. pp. 10-19.
Taupin, D. (1973). J. Appl. Crystallogr. **6**: 266-273.
Taylor, A. (1944). Philos. Mag. **35**: 215, 404.
Taylor, A. (1952). *An Introduction to X-ray Metallography.* London, Chapman & Hall.
Taylor, J. C. (1991). Powder Diffr. **6**: 2-9.
Taylor, J. C., Clapp, R. A. (1992). Adv. X-ray Anal. **35**: 49-55.
Taylor, J. C., Pecover, S. R. (1988). Aust. J. Phys. **41**: 323-335.
Thedchanamoorthy, A., McCarthy, G. (1989). Adv. X-ray Anal. **32**: 569-576.
Toraya, H. (1986). J. Appl. Crystallogr. **19**: 440-447.

Toraya, H., Yoshimura, M., Somiya, S. (1984). J. Am. Ceram. Soc. **67**: C119-C121.
Tournaire, M. (1954). J. Phys. Radium **15**: 11A-15A.
Uda, M. (1967). Z. Anorg. Allg. Chem. **350**: 105.
Umanski, J. L., Khidekel, S. S. (1941). Zh. Russ. Fiz.-Khim. Ova. **15**: 997.
Valvoda, V. (1986). Powder Diffr. **1**: 28-32.
Vasudevan, R., Ramesh, R., Pathiraj, B., Kolster, B. H. (1992). Powder Diffr. **7**: 77-82.
Villars, P., Calvert, L. D. (1985). *Pearson's Handbook of Crystallographic Data for Intermetallic Phases.*, Metals Park, Ohio, ASM.
Volkov, O. S., Kheiker, D. M. (1963). Proceedings of Niiasbestcement. no. 16. p. 43-55. Moscow, Gosstroyizdat.
Vonk, C. G. (1973). J. Appl. Crystallogr. **6**: 148-152.
Vonk, C. G., Fagherazzi, G. (1983). J. Appl. Crystallogr. **16**: 274-276.
Vonk, G. C. (1961). Norelco Reporter **8**: 92.
Voskamp, A. P. (1974). Adv. X-ray Anal. **17**: 124-138.
Vozdrigenski, M. P., Sergeev, G. A. (1935). J. Zeitcsh. Techn. Phys., **5**: 784-795.
Wallace, R. M. (1960). J. Phys. Chem. **64**: 899-901.
Wallace, R. M., Katz, S. M. (1964). J. Phys. Chem. **68**: 3890-3892.
Warren, B. E. (1969). *X-ray Diffraction*. Reading, MA., Addison-Wesley.
Weber, K. (1967). Acta Crystallogr. **23**: 720-725.
Weiss, Z., Krajicek, J., Smrcok, L., Fiala, J. (1983). J. Appl. Crystallogr. **16**: 493-497.
Werner, P.-E., Ekström, T. (1980). Anal. Chem. **52**: 833-837.
Werner, P.-E., Salomé, S., Malmros, G., Thomas, J. O. (1979). J. Appl. Crystallogr. **12**: 107-109.
Whelan, G., Moroz, J. P., Jr. (1985). Adv. X-ray Anal. **28**: 315-320.
Whittaker, E. J. (1953). Acta Crystallogr. **6**: 222-223.
Wilchinsky, Z. W. (1951). Acta Crystallogr. **4**: 1-9.
Will, G., Parrish, W., Huang, T. C. (1983). J. Appl. Crystallogr. **16**: 611-622.
Williams, P. P. (1959). Anal. Chem. **31**: 1842-1844.
Wilson, A. J. C. (1963). *Mathematical Theory of X-ray Powder Diffraction*. Eindhoven, Centrex.
Wilson, A. J. C. (1970). *Elements of X-ray Crystallography*. Addison-Wesley.
Wilson, A. J. C. (1972). Sov. Phys.-Crystallog. **16**: 991-993.
Wilson, A. J. C. (1980). Accuracy in methods of lattice-parameter measurement. *Proceedings of Symposium on Accuracy in Powder Diffraction*, pp 361-382, National Bureau of Standards. special publication no.567.
Wölfel, E. R. (1983). J. Appl. Crystallogr. **16**: 341-348.
Young, R. A., ed. (1993). *The Rietveld Method*. Oxford, International Union of Crystallography, Oxford University Press.
Young, R. A., Wiles, D. B. (1982). J. Appl. Crystallogr. **15**: 430-438.
Zachariasen, W. H. (1945). *Theory of X-ray Diffraction in Crystals*. New York, John Wiley & Sons.
Zaslavsky, A. I. (1977). X-ray Analysis of Mineral Matter. no.11: pp. 4-11. Moscow, Nedra.
Zav'yalova, L. L., Ivoilov, A. S. (1965). Indust. Lab. **31**: 1225-1228.
Zav'yalova, L. L., Ivoilov, A. S. (1970). Apar. Metody Rentgenovskogo Anal., 56-61.
Zav'yalova, L. L., Ivoilov, A. S., Vasilyev, E. K. (1964). X-ray Anal. Miner. Raw Mater. pp. 190-194.
Zevin, L., Lach, S., Levi, Y., Pregerson, B. (1988). Int. J. Mines. Process **24**: 235-245.
Zevin, L. S. (1977). J. Appl. Crystallogr. **10**: 147-150.
Zevin, L. S. (1979). J. Appl. Crystallogr. **12**: 582-584.
Zevin, L. S. (1981). J. Phys. E: Sci. Instrum. **14**: 725-730.
Zevin, L. S. (1988). Powder Diffr. **3**: 25-29.
Zevin, L. S. (1990). Acta Crystallogr. **A46**: 730-734.
Zevin, L. S. (1992). Standardless method of quantitative X-ray diffractometry, *In*: Prince, E. Stalick, J. K. eds. *Proceedings of the International Conference on Accuracy in Powder Diffraction II*, Gaithersburg, MD, National Institute of Standards and Technology, Special Publication 846. pp. 17-24.
Zevin, L. S., Kheiker, D. M. (1962). Ind. Lab. USSR **28**: 609-616.

Zevin, L. S., Kheiker, D. M. (1963a). *Apparatus and Methods of X-ray Analysis.* Proceedings of the Research Institute "Asbestos-Cem.", volume 16, pp. 3-24.
Zevin, L. S., Kheiker, D. M. (1963b). Ind. Lab. USSR **29**: 168-173.
Zevin, L. S., Kheiker, D. M. (1965). *X-ray Methods in the Investigation of Building Materials.* 361 p. Moscow, Stroizdat.
Zevin, L. S., Kheiker, D. M. (1967). Ind. Lab. **33**: 1798-1803.
Zevin, L. S., Malinin, Y. S., Ryazin, V. P. (1972). Comparison of methods for phase analysis of Portland cement. In: *Technical Information of All-Union Institutes of Building Materials,* series "Cem. and Asbestos Cem. Industry", no. 6. p. 16-20.
Zevin, L. S., Margolina, G. S., Flanzbaum, I. M. (1977). Cem. Conc. Res. **7**: 103-104.
Zevin, L. S., Messalem, R. (1982). Polymer **23**: 601-604.
Zevin, L. S., Rozenak, P., Eliezer, D. (1984). J. Appl. Crystallogr. **17**: 18-21.
Zevin, L. S., Shneck, R., Zevin, I. M. (1981). Twelfth Int. Cong. of Crystal., Collected Abstracts 16-25 August, Ottawa. Przybylska, M. ed., pp. C-280-281.
Zevin, L. S., Umanski, M. M. (1964). Sov. Phys.-Crystallogr. **9**: 399-402.
Zevin, L. S., Umanski, M. N., Kheiker, D. M. (1963). Sov. Phys.-Crystallogr. **8**: 528-536.
Zevin, L. S., Viaene, O. (1990). Clay Miner. **25**: 401-418.
Zevin, L. S., Zav'yalova, L. L. (1974). *Quantitative X-ray Phase Analysis.* Moscow, Nedra.
Zevin, L. S., Zevin, I. M. (1987). Powder Diffr. **2**: 78-81.
Zevin, L. S., Zevin, S. L. (1989). Powder Diffr. **4**: 196-200.
Zorn, G., Hellstern, E., Göbel, H., Schultz, L. (1988). Adv. X-ray Anal. **30**: 483-491.

Index

MIX
Papier aus verantwortungsvollen Quellen
Paper from responsible sources
FSC® C105338

If you have any concerns about our products,
you can contact us on
ProductSafety@springernature.com

In case Publisher is established outside the EU,
the EU authorized representative is:
Springer Nature Customer Service Center GmbH
Europaplatz 3, 69115 Heidelberg, Germany

Printed by Libri Plureos GmbH
in Hamburg, Germany